Intercept 1961: The Birth of Soviet Missile Defense

Intercept 1961: The Birth of Soviet Missile Defense

Mike Gruntman
University of Southern California

Ned Allen, Editor-in-Chief
Lockheed Martin Corporation
Bethesda, Maryland

Published by
American Institute of Aeronautics and Astronautics, Inc.
1801 Alexander Bell Drive, Reston, VA 20191-4344

American Institute of Aeronautics and Astronautics, Inc., Reston, Virginia

1 2 3 4 5

Library of Congress Cataloging-in-Publication Data

Gruntman, Mike, 1954-
   Intercept 1961: the birth of soviet missile defense/
     Mike Gruntman.—1st edition.
     pages cm.—(Library of flight series)
   Includes bibliographical references and index.
   ISBN 978-1-62410-349-0 (print) – ISBN 978-1-62410-350-6 (.pdf)
  1. Ballistic missile defenses—Soviet Union—History.   I. Title. II. Title:
Little known first breakthrough in non-nuclear missile defense.
   UG745.S65G78 2015
   358.1'740947–dc23

2015019662

Copyright © 2015 by the American Institute of Aeronautics and Astronautics, Inc. All rights reserved. Printed in the United States of America. No part of this publication may be reproduced, distributed, or transmitted, in any form or by any means, or stored in a database or retrieval system, without the prior written permission of the publisher.

Data and information appearing in this book are for informational purposes only. AIAA is not responsible for any injury or damage resulting from use or reliance, nor does AIAA warrant that use or reliance will be free from privately owned rights.

A Note from the Editor-in-Chief

Nothing has had more influence or given more impetus to the development of advanced aerospace technology than the search for security and especially for defense against missile attack. In the 1980s, U.S. President Reagan was concerned that deterrence in its then-canonical form, mutual assured destruction (MAD, as it was grotesquely dubbed at the time)—the idea that we would launch a noncancelable nuclear attack on our adversaries when we detected the launch of an attack upon us while their missiles were in the air, before even the trajectory of their launch was clearly discernable—required a decision that was beyond what any one person could be expected to make with a clear conscience. He held the MAD strategy to be a "suicide pact." So at his insistence researchers began a long and expensive search for other solutions, the Holy Grail of which was a technology that could destroy an enemy missile harmlessly in flight. Antimissile development programs then proceeded through the Strategic Defense Initiative (informally known as "Star Wars"), followed by the creation of one federal agency after another charged with the search for that Holy Grail, until today we have the U.S. Missile Defense Agency.

The Soviets realized the importance of a Holy Grail solution years before the United States even considered the concept. What was wanted was the ability to sustain an attack, even if barely survived, and then take the upper hand in the aftermath. That required an effective antimissile technology to at least protect the government, surviving forces, and command capabilities. Thus the Soviets established their own missile defense agency and concept of operations long before Reagan even entered the U.S. White House. The Soviets achieved their first successful intercept, "hitting a bullet with a bullet," during the heyday of Russian rocketry in the early 1960s. Mike Gruntman tells the story of that program, little known or understood before now. That dearth of information and understanding is remarkable, because the advance of Soviet weaponry was such a strong driver in 20th-century American technological history. So the story of the successful Soviet antimissile program and its political context takes its rightful place in the Library of Flight for all to read.

The Library of Flight is part of the growing portfolio of information services from the American Institute of Aeronautics and Astronautics; it documents the crucial role of aerospace in enabling, facilitating, and accelerating global commerce, communication, and defense. Distinct from the AIAA Education and Progress in Astronautics and Aeronautics series, the Library of Flight authors often express opinions on matters of controversy; of course, these are not the opinions of AIAA. As new aerospace programs grow and change around the world, we plan for this series to host a wide array of international authors, each expressing their own point of view on aerospace visions, events, and issues. As the demands on the world's space systems grow to support new capabilities such as unmanned vehicles, international relief, agricultural management, and environmental monitoring, the series will seek to document landmark events, emerging trends, and new viewpoints.

Ned Allen
AIAA Fellow and Chief Scientist
Lockheed Martin
Bethesda, Maryland

Contents

Preface xi
Acknowledgements xvii

Chapter 1   Introduction: Protect or Avenge 1

Away from Public Eyes 1
Geopolitical Importance 3
Selective Virtue of Defense 5
Common Sense 8

Chapter 2   Special Bureau SB-1 13

Advanced Weapons 13
Communist Princeling 16
New Special Bureau 19
Kometa KS-1 24

Chapter 3   Stalin's Order 29

A Summons to the Kremlin 29
Design Bureau KB-1 32
A New "Empire" Emerges 37
Track-While-Scan Radar 43
Antiaircraft Missile V-300 49
The Death of Stalin 56

Chapter 4   Air Defense System of Moscow 61

Early Warning Radars 61
Two Rings of Fire 65
The S-25 Site 66

S-25 Operational 71
U-2 Aircraft over Moscow 73
Two Bears in One Lair 83

CHAPTER 5   BEGINNING OF MISSILE DEFENSE 91

In Response to a New Threat 91
A Time of Changes in the Military-Industrial Complex 95
Missile Defense Challenges 97
Meeting at TGU 98
Development Leaders 101
System A Authorized 107
Development Team 110

CHAPTER 6   SARYSHAGAN TEST SITE 117

Desert in Kazakhstan 117
Construction in the Desert 123
GNIIP-10 129
Test Site "Put on the Map" 132
Growing Installation 137
Site 2 142
Priozersk 150
Nuclear Explosions in Saryshagan's Skies 155

CHAPTER 7   EXPERIMENTAL SYSTEM A 161

System Concept 161
Long-Range Search Radar Dunai-2 163
Precise Tracking and Guidance Radar RTN 173
Interceptor Missile Initial Guidance Radar RSVPR 179
Interceptor Missile V-1000 181
Data Transmission System 185
Central Computing Station 185

CHAPTER 8   INTERCEPTS 191

Autonomous and System Tests 191
Interceptor Warhead 194
The Success on 4 March 1961 198
Tests Continue 206
Battle Against Penetration Aids Begins 207

Chapter 9 Beyond Experiments 215

Toward Operational Missile Defense 215
Crisis in Missile Defense 227
Scientific-Industrial Association Vympel 231
Firing of Kisun'ko 232
A Gigantic Enterprise 234
Antisatellite Weapons 236
Ballistic Missile Early Warning 242
Soviet Princelings 244
Weapons in Space 248
Post-USSR Era 252

Appendix A First U.S. Missile Intercepts 255

Appendix B Acronyms and Abbreviations 265

Appendix C Pronunciation Guide 269

Appendix D List of Figures 281

Appendix E Selected Bibliography 285

Index 297

Supporting Materials 311

PREFACE

On 4 March 1961, a Soviet guided missile intercepted and destroyed the approaching warhead of an intermediate-range ballistic missile R-12 (SS-4) at the Saryshagan test site in the Kazakhstan desert. Several successful intercepts followed, paving the way for the emergence of a powerful political, military, scientific-technological, and industrial missile defense complex in the Soviet Union.

The spectacular nonnuclear destruction of a long-range ballistic missile clearly earned a place among the most important and consequential Soviet firsts in a rapidly advancing field of missiles and space. The other accomplishments of that era included the first intercontinental ballistic missile R-7 (SS-6) and artificial Earth satellite Sputnik in August and October, respectively, of 1957, and an orbital flight by the first cosmonaut, Yuri Gagarin,[1] in April 1961.

In contrast to Sputnik and Gagarin's flight, the history of the first intercept of a long-range ballistic missile remains poorly known and appreciated. The event stood out as especially impressive because it relied on advanced electronics, sophisticated radar systems, high-speed communications, real-time computing, and precise guidance and control. The Western public and media rarely viewed these areas of military technology as being particularly strong in the Soviet Union.

The missile defense achievement had its roots in the Soviet air defense establishment that was organizationally and technologically different and separate from those building strategic ballistic missiles and satellites and launching cosmonauts. The early missile defense programs also led to the development and emergence of new space-related areas such as tracking and cataloging of orbiting space objects, space object identification, detection from space of ballistic missile launches, and antisatellite weapons. Although abundant literature on ballistic missiles and space exploration fills bookshelves, very little is written about missile defense and first intercepts.

At the same time, the proliferation of ballistic missiles coupled with weapons of mass destruction, radical ideologies, and terrorism make missile defense increasingly important to the free world. Policy debates on the subject

---

[1] Yuri Alexeevich Gagarin, 1934–1968.

and media accounts are sometimes shallow in substance and not always informed, with ideologically driven narratives not uncommon.

After the collapse and disintegration of the Union of Soviet Socialist Republics (USSR) in the early 1990s, many leading defense research and development organizations in Russia published detailed corporate histories that described the introduction of major weapon, rocket, and space systems and associated research, development, and manufacturing efforts. In addition, a number of authors with varying degrees of expertise wrote accounts of some programs in defense areas. Much of this literature appeared in very limited editions and remains poorly known, especially outside Russia. The language barrier also plays an important restrictive role. Declassified U.S. government documents and reconnaissance imagery offer additional insight into the story, but they are not always conveniently accessible.

As a result, only a few, even among specialists, are familiar with the origins and history of the Soviet air defense establishment from which missile defense emerged and to which it remains closely linked to this day. Common sense calls for more attention to air defense systems, military units, and leading managers, scientists, and engineers because of their prominent and literally deadly role in contesting the policies of the United States and its allies in the free world.

Soviet air defense regiments, sometimes supported by scientists and engineers from development organizations, directly engaged U.S. fliers in a shooting war in Vietnam.[2] Such units also fought against Israel's pilots in Egypt for a few months in 1970.[3] These were real combat operations where the Soviets on one side and the Americans and Israelis on the other shed blood and took casualties. Providing antiaircraft missiles to anti-American regimes has been for decades and remains today a major tool in advancing Soviet, and now Russian, policies and interests.

A factually accurate sequence of events and the history of missile defense are not only indispensable for viewing the Cold War, but also essential for understanding the heritage, industrial capabilities, and relations among main defense "players" in the post-Soviet geopolitical space. Although some organizations changed their focus, merged, and reorganized, a number of key design bureaus and industrial plants survived through the turmoil of the 1990s. They now actively contribute to defense buildup and exports of weapons by the new assertive Russia.

The role of defense-related science and technology in the twentieth century offers insights into the current dynamics of the world. The historic background is thus critical for informed policy formulation, defense planning, threat evaluation, and counteracting the proliferation of weapons and sensitive technologies.

---

[2] e.g., Pervov, 2001, pp. 231–242; Semenov, 2008, pp. 237–255.
[3] e.g., Vinogradov et al., 1997; Pervov, 2001, pp. 243–249; Semenov, 2008, pp. 256–272.

This book focuses on the events that led to the first nonnuclear intercept of a long-range ballistic missile warhead by the Soviet Union in 1961. It introduces leading participants, largely forgotten now or unknown, and contains many technical characteristics of early air and missile defense systems, rarely found even in highly specialized publications. The latter details are not overbearing, and anyone interested in rocketry, space, and radar will navigate through the book without difficulty.

The book is for readers with diverse backgrounds and interests. It is for all those who are engaged in research, development, and operation of weapons systems in missile defense, air defense, military and national security space, and broad related fields of rocketry, missiles, satellites, space exploration and technology, and radar. It is also for science and engineering students who are considering careers in these areas.

The book will appeal to those involved in analyzing and formulating policies and engaged in planning in the areas of national security, defense, science, technology, and weapons proliferation. Missile defense has become an area of vital importance for the national security of a growing number of countries around the globe. Consequently, this publication will inform those interested in international relations and public policies. Students of history will find important, not-readily-available factual information.

In addition, the book is for anybody with a general interest in missiles, rocketry, space systems, space exploration, space applications, radar, military technology, and history and the Cold War.

A comparison between the Soviet and U.S. missile defense programs as well as antiballistic missile (ABM) and other arms control treaties, the evolution of national policies, and details of the development of missile defense systems after the first intercepts are, with a few exceptions, beyond the scope of this publication.

SOVIET BALLISTIC MISSILES AND AIR DEFENSE SYSTEMS

This book refers to a number of early Soviet ballistic missiles and air defense systems. Here, we consider the intermediate range (and intermediate-range ballistic missiles, or IRBMs) to be 1000–5500 km, as defined by the Intermediate-Range Nuclear Forces Treaty (INF Treaty) of 1987. [Sometimes, the intermediate range is further subdivided into medium-range ballistic missiles (MRBMs), with a range of 1000–3000 km, and IRBMs, with a range of 3000–5500 km.] Short-range ballistic missiles (SRBMs) are those with effective distances from 500 to 1000 km. Intercontinental ballistic missiles (ICBMs) are missiles with a range larger than 5500 km.

Tables 1 and 2 show the Soviet and Western nomenclatures of early ballistic missiles[4] and air defense systems.

---

[4]e.g., Gruntman, 2004, p. 289.

TABLE 1   EARLY SOVIET BALLISTIC MISSILES AND SPACE LAUNCHERS

USSR		Western		Type
R-1	8A11	SS-1	Scunner	
R-2	8Zh38	SS-2	Sibling	SRBM
R-5	8A62	SS-3	Shyster	IRBM
R-7	8K71	SS-6	Sapwood	ICBM
R-9A	8K75	SS-8	Sasin	ICBM
R-12	8K63	SS-4	Sandal	IRBM
R-14	8K65	SS-5	Skean	IRBM
R-16	8K64	SS-7	Saddler	ICBM
R-36	8K67	SS-9	Scarp	ICBM
Tsyklon-2	11K69	SL-11		Space launcher
Tsyklon-2A	11K67			Space launcher
UR-100	8K84	SS-11	Sego	ICBM
UR-200	8K81	SS-10	Scrag	ICBM

TABLE 2   EARLY SOVIET AIR DEFENSE SYSTEMS AND MISSILES

USSR	Western	
S-25 (Berkut)	SA-1	Guild
S-75	SA-2	Guideline
S-125	SA-3	Goa

UNITS

The fields of defense, rocketry, and space technology traditionally use a combination of metric and nonmetric units. Therefore, for convenience of the readers, many technical parameters are given simultaneously in the most widely used units. The unit ton with the symbol t means metric ton, that is, 1 t = 1000 kg = 2204.6 lb. We will use the same symbol t for both mass and weight. The unit n.m. stands for a nautical mile, with 1 n.m. = 1852.0 m. The unit lbf stands for pound-force, with 1 lbf = 4.448 N and 1 klbf = 1000 lbf.

TRANSLITERATION, STRESS, PRONUNCIATION, PATRONYMICS, AND GEOGRAPHIC NAMES

Several widely accepted systems transliterate Russian words, written using Cyrillic, into the English language.[5] In addition, journalists and government agencies used a traditional rendering of certain Russian words and names for many decades. Each system has advantages and disadvantages.

In this book I use transliteration (Romanization) that largely follows traditional approaches. For example, first names are rendered as Yuri, Grigorii, and

---

[5]e.g., Timberlake, 2004, pp. 24–27.

Vasilii, whereas the similar ending "iy" in last names is rendered as "y" (e.g., Kamenetsky). Examples of other conventions include the last name Korolev (sometimes rendered in literature as Korolyov) and the given names Petr (Pyotr) and Semen (Semyon). In addition, I do not use "y" when two separate Russian vowels appear together and belong to two syllables, as in Andreevich (sometimes spelled as Andreyevich and pronounced an-DRE-ye-vich) or in Nikolaevich (Nikolayevich).

The traditional spelling of the names of organizations adopted here are Energia (Energiya) and Poletami (Polyotami), the latter deriving from the word *polet* (polyot, or flight in English). The Russian names Alexander and Alexei and their derivatives are spelled using the letter "x" rather than the direct transliteration "ks"; the traditional rendering Alexander also differs slightly from the more accurate transliteration Alexandr. As widely used, the apostrophe (') in the middle or at the end of a word means softening of the preceding consonant. There are also two words with the double apostrophe (ob"ekt and ob"edinenie), which signifies separate pronunciation of syllables without softening of the preceding consonant.

The irregular stress patterns of Russian words present a challenge. Therefore, Appendix C provides a pronunciation guide to Russian words used in the book. It breaks down words into syllables and shows which syllables are stressed (capitalized syllables, as in an-DRE-ye-vich) to help with their proper pronunciation. Most of these words are personal names and names of organizations. In addition, the guide shows the pronunciation and stress of Russian acronyms and abbreviations.

According to the Russian tradition, the name of a person consists of the first or given name, the patronymic name, and the last name or surname (family name). The patronymic is a modified form of a genitive case of the person's father name. For example, consider the name of Grigorii V. (Vasil'evich) Kisun'ko, who led the effort resulting in the first successful long-range ballistic missile intercept. Here, Grigorii and Kisun'ko were the first (given) and the last (surname or family) names, respectively. The given name of Grigorii's father, Vasilii, resulted in the patronymic Vasil'evich.

Footnotes give full names, including patronymics, of individuals when they first appear in the narrative. They also show the years of birth and death, when known. If only the year of birth is given, then the person is either alive or the year of death is not known.

The geographic names are given in the form in which they existed at the time of the described events. Presently used new names are sometimes provided in parentheses when mentioned for the first time; examples include lake Balkhash (Balqash) and the towns of Leningrad (St. Petersburg) and Nikolaev (Mykolaiv).

I also spell the names of two major Soviet test sites, Tyuratam (Baikonur) and Saryshagan, as one word. Some quoted sources also spelled them Tyura Tam (and Tyura-Tam) and Sary Shagan (Sary-Shagan).

FIGURES AND RECONNAISSANCE PHOTOGRAPHS

The book has more than 120 figures, including a number of photographs never published outside Russia and many others appearing for the first time ever (at least for the first time in open literature).

Figure captions identify cameras and optical systems for photographs obtained by U.S. U-2 aircraft and reconnaissance satellites.

The official CIA history of the U-2 aerial reconnaissance program details its optical systems and missions.[6] The first space optical intelligence system, Corona, used KH-1, KH-2, KH-3, and KH-4 cameras with gradually improving capabilities from 1960 to 1972; KH-5 (Argon) mapping cameras took photographs of large areas with relatively low resolution; and KH-7 (Gambit-1) cameras produced images from 1963 to 1967 that were smaller than KH-4 swaths but with higher resolution on the ground.[7]

Many reconnaissance photographs were taken at oblique angles. Because they were not corrected, shown scales and compass directions are approximate.

APPENDICES AND INDEX

The book concludes with five appendices and the index.

- *Appendix A* briefly summarizes U.S. intercepts of smaller missiles *prior* to the Soviet feat in 1961.
- *Appendix B* provides a list of acronyms and abbreviations.
- *Appendix C* provides a pronunciation guide for Russian words and names.
- *Appendix D* contains a list of figures with abridged captions.
- *Appendix E* contains a selected bibliography with more than 200 items. All referred to and quoted U.S. government documents are declassified.

The opinions expressed in the book are solely mine and are not necessarily shared by the acknowledged individuals and organizations. Needless to say, I take the responsibility for all errors.

Mike Gruntman
August 2015

---

[6]Pedlow and Welzenbach, 1998.
[7]e.g., Perry, 2012; Oder et al., 1991; Ruffner, 1995; Day et al., 1998; Gruntman, 2004, Chapter 16; Berkowitz, 2011.

ACKNOWLEDGEMENTS

A number of individuals and organizations kindly helped me in the preparation of this publication. I would like to thank each and every one of them. I am especially grateful to (alphabetically)

Alexander Yu. Bonchkovsky, Minsk, Belarus
borbor, user, Google's Panoramio
Bob Brodsky, Claremont, California
Alexander Degtyarev, Yuzhnoe Design Bureau (Office), Dnepropetrovsk, Ukraine
Freedom of Information Act program, Central Intelligence Agency
Dmitrii Gordon, Kiev, Ukraine
Sergei A. Gruntman, Moscow, Russia
ITAR-TASS News Agency, Moscow, Russia
Larry Kaplan, U.S. Missile Defense Agency, Huntsville, Alabama
Mikhail Khodarenok, editor-in-chief, *Vozdushno-Kosmicheskaya Oborona* (journal), Moscow, Russia
Sergei N. Khrushchev, Rhode Island
Sharon Lang, U.S. Army Space and Missile Defense Command, Huntsville, Alabama
Jerry Luchansky, National Archives and Records Administration, Maryland
Doyle Piland, White Sands Missile Range Museum, New Mexico
Leonid Sokolovsky, Moscow, Russia
Oleg Veideman, Tallinn, Estonia

Chapter 1

INTRODUCTION: PROTECT OR AVENGE

AWAY FROM PUBLIC EYES

Launched eastward from a base north of the Caspian Sea, an R-12 intermediate-range ballistic missile had separated its warhead. Then, an interceptor missile dashed towards them from the Saryshagan test range in Kazakhstan.[1] The date was 4 March 1961.

The interceptor detonated a fraction of a second before passing the target warhead at a distance of 104 ft (31.8 m) and released thousands of spherical balls, each containing a high-explosive charge, which destroyed the approaching warhead on impact. The direct, head-on, hit-to-kill collision of today's kinetic kill vehicle (KKV) intercept was still in the far-distant future. The first Soviet experimental missile defense System A, or Sistema A in Russian, did not rely on a nuclear explosion to stop the approaching warhead. Several successful intercepts followed, which laid the foundation for a vigorous missile defense effort in the Union of Soviet Socialist Republics (USSR).

Three years prior to the event in Kazakhstan, Soviet leader Nikita S. Khrushchev[2] had spoken to graduates of military academies. Khrushchev described, in his memorably folksy way, the new age of automated computer-controlled weapon systems: "It is sufficient now only to press a button and not only [military] airfields and means of communications of various headquarters will [be hit and] fly into the air but entire cities. Entire countries can be destroyed."[3] The missile intercept perfectly illustrated these emerging push-button futuristic weapons.

The spectacular first nonnuclear destruction of an intermediate-range ballistic missile (IRBM) warhead remained poorly known in the USSR and

---
[1] Golubev et al., 1992, 1994; Yakovlev, 1999; Pervov, 2003; Zavaly, 2003; Kulakov, 2006; Belous et al., 2009; Khodarenok, 2010.
[2] Nikita Sergeevich Khrushchev, 1894–1971.
[3] N.S. Khrushchev, 1958, p. 29.

abroad, in contrast to the highly publicized achievements of the first intercontinental ballistic missile (ICBM), the first satellite, and an orbital flight by the first cosmonaut. In the secrecy-obsessed Soviet Union, a limited number of officials and specialists knew about the accomplishment in Kazakhstan in March 1961. As leading Soviet missile defense specialists wrote many years later,

> The fact of the intercept and destruction of ballistic targets had not only a big military-technological but also political meaning. At that time [in the late 1950s and early 1960s] even a single ballistic [missile] target was considered an absolute weapon. However the event [of the warhead intercept] had not received appropriate attention in the country. There were sufficient reasons for that. It happened so, in part, because the air defense work was treated as top secret and because leaders of the military-industrial complex were concerned at that time with preparation for launch of the first artificial satellite of the Earth with a man on board. The latter launch [of Yuri Gagarin] occurred [several weeks after the first warhead intercept] on April 12, 1961, and left in the shadows the outstanding event in the history of missile defense.[4]

Outside Russia, not many are familiar today with the history of the first intercept. The U.S. Army's Center of Military History recently published a comparative history of Soviet and U.S. strategic air and ballistic missile defenses. The 650-page publication does not even mention the intercept.[5] A monthly publication of the American Institute of Aeronautics and Astronautics (AIAA), *Aerospace America*, lists major historical events in the world of aerospace 25, 50, 75, and 100 years ago in a special section. It did not include the first intercept in the 50-years-ago category in its March 2011 issue.[6]

First missile intercepts at Saryshagan eventually led to a strategic missile defense system for the Soviet capital, Moscow. (Strategic missile defense refers to the protection of large areas or important administrative, military, and industrial centers in the homeland against strategic intercontinental and intermediate-range ballistic missiles. Theater, or tactical, missile defense protects specific sites and deployed troops from smaller, shorter range missiles.) A huge and exceptionally powerful military and industrial missile defense complex emerged. Budgetary estimates put the Soviet Union's expenditures on strategic air and missile defense at between 50% and 100% of the funds spent on development of strategic offense capabilities in the years 1955–1972.[7]

---

[4]Golubev et al., 1992, p. 174.
[5]U.S. Army, 2009, Vol. 2, pp. 270, 271, 347–351.
[6]Winter and van der Linden, 2011.
[7]U.S. Army, 2009, Vol. 2, p. 121.

The continuously expanding missile defense establishment included numerous scientific research institutes, design bureaus, industrial plants, and various military units. The development culminated in deployment of operational missile defense systems protecting Moscow with increasing capabilities, first the A-35 in 1971–1974, then its upgrade the A-35M in 1978, and finally the A-135 in the mid-1990s. The latter stands on combat duty today.

This national effort also branched off into other areas essential for defense and for space operations, such as optical and radar monitoring and cataloging of orbiting space objects (the approximate equivalent of space situational awareness, or SSA, in U.S. terminology); early warning of ballistic missile attack by sensors on satellites and by above-the-horizon and over-the-horizon radars; prediction and warning of overflights by U.S. reconnaissance satellites; and space defense, including antisatellite weapons and space-based weapon systems.

The public did not know much about this vast enterprise, in spite of occasional publications[8] and statements by Soviet leaders. The official Soviet propaganda rarely highlighted missile defense. General Yuri V. Votintsev[9] served as the first commander, from 1967 to 1986, of the Missile and Space Defense Forces. After the demise of the Soviet Union he published his recollections under the title *Forgotten Troops of the Vanished Superpower*. Votintsev observed that the achievements in missile defense "were ignored for a long time and did not get the proper recognition."[10]

GEOPOLITICAL IMPORTANCE

Soviet leaders understood the importance of the accomplishments in missile defense. They publicized them sparingly in a precisely targeted manner, in contrast to the feats of offensive ballistic missiles and space exploration, which continuously filled communist propaganda. Seven months after the first successful warhead intercept at Saryshagan, Minister of Defense Marshal Rodion Y. Malinovsky[11] reported to the 22nd Congress of the Communist Party of the Soviet Union (CPSU), "The Central Committee of the [Communist] Party has been and is showing particular care about air and missile defenses of the country.... I have to specially report [to the Congress] that the problem of destroying rockets in flight has been successfully solved."[12]

The Soviet military doctrine combined strategic offense and defense, with the latter emphasizing protection against missiles. In September 1962, the USSR Ministry of Defense published 20,000 copies of an authoritative open

---

[8]e.g., Talensky, 1964.
[9]Yuri Vsevolodovich Votintsev, 1919–2005.
[10]Votintsev, 1993, Part 2, p. 31.
[11]Rodion Yakovlevich Malinovsky, 1898–1967.
[12]*Pravda*, 1961, p. 4.

treatise on military strategy, edited by Marshal Vasilii D. Sokolovsky.[13] Sokolovsky had served as chief of the general staff from 1952 to 1960 and played a key role in the birth of missile defense in the USSR. The treatise recognized that "in a future world war, massive nuclear-missile strikes will be of decisive importance" and that "the main problem in Soviet military strategy is thus development of methods of reliably *repelling a sudden nuclear attack by an aggressor.*"[14]

Sokolovsky's publication neither explicitly mentioned recent intercepts at Saryshagan nor noted vigorous development of a follow-on operational missile defense system. It stated, however, that "in our country, the problem of eliminating rockets in flight *has* been successfully solved by Soviet science and technology." The treatise optimistically added that "the creation of an invulnerable antimissile defense system has become quite possible."[15] It also emphasized the imperative that "in the future, this form of defense [against missiles] must be perfected."[16]

In July 1962, Soviet leader Nikita Khrushchev met with a group of visiting U.S. newspaper editors led by Lee Hills.[17] "I know what antimissile systems are since we have them," said Khrushchev with confidence, "our missile, one can say, hits a fly in space."[18]

The Soviet government considered missile defense so important for national security that real or imaginary U.S. efforts in this field helped to "sell" space reconnaissance in the USSR. Most Soviet people knew Sergei P. Korolev and Mstislav V. Keldysh[19] as enigmatic chief designer and chief theoretician, respectively, of the early Soviet space program.[20] Korolev led the building and launching of the first ICBM and artificial satellites and the sending of the first cosmonauts into space; Keldysh served as president of the USSR Academy of Sciences from 1961 to 1975.

These two Soviet space leaders had been arguing for development of heavy reconnaissance satellites since (at least) April 1957, prior to the launch of Sputnik. Korolev and Keldysh emphasized three priorities for space reconnaissance, which included "conducting radio reconnaissance of the means of missile defense of a possible adversary with aid of special receiving and recording devices onboard a satellite."[21] They finally got the authorization to proceed in early 1959.

---

[13]Vasilii Danilovich Sokolovsky, 1897–1968.
[14]Sokolovsky, 1963, pp. 196, 204 (emphasis by Sokolovsky).
[15]Sokolovsky, 1963, p. 297 (emphasis by Sokolovsky).
[16]Sokolovsky, 1963, p. 231.
[17]Lee Hills, 1906–2000.
[18]*Pravda*, 1962, p. 2.
[19]Sergei Pavlovich Korolev, 1906–1966; Mstislav Vsevolodovich Keldysh, 1911–1978.
[20]Gruntman, 2004, p. 338.
[21]Raushenbakh, 1998, pp. 264–268.

Sokolovsky's strategy treatise stated that "in principle, the technical solution of this problem [of missile defense] has now been found [in the Soviet Union]" and noted that at the same time "the problem of antimissile defense is far from being solved in the West."[22] Since those early days, the Soviet Union steadily advanced science and technology essential for missile defense and acquired invaluable and unique experience by fielding operational systems.

Many years later, U.S. physicist Edward Teller,[23] who had been playing a major role in research and development in critical defense areas, assessed the leading position of the USSR in missile defense. Teller wrote that by that time (in 1987), the missile defense technology was "a quarter of a century old. As the leaders in defensive technology, they [the Soviets] have nothing to learn from others." He also observed that "whenever we [in the United States] find 'new' possibilities [in missile defense], we have indications that, for the Soviets, they are quite old."[24]

SELECTIVE VIRTUE OF DEFENSE

The Soviet Union always emphasized the value of both offensive and defensive systems in the arms race. Chairman of the USSR Council of Ministers Alexei N. Kosygin[25] notably reiterated the importance of defense in 1967:

> I think that defensive systems that preempt assault are not the cause of the arms race. They represent a means of preventing death of people. Some pose a question: what is cheaper—to have attack weapons capable of annihilating towns and entire states or to have defensive weapons which may prevent such annihilation? ... Perhaps, a missile defense system costs more than an attack system but it is aimed not at killing people but at saving human lives.[26]

Remarkably, this praise of saving human lives came from a leader of a Marxist regime that had exterminated many millions of its own citizens on a 50-year march to socialist paradise.[27]

The Soviet approval of defensive technologies did not extend, however, to the U.S. missile defense effort, stepped up by President Ronald Reagan[28] in 1983. Teller observed, "That the shield is better than the sword is consistent both with what the Soviets used to say and with Russian behavior through the centuries. Only after President Reagan agreed with them [by announcing the

---

[22]Sokolovsky, 1963, pp. 231, 297.
[23]Edward Teller, 1908–2003.
[24]Teller, 1987, pp. 22, 38.
[25]Alexei Nikolaevich Kosygin, 1904–1980.
[26]*Pravda*, 1967, p. 3.
[27]Courtois et al., 1999.
[28]Ronald Reagan, 1911–2004.

Strategic Defense Initiative] did they start denying it."[29] As a U.S. National Intelligence Estimate (NIE) predicted in 1982, "the Soviets would stress the defensive nature of... [their antiballistic missile] system and try to use Western public opinion to constrain the freedom of action of Western governments."[30]

In March 1983, President Reagan challenged the nation to develop missile defense. Following quick denunciation by Soviet leader Yuri V. Andropov,[31] the USSR Academy of Sciences harshly criticized the U.S. Strategic Defense Initiative (SDI).[32] Then, in the words of Teller, "that cue was picked up by many American academics and, throughout the free democracies, by all parties to the left of center."[33] Missile defense policies and associated politics and debates remain highly charged ideologically to this day.

Russian publications proudly and unanimously have emphasized that the United States took 23 years to duplicate the 1961 nonnuclear intercept of a ballistic missile.[34] One Soviet missile defense veteran gleefully wrote, for example, that they had "won in a most difficult competition against the U.S., and the boastful American military science could do [repeat] it only 23 years later, in 1984."[35]

STRATEGIC DEFENSE

... I have become more and more deeply convinced that the human spirit must be capable of rising above dealing with other nations and human beings by threatening their existence. ...

Let us turn to the very strengths in technology that spawned our great industrial base and that have given us the quality of life we enjoy today.

What if free people could live secure in the knowledge that their security did not rest upon the threat of instant U.S. retaliation to deter a Soviet attack, that we could intercept and destroy strategic ballistic missiles before they have reached our own soil or that of our allies?

... I call upon the scientific community in our country, those who gave us nuclear weapons, to turn their great talents now to the cause of mankind and world peace, to give us the means of rendering these nuclear weapons impotent and obsolete.

—Ronald Reagan, address to the Nation on Defense and National Security, 23 March 1983 (Reagan, 1984, pp. 442, 443)

---

[29]Teller, 1987, p. 22.
[30]Central Intelligence Agency, 1982, p. 9.
[31]Yuri Vladimirovich Andropov, 1914–1984.
[32]*Pravda*, 1983.
[33]Teller, 1987, p. 7.
[34]e.g., Votintsev, 1993, Part 2, p. 31; Kisun'ko, 1996, p. 6; Chertok, 1997, p. 140; "Chronicle...", 1998; Drozdov, 1998; Yakovlev, 1999, p. 179; Gubenko, 2003, p. 680; Zhadeiko, 2005, p. 639; Semenov, 2008, p. 173; Belous et al., 2009, pp. 14, 177, 179, 181; Egorov, 2009, p. 3.
[35]Belous et al., 2009, p. 179.

They referred to the U.S. Homing Overlay Experiment (HOE) program that demonstrated, in 1984, a kill of a target warhead by a direct collision.[36] The HOE interceptor did not use a blast fragmentation warhead but relied on a lethality enhancement device with 36 deployed arms, each carrying three rods. After "the interceptor arrived in the designated area in space and its infrared sensor acquired the target, onboard systems guided the interceptor into the path of the target." It then deployed "a 13-foot [4-m] diameter, aluminum-ribbed net, laced with steel fragments" to enhance the interceptor effective area.[37]

Although the Soviet Union was indeed the first to intercept, without a nuclear charge, a long-range ballistic missile, their proud narrative requires some qualification. The U.S. Army had intercepted and destroyed four shorter range missiles (Honest John, two Corporals, and one Nike-Hercules) at White Sands Missile Range in 1960 *prior* to those Soviet IRBM warheads shot down at Saryshagan. (See Appendix A.) After two more intercepts in 1961 and 1962, the U.S. Army successfully demonstrated a simulated nuclear intercept of an ICBM warhead in 1962.

A direct collision with an approaching warhead often does not require high explosives for its destruction. Such a head-on, hit-to-kill interception relies on precise guidance to place an interceptor in the path of the approaching target warhead. A body of a certain mass moving with a velocity of approximately 2 miles/s (3 km/s) possesses kinetic energy equivalent to the energy produced by detonating the same mass of high-explosive trinitrotoluene (TNT). Significantly higher relative velocities of warheads of strategic ballistic missiles result in three to five times larger kinetic energies per unit mass of the interceptor released in head-on collisions without any explosives.

In contrast, a nuclear intercept requires less precise guidance and relies on detonating an optimized nuclear charge at the right moment of time at some distance from the target warhead. Even before the successful nonnuclear intercepts at Saryshagan in 1961, the Soviet Union had already been working for three years on an operational missile defense system based on nuclear-armed interceptors. All deployed Soviet strategic missile defense systems as well as the U.S. Safeguard, briefly operational in the mid-1970s, would rely on nuclear interceptors. The state of technology prior to the 1980s made this approach simply unavoidable by both the Soviet Union and the United States.

The early Soviet missile defense programs also led to initiation in the 1970s of a major effort in space-based weapons to destroy satellites and intercept ballistic missiles.[38] It culminated in a failed attempt to orbit a prototype of a secret laser space battle station, Polyus,[39] in May 1987. (See Chapter 9.)

---

[36]Government Accounting Office, 1994; Lloyd, 2001, pp. 14, 15, 306; Army Space and Missile Defense Command, 2007.
[37]Government Accounting Office, 1994, pp. 17, 22.
[38]e.g., Semenov, 1996, pp. 419, 420.
[39]Kornilov, 1992; Semenov, 1996, p. 370; Gubanov, 1998, Chapter 34; Lantratov, 2005.

It is ironic that it was Soviet leader Mikhail S. Gorbachev,[40] so much adored by so many opponents of missile defense in the free world, who authorized—although somewhat reluctantly—the launch into orbit of the gigantic 80-metric-ton Polyus. The development of this battle station had predated President Reagan's Strategic Defense Initiative (SDI). At the same time, Gorbachev lectured the world about restraining the arms race, particularly targeting the U.S. SDI.

This inconvenient, for some, historical, event remains almost never mentioned, with rare exceptions,[41] in political discourse, by media commentators, and in publications. It could not even be found in directly related specialized articles[42] where one would expect the important Polyus program to be the sine qua non. Such peculiar factual selectivity illustrates ideological polarization and politicization of missile defense in spite of its utmost importance and consequence.

COMMON SENSE

Today, as during the past 60 years, the concept of mutually assured destruction, or deterrence based on unavoidable deadly retaliation, remains acceptable for many in the United States. They essentially share a view that "the dilemma of steadily increasing military power and [consequently] decreasing national security ... has no technical solution."[43] Some adhere to the "dogma, equating stability with mutual vulnerability, [which] has the consequence that deterrence and defense are regarded as incompatible."[44] At the same time, as was true 60 years ago, others call for and strive to find technical solutions to this national security challenge.

Many place unbounded faith in the power of international treaties. For example, the Comprehensive Test Ban Treaty (CTBT) of 1996 "ended" all nuclear tests. A respected physicist who had been advising the U.S. government on defense issues for many years highly praised the decisive role of "appropriate sanctions" that would be "applied for noncompliance." He argued that the treaty would thus "force rogue states seeking nuclear capability to place confidence in untested bombs."[45]

The real world clearly shows, however, that exclusive reliance on international treaties to contain existential threats fails. An editorial in a leading national newspaper labeled similar wishful thinking and expectations "the arms control illusion."[46] Some countries simply ignored sanctions and treaties

---

[40]Mikhail Sergeevich Gorbachev, b. 1931.
[41]e.g., Hoffman, 2009, pp. 286–288.
[42]e.g., von Hippel, 2013.
[43]Wiesner and York, 1964, p. 35.
[44]Dyson, 1964, p. 16.
[45]Drell, 1999, p. S465.
[46]*Wall Street Journal*, 2014a.

such as CTBT. Both Pakistan and the Democratic People's Republic of Korea (North Korea) tested their atomic bombs after activation of the treaty. In addition, all sanctions and numerous United Nations resolutions have proved impotent to stop Iran on its march to creating nuclear weapons. In the area of missiles, North Korea especially clearly and repeatedly demonstrated that the United Nations could be safely brushed aside by steadily advancing its offensive ballistic missile capabilities, culminating with a successful satellite launch in December 2012.

The evolution of world affairs keeps the pursuit of effective and efficient defenses against missiles, both strategic and tactical, vitally important for national survival. A number of countries today, such as Japan, South Korea, India, and Turkey, have embarked on or are considering development of missile defenses. Cooperation in this area grows among North Atlantic Treaty Organization (NATO) members. A few countries have purchased and deployed weapons with some theater missile defense capabilities. The People's Republic of China will inevitably build strategic missile defense in addition to its energetic program in antisatellite weapons, demonstrated in a test[47] in 2007.

Thirty-five years ago fewer than 10 countries possessed ballistic missiles. Today, more than two dozen countries boast such weapons, some combining them with radical political views or fanatical religious ideologies—with inherent disregard of human life and hatred of the free world—and pursuits of weapons of mass destruction. In addition, extremist militant groups in the Middle East have access to and even manufacture missiles with increasing capabilities.

Clearly, the doctrine of mutually assured destruction does not provide deterrence against certain countries and especially nonstate actors preaching martyrdom. At the same time, many governments and political leaders prefer appeasement and choose, for various reasons, to look the other way. Such lack of action enables and sometimes encourages violence, including missile attacks, as well as defiance of feel-good but inherently ineffective treaties.

This reality puts a spotlight on the main question, succinctly formulated in 1983 by then–U.S. Chief of Naval Operations Admiral James Watkins,[48] "Wouldn't it be better to protect the American people rather than avenge them?"[49]

The United States is not the only country facing this protect-or-avenge dilemma. Israel has become the most threatened country in the world. With tacit acceptance by numerous governments, supported by what a newspaper editorial[50] described as "eruptions of anti-Semitic venom on Europe's streets,"

---

[47]Johnson et al., 2008.
[48]James D. Watkins, 1927–2012.
[49]Baucom, 1992, p. 193.
[50]*Wall Street Journal*, 2014b.

Fig. 1.1 Fragment of an Iraqi Scud ballistic missile on display in the Air Force Museum of the Israel Defense Forces near Beer Sheba, Israel. Iraqi Scuds targeted Israeli population centers and U.S. forces deployed in the region during the First Gulf War in 1990–1991. Photo (2004) courtesy of Mike Gruntman.

for many years emboldened Hamas and Hezbollah militants have been firing missiles into Israel's population centers (see Fig. 1.1).

Had Israeli leaders relied on the international community and treaties for defense, many thousands of lives would have been lost. Instead, the country made and continues to make timely major advances, with the support of the United States, in the development of missile defense layers Iron Dome, David's Sling, Arrow 2, and Arrow 3. "During [the] November 2012 Gaza conflict, Hamas fired a total of 1,500 rockets in [an] attempt to overwhelm Israel's defenses. But Iron Dome shot down 421 of roughly 500 it deemed to be threats...."[51] In 2014, Israel's system demonstrated its effectiveness against a new Hamas onslaught, intercepting 735 rockets and mortar bombs launched at Jewish population centers.[52]

Logic and common sense call for fielding a combination of various means of missile defenses as long as evil exists in the world. No defensive system would ever be 100% reliable and effective. History teaches, however, that even partial protection helps to save lives and provide deterrence by undermining confidence in offensive weapons. Waiting for the silver bullet and delaying or limiting introduction of imperfect defenses are as wise as denying police officers bulletproof vests based on the theory that a criminal may shoot the officer in the face, which would be unprotected.

---

[51]Selinger, 2013, p. 28.
[52]Ben David, 2014.

For science and technology, missile defense poses an enormous challenge often described as hitting a bullet with another bullet. The first successful ballistic missile intercepts by the United States and the Soviet Union demonstrated more than 50 years ago that this challenge could be met.

The seeds for the Soviet accomplishment in 1961 had been planted in a new special design bureau created in Moscow in 1947 as part of the rapidly expanding weapons development complex.

Chapter 2

SPECIAL BUREAU SB-1

ADVANCED WEAPONS

While recovering from the devastation of World War II, the Union of Soviet Socialist Republics (USSR) poured enormous resources into the development of nuclear weapons, ballistic and guided missiles, jet aviation, air defense systems, and electronics. The American Ambassador to Moscow from 1946 to 1949, Walter Bedell Smith,[1] observed that "the total [Soviet] effort in all fields which contribute to military strength obviously is much greater than should be expected of a nation which for the first time in history is without any strong neighbor on the entire Eurasian land mass."[2]

The Soviet ideological outlook provided the foundation for "the irreconcilable hostility of communism toward the capitalist world."[3] As Moscow dictator Joseph (Iosif) V. Stalin[4] (Fig. 2.1) emphasized in an important speech in February 1946, the recently ended world war "was the inevitable result of development of world economic and political forces based on present day monopolistic capitalism." He asserted a common Marxist view that "the capitalist system ... [inherently] contains elements of a general crisis and military clashes."[5]

For many years the Cold War was dangerously close to turning hot. After World War II, the Soviet Union forced communist regimes on Eastern Europe, harshly suppressing dissent. In Western Europe, national communist parties, which were subservient to the Kremlin, fought against democratic restoration and revitalization of their countries' economies by the U.S. Marshall plan. The Soviet Army blockaded Berlin in the summer of 1948 and tried to strangle the city. The U.S. and British airlift brought supplies, which broke the blockade.

---

[1] Walter Bedell Smith, 1895–1961.
[2] Smith, 1950, p. 320.
[3] Smith, 1950, p. 320.
[4] Iosif Vissarionovich Stalin, 1878–1953.
[5] *Pravda*, 1946, p. 1.

Fig. 2.1 Soviet dictator Joseph (Iosif) V. Stalin during the Potsdam Conference in Germany, 20 July 1945. Photo courtesy of National Archives and Records Administration.

In 1956, Soviet soldiers crushed the insurrection in Hungary. That same year, Polish communists used bullets to quell unrest in their country. New tensions in the divided Germany culminated in the erection of the Berlin Wall in 1961.

In Asia, mainland China transformed into a brutal Marxist regime under Mao Zedong[6] following defeat of the nationalists led by Chiang Kai-shek[7] in the civil war. Chinese communists not only relied on backing by the Soviet Union, but also enjoyed important support from communists and fellow travelers in Western countries, especially the United States. The latter engaged in a coordinated propaganda war that effectively isolated the nationalist government of the Republic of China.[8] Then, communist troops attacked South Korea in 1950, leading to a protracted bloody conflict there. The war in Vietnam followed in the 1960s in response to a seemingly unstoppable communist expansion. Military tensions between the People's Republic of China (PRC) and the Republic of China (ROC) in Taiwan continued for many years, with communists periodically shelling the island of Quemoy (Kinmen) from the early 1950s into the 1970s.

During and shortly after World War II, the Soviet government created an extensive network of organizations for the pursuit of advanced weapons.

---

[6]Mao Zedong (Mao Tse-tung), 1893–1976.
[7]Chiang Kai-shek (Jian Zhongzheng), 1887–1975.
[8]Chiang Kai-shek, 1967.

These new establishments and their successors provided the foundation for the missile defense effort of the 1950s.

Formed in 1943, the Council for Radars directed development of radar systems and related electronics. Another especially important government body, Committee No. 1, oversaw the atomic bomb program. On 13 May 1946, the USSR Council of Ministers issued a special decree, No. 1017-419ss, Matters of the Rocket Weapons, signed by Stalin.[9] The decree formed essential organizational elements and directed ballistic missile and space establishments for many years. The government also activated a new Special Committee on rocket weapons, later reorganized into Committee No. 2.

In June 1946, another government decree, No. 12866-525, formed Design Bureau No. 11 (KB-11) at a secret location, Arzamas-16 (also known as Sarov), to accelerate the development of atomic weapons. KB-11 and its rival, Scientific-Research Institute No. 1101 (NII-1011), which opened later in Chelyabinsk-70 in the southern Ural mountains, would design and build nuclear warheads, including for air and missile defenses.[10] One month later, in July 1946, the government approved an ambitious three-year plan for introducing a new generation of radar.

The 1946 decree Matters of the Rocket Weapons provided a comprehensive plan for the development of rocket and missile technology, which marshaled enormous resources of the totalitarian state. The government specifically identified "the top priority goal to reconstruct, using domestically available materials, the [World War II German] rockets V-2 (long-range guided rocket) and Wasserfall (antiaircraft guided missile)."

The decree also activated a new missile test site, the State Central Test Range No. 4, which was near the settlement Kapustin Yar on the eastern side of the Volga river, 70 miles (110 km) southeast of Stalingrad (Volgograd). Initially, the test site, or poligon in Russian, supported work on ballistic missiles. With time, activities at Kap Yar, as those in the know commonly called the proving ground, expanded to development of air defense systems and various tactical and guided missiles. Trials of bigger intercontinental ballistic missiles transferred to a new range, Scientific-Research Test Range No. 5 (Nauchno-Issledovatel'skii Poligon No. 5, or NIIP-5), also known as the Tyuratam missile range (later called Baikonur), in Kazakhstan in the mid-1950s.

The government ordered various ministries to assign their research, development, and manufacturing facilities to a new massive effort in rocketry. The decree also provided "increased salaries for specially qualified specialists." In addition, the authorities directed institutions of higher learning to create dedicated programs to educate and train engineers and scientists for the new field.

---

[9] e.g., Gruntman, 2004, pp. 275–277; Baturin, 2008, pp. 30–36.
[10] KB and NII stand for konstruktorskoe byuro (design bureau) and nauchno-issledovatel'skii institut (scientific-research institute), respectively, in Russian. See the list of acronyms and abbreviations in Appendix B and their pronunciation in Appendix C.

As ordered by the rocket weapons decree, the military activated "a special artillery unit to learn and master the preparations and launching of the V-2 rockets." The strategic rocket forces would grow from this missile unit. As was common for a totally controlled socialist society, the decree of the Council of Ministers also dealt with numerous minute items such as allocations of food rations, transport, and living quarters.

COMMUNIST PRINCELING

Sergo Beria[11] (Fig. 2.2) graduated from the S.M. Budenny Military Academy of Communications in Leningrad (today's Saint Petersburg) in 1947.[12] The 23-year-old Sergo (or Sergei, in the Russified form of his first name) already held the rank of Soviet Army captain. He was the son of Lavrentii P. Beria,[13] Stalin's trusted henchman and the feared head of the notorious secret service KGB from 1938 to 1945.[14]

Fig. 2.2 Sergo Beria, the son of Lavrentii P. Beria, in the 1940s. Photograph from http://www.gordon.com.ua/books/heroes/beriya/; (accessed 6 September 2011); courtesy of Dmitrii Gordon, Kiev, Ukraine.

---

[11]Sergei (Sergo) Lavrentievich Beria, 1924–2000, also known as Sergei Alexeevich Gegechkori. The last name Beria is sometimes transliterated as Beriya.
[12]Beria, 1994, p. 429.
[13]Lavrentii Pavlovich Beria, 1899–1953; like Stalin, Beria was an ethnic Georgian from the Caucasus.
[14]Komitet Gosudarstvennoi Bezopasnosti (Committee for State Security), or KGB, was the main organization in the Soviet Union responsible for various matters of state security and repression. The name and composition of the organization evolved with time. It was known as VChK (1917–1922), GPU and OGPU (1922–1934), NKVD (1934–1941), NKVD-NKGB (1941), NKVD (1941–1943), NKGB (1943–1946), MGB (1946–1953), MVD (1953–1954), and finally KGB starting in 1954 (Kokurin and Petrov, 1997). I use the name KGB throughout this publication unless it is a direct quote with the contemporary name of the organization.

U.S. Ambassador to Moscow W.B. Smith noted that "[Lavrentii] Beria renovated and expanded the agencies of state security beyond anything that had been known before, and under his leadership they have become increasingly efficient." In the Soviet Union, explained Smith, "the agencies of state security have complete supervision of every phase of national life."[15]

Before his transfer to Moscow in 1938, Lavrentii Beria worked in the state security apparatus in the Soviet republics of Azerbaijan and Georgia in the Caucasus from 1921 to 1931. He then served as the Communist Party chief in his native Georgia from 1931 to 1938. For 12 years, from 1941 to 1953, Beria was deputy chairman of the USSR Council of Ministers. After World War II, he directed expansion of the Soviet weapons programs, including crash development of the atomic bomb, ballistic missiles, and air defenses.

Sergo Beria's diploma project (equivalent to a master's thesis) at the Military Academy of Communications focused on design of an air-launched antiship missile, controlled by radio. An experienced radio engineer, Pavel N. Kuksenko (Fig. 2.3),[16] guided the student as his advisor.

The future chief designer of the first Soviet missile defense system, Grigorii V. Kisun'ko,[17] at that time served on the staff of the Academy. He recalled details of the master's thesis defense by the young Beria.[18]

Fig. 2.3 Pavel N. Kuksenko in 1947. Photograph from K.S. Al'perovich, *Gody Raboty nad Sistemoi PVO Moskvy* **... , 2003, p. 10.**

---

[15] Smith, 1947, p. 126.
[16] Pavel Nikolaevich Kuksenko, 1896–1980.
[17] Grigorii Vasil'evich Kisun'ko, 1918–1998.
[18] Kisun'ko, 1996, pp. 187–189.

After completing his presentation to the committee, Sergo stepped out of the room. At that point, a respected academy professor and one-star general, who was a leading specialist in radio transmitters, suddenly commented that

> everything had been thought through and decided about this mother's darling. ... Isn't it clear that the thesis [work] was done and [the thesis itself] written by dozens if not one hundred specialists. ... Let us recognize this project as outstanding and award the author [Sergo Beria] the [master's] degree diploma with the [highest] grade of excellence and thus finish this comedy.

An awkward dead silence fell among the committee members. The head of the academy, Konstantin K. Murav'ev, who conducted the defense proceedings, tried to find a safe exit from a dangerous situation by characterizing the comment as "an inappropriate joke."

The committee unanimously approved the thesis and gave it the highest grade. In a few months, the Communist Party kicked out the offending general from its ranks, calling him a Trotskyite, and discharged him from the Armed Forces without a retirement pension. Several instructors at the academy were arrested on charges of being spies. The Army also sent academy head Murav'ev into early retirement.

In the late 1940s, Boris E. Chertok[19] worked in Scientific-Research Institute No. 88 (Nauchno-Issledovatel'skii Institut No. 88, or NII-88) established in Podlipki, near Moscow, by the Matters of the Rocket Weapons decree. This area, later known as the town of Kaliningrad and now Korolev, would accommodate a number of leading research and development organizations and grow to a special prominence in Soviet ballistic missile, rocket, and space establishments.[20] The suburban railroad station retains its original name, Podlipki, to this day. Those who worked in defense, rocketry, and space continued to commonly call the location Podlipki instead of the new town name.

The NII-88 institute became the home of the initial Soviet effort to build analogs of German World War II ballistic and antiaircraft missiles. Although work on the R-1, the Soviet variant of the V-2 ballistic missile, received most attention in historic publications, at that time NII-88 also engaged in a large-scale effort in guided and unguided missiles for air defense. Chertok would serve for many years as a deputy to Sergei P. Korolev, the chief designer who achieved the first intercontinental ballistic missile (ICBM); launched the first artificial satellite, Sputnik; and put the first man into space.

---

[19]Boris Evseevich Chertok, 1912–2011.
[20]e.g., Gruntman, 2004, pp. 279–283.

Boris Chertok recalled that all-powerful and feared Lavrentii Beria had telephoned the People's Commissar (or Minister in later terminology) of Armaments, Dmitrii F. Ustinov,[21]

> and "asked" him to gather specialists to listen to the proposal [of the antiship missile put together by his son Sergo Beria]. [The meeting] was called, not for an evaluation of the [merits of the] proposal, but to decide where to implement it. There was no question at all of assessing the proposal in the sense whether it was worthy of pursuing.

The proposed missile, noted Chertok, had "much in common with the [German] Wasserfall [antiaircraft missile concept]. ... A seasoned expert immediately saw many naive and childish suggestions as well as earlier rejected approaches."[22] The similarity with Wasserfall, noted by Chertok, was apparently in the proposed semi-active homing head of the missile.

Chertok represented NII-88 at this meeting. He smartly emphasized that his institute lacked the expertise in turbojets and radar required for the project. In contrast, Mikhail L. Sliozberg,[23] the chief engineer of a leading radar institute of the Ministry of Armaments, Scientific-Research Institute No. 20 (NII-20) in Moscow, enthusiastically offered his organization's support. He "argued that his institute had all the capabilities for implementation of the radar part of the proposed program."[24]

NEW SPECIAL BUREAU

On 8 September 1947, Joseph Stalin signed the decree of the USSR Council of Ministers No. 3140-1028, establishing a new independent organization in the Ministry of Armaments, Special Bureau No. 1 (Spetsial'noe Byuro No. 1, or SB-1). The decree tasked SB-1 with developing the proposed air-launched antiship missile and assigned the codename *Kometa* to the program (*kometa* means *comet*). The thesis advisor of the young Beria, Pavel Kuksenko, and Sergo himself became director and chief engineer, respectively, of the new bureau. The SB-1 and its successor organizations would later lead development of Soviet air and missile defenses.

Not surprisingly, SB-1 promptly took over parts of the buildings and territory of Sliozberg's NII-20. The institute was conveniently located 5 miles (8 km) from the center of Moscow at a road fork where the Volokolamsk highway (Volokolamskoe Shosse) branched out from the Leningrad highway (Leningradskoe Shosse). In fact, Sergo Beria had already become head of a new radar department in this institute prior to Stalin's decree.

---

[21] Dmitrii Fedorovich Ustinov, 1908–1984.
[22] Chertok, 1994, p. 271.
[23] Mikhail Lvovich Sliozberg, 1906–1970.
[24] Chertok, 1994, p. 271.

Minister of Armaments Ustinov ordered NII-20 and its supporting Plant No. 465 to treat all design, machining, procurement, and personnel requests of SB-1 as top priority. He also directly assigned 40 of their engineers and technicians to the new special design bureau.[25]

Three years later, in 1950, the growing SB-1 absorbed all the remaining facilities and buildings of NII-20, whose leaders had so unwisely volunteered their organization's expertise to Sergo Beria. Then, the government finally relocated the institute to a Moscow suburb.

SCIENTIFIC-RESEARCH INSTITUTE NO. 20 (NII-20)

The NII-20 institute of the Ministry of Armaments emerged from the World War II effort to build gun-laying (pointing) radar for air defense fire control systems.

In February 1942, the Soviet government established a new Plant No. 465 in Moscow. By the fall of that year, the plant had more than 1000 employees, including many scientists, engineers, and technicians. They reverse-engineered and produced the SON-2ot radar, the analog of the British gun-laying radar GL MkII supplied to the Soviet Union through the lend-lease program.

The new organization played an important role in building up the Soviet radio industry. In 1943, the plant contributed a core group of specialists for the new Scientific-Research Institute No. 108 (NII-108), which would play a major role in advancing radar science and engineering. Then, 100 engineers transferred to the town of Fryazino, 5 miles (8 km) from Podlipki, to expand a local plant and establish the new Scientific-Research Institute No. 160 (NII-160). The institute, later known as NII Istok, focused on the science and engineering of radio vacuum tubes.

In 1945, government orders established a new Central Design Bureau No. 20 (TsKB-20), soon collocated with Plant No. 465. In the middle of 1946, the design bureau was reorganized into Scientific-Research Institute No. 20 (NII-20), with Plant No. 465 supporting the institute's pilot production. The institute took many scientists and engineers from the plant. Initially, Mikhail L. Sliozberg headed both organizations. In 1958, NII-20 formally absorbed the plant, which became its manufacturing development arm.

NII-20 introduced many air defense and tactical missile defense systems, primarily for army field units, such as SA-4 (Fig. 2.4), SA-8,

---

[25]Davydov, 2009, pp. 86–88.

SA-12a, SA-12b, SA-15, and SA-23. It was also among the first to study the feasibility of strategic missile defense in the late 1940s. In 1966, the government reorganized NII-20 into the Scientific-Research Electromechanical Institute, or Nauchno-Issledovatel'skii Elektromekhanicheskii Institut (NIEMI). Today NIEMI is part of the Almaz-Antei conglomerate, the main development organization of Russian air and missile defenses.

Fig. 2.4 Launch unit 2P24 with antiaircraft missiles 3M8 of the Army field air defense system Krug (SA-4; deployed in 1965) near the main building of NIEMI. Photograph (2008) courtesy of *borbor* (Google's Panoramio).

The new secret SB-1, also known as Organization Post Box 1323,[26] rapidly expanded. Government orders transferred many experienced engineers from other organizations to the new program and assigned young graduating engineers and military officers to the effort.

In addition, the bureau employed two other special categories of highly skilled individuals "supplied" by the predecessor of the KGB. One group was composed of German specialists forcibly brought to the Soviet Union after World War II. In 1946, Soviet secret services in the occupied East Germany rounded up a large number of German scientists and engineers with expertise in air defense, ballistic missiles, and other military technologies.[27] Special trains took them and their families under guard to the Soviet Union to work in

---

[26]The government assigned special post box codes (pochtovyi yashchik, or literally a post box) to organizations involved in secret work.

[27]Groettrup, 1959; Albring and Vinke, 1991; Magnus, 1993; Chertok, 1994; Riehl and Seitz, 1996; Gruntman, 2004.

various weapon-related industries. (Several nuclear physicists and chemists had been brought to the USSR earlier to aid in the enrichment and processing of uranium.) In contrast, the United States (through operation Paperclip), Great Britain, and France utilized German experts through voluntary contracts.[28]

Future leading specialist in interceptor warheads for missile defense Yuri A. Kamensky[29] joined SB-1 in the early 1950s. He recalled that every day buses brought German engineers to SB-1 from the town of Khimki[30] 10 miles away. "One could easily recognize the Germans among our [SB-1] engineers," wrote Kamensky. "They always wore white shirts and had a walk in the yard during lunch breaks."[31] Some Germans also resided in a special settlement in Tushino on the outskirts of Moscow.[32]

The "imported" Germans usually lived in isolation from the Soviet population. By the mid-1950s, after draining off all their knowledge, the authorities sent them back to the Soviet-controlled East Germany. Although some stayed in the communist part of Germany, others promptly escaped to the West.[33] Debriefing of the latter by U.S. and British intelligence provided important information about post–World War II early development of ballistic missiles, air defense, atomic bombs, and other weapons in the Soviet Union.[34]

In addition to the Germans, every day KGB guards delivered another unusual category of specialists to SB-1: Soviet prisoners with technical expertise. During internal purges in the 1930s and 1940s, the authorities arrested numerous scientists and engineers. Beatings and harsh treatment extracted "confessions," with many thousands being shot, banished, or imprisoned.[35]

Those specialists who survived ended up in special prisons—sharashkas. A sharashka was a prison that was turned into a research and design establishment. Thousands of imprisoned scientists and engineers worked in various sharashkas,[36] which at least provided them some hope of survival. A very different fate awaited those millions sent straight to GULAG labor camps.[37] There, malnutrition, excessive hard work, savage living conditions, starvation, and abuse by the guards took a tremendous toll. Many millions perished without a trace on this march to a socialist paradise.

---

[28]Lasby, 1971; Gruntman, 2004, pp. 157–163.
[29]Yuri Alexandrovich Kamensky, b. 1926.
[30]Similar to Podlipki, Khimki boasted a growing cluster of missile-related research, development, and industrial establishments (Gruntman, 2004, p. 279).
[31]Kamensky, 2002.
[32]Falichev and Sukharev, 2012, p. 3.
[33]Groettrup, 1959; Albring and Vinke, 1991; Magnus, 1993; Riehl and Seitz, 1996; Gruntman, 2004, pp. 163, 164.
[34]e.g., Central Intelligence Agency, 1959, p. 7.
[35]Solzhenitsyn, 1973; Courtois et al, 1999; Gruntman, 2004, pp. 273–275; Gruntman, 2007, pp. 42, 43.
[36]e.g., Saukke, 2006; Kachur and Glushko, 2008, Chapter 5; Kerber, 2008–2010.
[37]Solzhenitsyn, 1973; Courtois et al, 1999.

This book includes a number of scientists, engineers, and managers in Soviet air and missile defenses, guided and ballistic missiles, and space who went through sharashkas, survived, and rose to prominent positions. Among them were Aksel' I. Berg, Valentin P. Glushko, Korolev, Kuksenko, Alexander L. Mints, Dominik D. Sevruk, Dmitrii L. Tomashevich, and Andrei N. Tupolev.[38]

Oleg V. Golubev,[39] a future specialist in the guidance of missile defense interceptors, joined design bureau KB-1 as a young engineer in January 1951. (SB-1 changed its name to KB-1 in 1950.) Director Kuksenko assigned him to a theoretical group working on the guidance of air defense missiles. Golubev later recalled that he had been amazed and terrified to find out that the head of his laboratory[40] was a prisoner:

> He did not have a surname, just a number, 0-42, and the first name and the patronymic name, Sergei Mikhailovich. He was ... a middle age man dressed, as all his prisoner-colleagues, in a standard gray suit with a gray shirt and a gray tie. We were strictly forbidden to have any contacts with him which were not job related. Naturally, we were stressed by such a situation, especially in the beginning. Later, we got accustomed to this environment.
>
> Sergei Mikhailovich happened to be, as we realized later, a brilliant engineer and scientist. ... [He] taught us to be engineers and, in fact, he created a [scientific] school of control and guidance of interceptor missiles in KB-1. ...
>
> Guards daily brought Sergei Mikhailovich, as all other prisoners, [to KB-1]. Once, he did not show up with everybody else.... He appeared two hours later, with a pale face ... "I was released from being a prisoner," he said with a trembling voice, "and was awarded the Order of the Red Banner."[41]... We were astonished. ... Soon, the surname of Sergei Mikhailovich became known—Smirnov.[42]

Many accomplished Soviet scientists toiled as prisoners in KB-1. They included,[43] for example, physicist Lev A. Sena, mathematician and corresponding member of the USSR Academy of Sciences Nikolai S. Koshlyakov,

---

[38] Aksel' Ivanovich (Ioganovich) Berg, 1893–1979; Valentin Petrovich Glushko, 1908–1989; Alexander Lvovich Mints, 1895–1974; Dominik Dominikovich Sevruk, 1908–1994; Dmitrii Lyudvigovich Tomashevich, 1899–1974; Andrei Nikolaevich Tupolev, 1988–1972.

[39] Oleg Vasil'evich Golubev, b. 1924.

[40] Laboratories as well as departments were common administrative units in many Soviet research and development organizations.

[41] One of the highest noncombat state decorations in the Soviet Union.

[42] Golubev, 2003, p. 572.

[43] Kisun'ko, 1996, pp. 208, 209; Semenov, 2008, p. 427.

and dynamicist Georgii V. Korenev.[44] Some specialists continued to work there after being released from incarceration.

Golubev recalled that "ascetic" Korenev "always wore a weathered leather jacket and aviator helm."[45] Two decades after those events, the author of this book, a freshman student, took a course in theoretical mechanics taught by Korenev, who had not changed his habits. He always appeared and lectured in a well-weathered leather jacket.

This unique combination of Soviet scientists, engineers, and technicians as well as enemy-of-the-state prisoners and forcibly imported Germans, all overseen by the omniscient and omnipotent KGB and guided by the Communist Party, toiled on new weapon systems in the design bureau headed by former prisoner Kuksenko and communist princeling Sergo Beria.

KOMETA KS-1

The special bureau SB-1 led the development of the entire Kometa antiship missile system. Several other major organizations contributed as principal contractors. In particular, the aviation design bureau of Artem I. Mikoyan (Fig. 2.5) and Mikhail I. Gurevich,[46] who gave the letters *M* and *G* to the name of the MiG aircraft family, designed the missile airframe.

Andrei Tupolev's Tu-4 bomber served as a launch platform for the Kometa missiles. During World War II, three damaged U.S. B-29 bombers landed in the Soviet Far East after raids on Japan and Japan-occupied Manchuria in 1944.[47] (The fourth Superfortress crashed in the wilderness of Siberia.) The Soviets never returned the bombers to their U.S. ally. On direct orders from Stalin, Tupolev reverse engineered the plane in two years,[48] producing the strategic bomber Tu-4. The airplane, known in the West as Bull, made its first surprise appearance at an air demonstration in 1947.

A Tu-4 bomber could carry two Kometa missiles under its wings (Fig. 2.6). The plane released the missile at an altitude of 10,000–13,000 ft (3–4 km). Separated from the plane, the missile first dropped 300–400 ft (100–130 m) in altitude, accelerated, and then overtook the plane.[49] Then, the onboard operator guided it to a target ship up to 60 miles (100 km) away. Initial missile capture into the narrow guiding radio beam presented a major technical challenge.

---

[44]Lev Aronovich Sena, 1907–1996; Nikolai Sergeevich Koshlyakov, 1891–1958; Georgii Vasil'evich Korenev, 1902–1980.
[45]Golubev, 2003, p. 572.
[46]Artem Ivanovich Mikoyan, 1905–1970 (his older brother, Anastas I. Mikoyan, would serve as chairman of the Presidium of the Supreme Soviet of the USSR from 1964 to 1965); Mikhail Iosifovich Gurevich, 1893–1976.
[47]Hays, 1990.
[48]e.g., Saukke, 2006, p. 143.
[49]Vlasko-Vlasov, 2002, p. 26.

Fig. 2.5 Leading Soviet aircraft designers at the airfield of the V.P. Chkalov Central Air Club in Tushino (Moscow) in July 1947. Left to right: Alexander S. Yakovlev (Yak family of aircraft), Andrei N. Tupolev (Tu family), Semen A. Lavochkin (La family), and Artem I. Mikoyan (the letter "M" in the MiG family). The design bureaus of Tupolev and Mikoyan built, respectively, the airplane platform for launching Kometa missiles and the missile airframes. In the 1950s, Lavochkin played an important role in development of the first generations of Soviet antiaircraft missiles. Photograph courtesy of ITAR-TASS.

A radar system on the Tu-4, Kobalt-M,[50] resembled its prototype, the U.S. AN/APQ-13 installed on B-29 bombers. It continuously illuminated the target at the 3-cm wavelength (10-GHz frequency). Usually the missile flew at a cruise altitude of 1300 ft (400 m) over the water surface. For terminal guidance at distances of 10 miles (16 km) from the target, the Kometa's sensor picked up radar signals reflected from the ship. The missile then homed in on the illuminated target autonomously.

The SB-1 built air-to-surface, ship-to-ship, and surface-to-surface variants of the subsonic Kometa. The government designated the air-to-surface missile KS-1 (krylatyi snaryad, or winged projectile).

A human test pilot squeezed into a specially built small cabin and manually flew the first Kometa missiles. After 150 such flights, trials of pilotless missiles began in 1952, targeting the old cruiser *Krasnyi Kavkaz*. The cruiser automatically circled in the designated area, with its crew temporarily evacuated from the ship by accompanying torpedo boats.[51] In the final test on 21 November 1952, the fully armed Kometa sank the cruiser. The 27-ft (8.3-m) long missile

---

[50]Shirokorad, 2003, p. 335.
[51]Vlasko-Vlasov, 2002, p. 25.

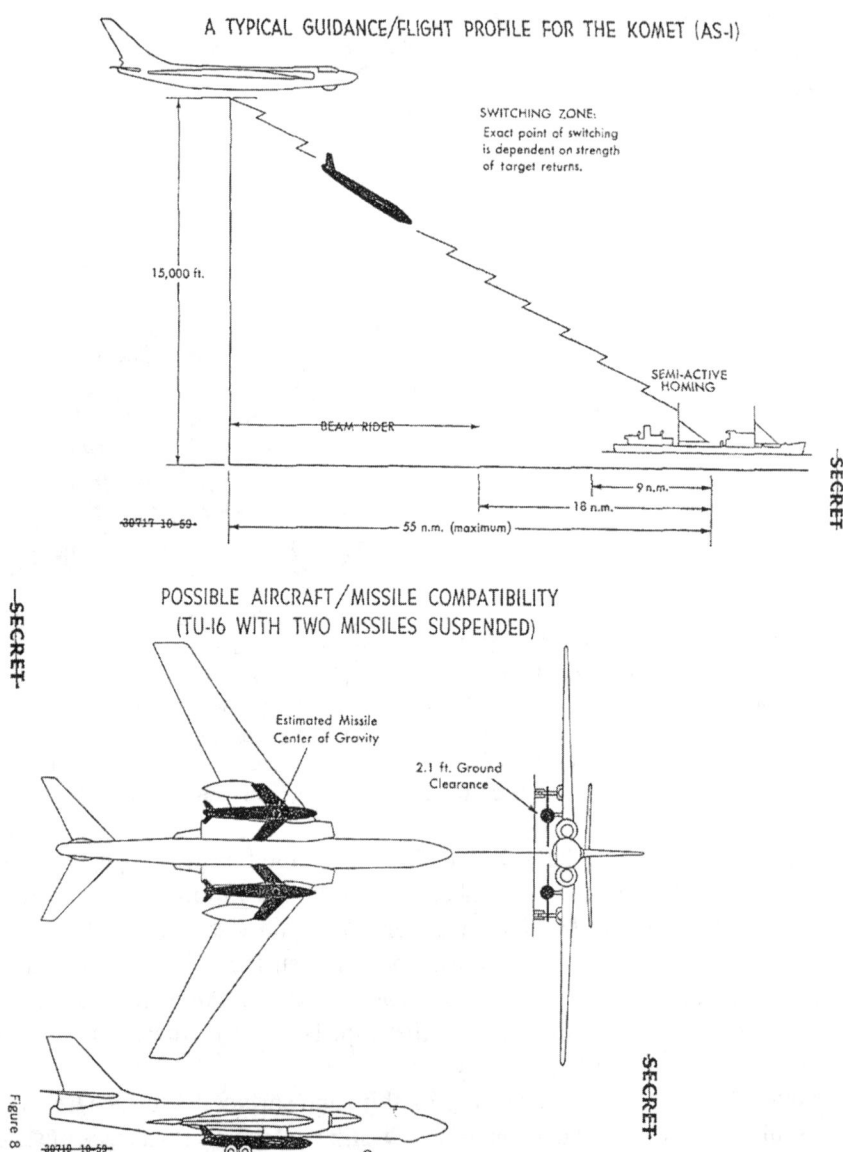

Fig. 2.6 Typical guidance profile (top) and missile mounting on an aircraft (bottom) of the air-launched antiship guided missile Kometa KS-1 (AS-1) developed by SB-1. Figures 7 and 8 (following page 11) from Central Intelligence Agency, 1959.

had a mass of 3 t and carried a 2200-lb (1000-kg) warhead. Its wings swept back 55 deg with a span of 15.7 ft (4.8 m).

Kometa became the first Soviet air-to-sea guided missile system. The government decorated both Kuksenko and Sergo Beria with its highest award, the Order of Lenin. Many leading specialists and managers received the prestigious Stalin Prize, which also included significant monetary bonuses. Operational deployment of Kometa missiles on Tu-4 bombers followed in September 1953. By that time, Kuksenko had been demoted and both Berias, the father and the son, were arrested, awaiting execution and banishment, respectively, in an internal Kremlin struggle for power following the death of Stalin.

The KS-1 Kometa missile became known in the West as the AS-1 Kennel. The CIA's National Intelligence Estimate noted in March 1957 that "a missile system of the Kometa type, or other air-to-surface system with similar characteristics, has reached at least final flight test stage."[52] Two years later, a new NIE more accurately stated that

> [t]he USSR has had operationally available since 1956–1957 a subsonic antiship system (AS-1) with a maximum range of about 55 n.m. It achieves a speed of Mach 0.8 and can carry a nuclear or possibly HE [high explosive] warhead of about 3,000 pounds, with a CEP [circular error probable] of about 150 feet against well-defined radar targets. It uses a guidance system known as Komet (a beam rider with semiactive homing ...), the characteristics of which limit its employment almost exclusively to ships at sea.[53]

During its lifetime, the KS-1 Kometa went through many upgrades. In 1955 it was adapted to a new launch platform, the new advanced Tupolev bomber Tu-16 (Badger).[54] The missile retired from service in the late 1960s. The Soviet Union also sold KS-1 missiles to Egypt and Indonesia. The successful development of Kometa provided experience and laid the foundation for designing and building many other air-to-sea, air-to-surface, air-to-air, and surface-to-surface missiles by the successor organization of the SB-1.[55]

By the time of the first trials of Kometa in 1952, the special design bureau SB-1 already was focusing its main effort on a much more challenging and bigger task—building an air defense system for the Soviet capital, Moscow.

---

[52]Central Intelligence Agency, 1957, p. 17.
[53]Central Intelligence Agency, 1959, p. 11.
[54]Gordon and Rigmant, 2004, pp. 37–38.
[55]Vlasko-Vlasov, 2002.

Chapter 3

STALIN'S ORDER

A SUMMONS TO THE KREMLIN

One night in 1950, Soviet dictator Joseph Stalin called director of SB-1 Pavel Kuksenko to the Kremlin. Stalin wished to discuss a matter of utmost state importance. A few more meetings followed. Then in August 1950, Stalin gave Kuksenko's special bureau a challenging task: to build an "impenetrable" air defense system for the Soviet capital, Moscow.[1] The new system was to defend the city against a massive raid of up to 1000 bombers. The government named the program Berkut (a golden eagle).

That year, 1950, witnessed a sharp military confrontation between the free world and totalitarian socialist countries. In Korea, communist troops attacked the Republic of Korea in late June, starting a bloody war. The first U.S. units joined fighting in July, suffering significant casualties. There was concern that the conflict could escalate beyond the region. Rushing development of Berkut and improving the USSR's air defenses would remain among the highest priorities of the Kremlin for a number of years.

A 1958 study by the CIA's Office of Scientific Intelligence put this effort into perspective:

> Soviet leaders seem to have been impressed with three important facts at the end of World War II, i.e., the role of airpower during the war, U.S. possession of the atomic bomb and means for its delivery, and the absence of air defenses of the USSR. Faced with these facts, the USSR undertook an ambitious program to provide air defenses. It is believed that the highest priority was given to their development. It is difficult to assess the relative priority of Soviet atomic weapon research in comparison to air defense, but clearly their bomber development to provide

---

[1]Kisun'ko, 1996, pp. 196–198; Belous et al., 2009, pp. 50, 51.

a delivery capability did not enjoy near the priority accorded air defense. Perhaps the first evidence of this large effort was the rapid growth of their jet fighter strength which had reached significant proportions prior to 1950. By 1951, the results of their efforts in electronics for air defense began to appear and important advances have marked each year since that time.[2]

After World War II, the NII-88 institute in Podlipki became the home to a large portion of the Soviet work on air defense systems. There, three departments concentrated on reconstruction of the German antiaircraft guided missiles Wasserfall (the Soviet copy designated R-101) and Schmetterling (R-105) and the unguided solid-propellant Taifun (R-110). An effort to build another German missile, Rheintochter, was limited.

The first variants of the R-101 missile reached the Kapustin Yar test range in 1948, followed by the R-105 and R-110 in 1949. In 1951, the government focused on development of the highest priority Berkut, restructured its entire air defense effort, and terminated work on antiaircraft missiles in NII-88.[3] The derivatives of Taifun prototypes would merge into designs of unguided missiles by other organizations later in the 1950s.

The beginning of Berkut coincided with a reorganization of the massive armaments complex in the country. Of particular importance, Stalin established a new Special Committee to coordinate the development of advanced weapons, with his close associate, Lavrentii Beria, as its head (Fig. 3.1). Under this arrangement, Beria exercised enormous power directing postwar programs in nuclear weapons, long-range ballistic missiles, and air defense systems. Following Stalin's death in March 1953, the Soviet government promptly reconfirmed Beria's responsibilities in weapon development by a new decree, No. 697-355ss/op, On the Direction of Special Works.[4] He would soon be arrested, however, and then executed during an internal struggle for power in the USSR.

The First Chief Directorate of Beria's Special Committee oversaw work on nuclear weapons. The top secret decree of the USSR Council of Ministers, No. 307-144ss from 3 February 1951, soon activated another Third Chief Directorate (Tret'e Glavnoe Upravlenie, or TGU) of the special committee with a particular task of "development, design, and production of the means of the air-defense system 'Berkut.'"[5] The Council of Ministers appointed Vasilii M. Ryabikov to direct the TGU; Sergei I. Vetoshkin became his first

---

[2]Central Intelligence Agency, Office of Scientific Intelligence, 1958, p. 5.
[3]Evtif'ev, 2000, pp. 25–48; Pervov, 2001, pp. 35–47; Shirokorad, 2003, pp. 107–109, 248–252.
[4]Ivkin and Sukhina, 2010, pp. 321–323.
[5]Ivkin and Sukhina, 2010, pp. 211–213.

Fig. 3.1 Lavrentii P. Beria (left) with the revered communist dictator, "the father of the peoples," Joseph Stalin (right) in the Kremlin in 1936. A long-time close associate of Stalin, Beria headed the Soviet state security apparatus for many years. He also directed development of nuclear, ballistic missile, and air defense weapons after World War II. Beria's son Sergo initiated and led work on the air-launched antiship missile Kometa at SB-1. Photo courtesy of ITAR-TASS.

deputy, and Alexander N. Shchukin assumed the role of deputy responsible for science and engineering.[6] The TGU quickly established a specialized organization in the Ministry of Defense for quality control and acceptance of weapons procured by the directorate, which would grow to more than 3000 people in a mere half a year.[7]

---

[6]Vasilii Mikhailovich Ryabikov, 1907–1974; Sergei Ivanovich Vetoshkin, 1905–1985; Alexander Nikolaevich Shchukin, 1900–1990.
[7]Semenov, 2008, p. 411.

DECREE OF THE USSR COUNCIL OF MINISTERS NO. 2837-1349[8] PROBLEMS OF ROCKET WEAPONS

August 4, 1951
Top Secret
Special Importance

The Council of Ministers of the Union of SSR decrees:

1. Because development of long-range rockets R-1 [SS-1], R-2 [SS-2], and R-3 and organization of series manufacturing of the rocket R-1 are related to work on [systems] "Berkut" and "Kometa," assign the task of overseeing the work of ministries and organizations on these rockets to Deputy Chairman of the USSR Council of Ministers Comrade L.P. Beria. ...

Chairman of the Council of Ministers of the USSR   J. Stalin
Administrator of the Council of Ministers of the USSR   M. Pomaznev

DESIGN BUREAU KB-1

On 9 August 1950, the USSR Council of Ministers issued decree No. 3389-142bss/op to implement Stalin's order to Kuksenko to develop Berkut. The wheels of state bureaucracy began to turn. Three days later, Minister of Armaments Dmitrii F. Ustinov followed with order No. 427, reorganizing and expanding Special Bureau SB-1 into Design Bureau No. 1 (Konstruktorskoe Byuro No. 1, or KB-1) as the lead organization to build air defenses for the Soviet capital. Three more days went by and Ustinov ordered the scientific-research institute NII-20 and its supporting Plant No. 465 to hand over all their remaining premises to the new KB-1 and relocate, within 10 days, to new quarters in the Moscow suburb of Kuntsevo 7 miles (11 km) away.

This surprise transfer delivered a major disruptive blow to the institute and its specialists. The new facilities were smaller, although the area provided space for growth in the future. Its previous tenant, another research institute of the Ministry of Armaments, NII-6, which specialized in gunpowder and explosives, had just been evicted. Both organizations, NII-6 and NII-20, would later contribute to the first ballistic missile intercept.

After relocation of NII-20, commute time for many of its employees increased dramatically. At that time, people could not resign their positions in the defense industry at will and seek other employment. In the Soviet Union,

---

[8]Ivkin and Sukhina, 2010, pp. 247, 248.

the omnipotent state controlled the lives of its subjects and did not allow such freedom to many. Instead, the government allocated six buses to the institute to help its workers to reach public transport connections.[9]

The new KB-1 also took over most of the equipment and many specialists from NII-20 as well as two thirds of employees and all the equipment of the supporting Plant No. 465. As the institute's veterans described it later, they had been "evicted ... in undershorts only."[10] Minister Ustinov promised, however, to rebuild the organization at a new location. With his help, NII-20 significantly expanded as time went by. Under its new name since 1966, the Scientific-Research Electromechanical Institute NIEMI, it designed and built a number of radar systems primarily for army field units and naval ships, including SA-4, SA-8, and SA-15.

Two years after beginning work on Berkut and forming KB-1, the Central Intelligence Agency's (CIA's) analysts accurately observed in a special intelligence estimate (SIE) that the "USSR is carrying out an intensive program for the improvement of its air defense system, with the priority which is probably second only to the Soviet atomic weapons program."[11]

Dmitrii Ustinov, whose ministry oversaw SB-1 (and then KB-1), continuously played a prominent role in the Soviet military industrial complex from the critical days of World War II until his death in 1984. Stalin appointed him the armaments minister during the war when he was only 33 years old. With time, Ustinov's influence reached the highest levels of power. He actively participated in a "palace coup" that ousted Khrushchev from the Kremlin in 1964 and elevated Leonid I. Brezhnev[12] to the top of the Soviet empire. Ustinov served as the USSR Minister of Defense from 1976 to 1984.

While organizing the work on air defense, the government appointed two chief designers for Berkut, Kuksenko and the young Sergo Beria. This was an arrangement similar to the concurrent development of Kometa at KB-1. In August 1950, the same month it established KB-1, the government ordered the transfer of Alexander A. Raspletin[13] (Fig. 3.2) to this new design bureau, making him deputy chief designer of Berkut and head of the radar department.[14]

Prior to the new appointment, Raspletin was in charge of a main radar unit in the Central Scientific-Research Institute No. 108 (Tsentral'nyi Nauchno-Issledovatel'skii Institut No. 108, or TsNII-108), also known as NII-108, as well as the Central Scientific-Research Radar Institute. The institute's director Aksel' I. Berg (Fig. 3.3) could not prevent the departure of his key specialist.

---

[9]Davydov, 2009, p. 98.
[10]Davydov, 2009, p. 96.
[11]Central Intelligence Agency, 1952, p. 2.
[12]Leonid Ilyich Brezhnev, 1906–1982.
[13]Alexander Andreevich Raspletin, 1908–1967.
[14]Al'perovich, 2003, p. 9.

Fig. 3.2 Alexander A. Raspletin, 1908–1967, served as first deputy chief designer for Berkut; after Stalin's death and the arrest and banishment of Sergo Beria, he became chief designer. He led the development of the main Soviet air defense systems in the 1950s and 1960s. Photo courtesy of ITAR-TASS.

Fig. 3.3 Aksel' I. Berg, 1893–1979, played a prominent role in the development of Soviet radar during and after World War II. He served as Deputy Minister of Defense from 1953 to 1957. Photo (1963) courtesy of ITAR-TASS.

The Soviet government established Berg's TsNII-108 on 4 July 1943, simultaneous with the overseeing Council for Radars. Located a couple of miles from the Kremlin, TsNII-108 grew into a leading research institution in radar. From its early days, the organization also engaged in work on electronic warfare, particularly radar jamming, and also initiated development, in 1958, of the first penetration aids against missile defenses.

Vice Admiral Aksel' Berg had begun his military career in the czarist Navy and then continued to serve, after the Bolshevik revolution, in the Soviet Navy. He got involved in the military applications of radio in the late 1920s and soon became director of the Navy's research radio institute. At one time a prisoner during internal purges of the late 1930s, Berg would rise to serve as Deputy Minister of Defense from 1953 to 1957.

Berg's personal meeting with Stalin[15] paved the way for the creation of the Council for Radars by the Ordinance of the State Committee for Defense No. 3683ss on 4 July 1943. An influential Communist Party apparatchik and Stalin's lieutenant, Georgii M. Malenkov,[16] chaired the council. It was at this time that the word "radiolokatsiya," literally meaning "radio location" or "location by radio waves," formally replaced the older "radioobnaruzhenie," or "radio finding," as a term for radar in the Russian language. Reorganized into the committee in 1947, the council directed the development of Soviet radar until its dissolution in August 1949.[17]

The Council for Radars served as a focal point for gathering and assessing information on development of radar in the Soviet Union and abroad. In addition to Berg, Shchukin, the head of the council's science department (who would oversee science in the future Third Chief Directorate, or TGU), and Alexander I. Shokin[18] played the most important roles in advancing science and technology in this new field. (Shokin would direct the Soviet electronics industry from 1961 to 1985 and spearhead establishing the Soviet "Silicon Valley," a research and development cluster in Zelenograd near Moscow.) Shokin also headed a group of Soviet engineers, including Raspletin, who in 1945 went to the defeated Germany to examine German aircraft and air defense radar. The group secured the full set of technical documentation for the Wuerzburg radar and familiarized itself with the air defense system of Berlin.[19]

When the Council for Radars formed in 1943, the government appointed Berg a deputy to the chairman of this powerful coordinating body as well as director of a new radar research institute, TsNII-108. (Some specialists from the above-mentioned Plant No. 465 joined the institute at the time of its birth.)

---

[15]e.g., Pervov, 2001, pp. 26, 27; Erofeev, 2007, pp. 96, 100–102.
[16]Georgii Maximilianovich Malenkov, 1902–1988.
[17]Lobanov, 1982, pp. 155, 159, 160; Erofeev, 2007.
[18]Alexander Ivanovich Shokin, 1909–1988.
[19]Semenov, 2008; pp. 69–70.

In addition, he assumed the position of Deputy Minister of the Electronics Industry overseeing radars.

A scientific seminar in Berg's institute offered an important venue for discussing and evaluating new concepts and developments in this emerging area of science and technology. Many leading specialists, including Kuksenko and Kisun'ko, defended their theses for second advanced science (D.Sc.) degrees[20] at the TsNII-108's Scientific Council, presided over by Berg.[21] Throughout its history, several branches of the institute evolved into independent research and development organizations. The A.I. Berg Central Scientific-Research Radiotechnical Institute, as it is known today, focuses on electronic warfare, surveillance, and information protection.

Within two months of his transfer to KB-1, Alexander Raspletin "stole" four former colleagues from Berg's TsNII-108: Karl S. Al'perovich, Boris V. Bunkin, Iliya L. Burshtein, and Mikhail B. Zakson.[22] They joined Raspletin in the newly formed design bureau and would play important roles there. Al'perovich remained close to Raspletin, both professionally and personally, for many years. In 1999 and 2003 he published memoirs about the early years of KB-1 and development of the first air defense systems.[23] Bunkin succeeded Raspletin after his death in 1967 and served from 1968 to 1998 as general designer of the Scientific-Production Association, or NPO, Almaz, as KB-1 became known in the 1970s.[24]

Al'perovich provided details of their transfer from TsNII-108 to KB-1. Raspletin and science deputy head of the overseeing TGU Shchukin chose at least two of these transferred specialists, Al'perovich and Zakson, based on the desire to utilize their expertise in spite of the growing anti-Semitic campaign in the USSR.[25] Anti-Semitism had faithfully served the rulers in Russia, czars and Marxist commissars alike, for almost two centuries. Now Stalin initiated a new convenient campaign. In the late 1940s and early 1950s many Soviet scientists, engineers, administrators, doctors, writers, and other professionals lost their jobs simply because they were Jews.[26] The socialist state stepped up their persecution under the guidance and coordination of Communist Party apparatchiks and the security services. The high profile of KB-1 and the importance of its task provided some flexibility in employing technically accomplished and practically useful but otherwise undesirable employees.

---

[20]See Chapter 4 for an explanation of the advanced degrees in the USSR.
[21]Erofeev, 2007, pp. 133, 134.
[22]Karl Samuilovich Al'perovich, b. 1922; Boris Vasil'evich Bunkin, 1922–2007; Iliya Lvovich Burshtein; Mikhail Borisovich Zakson, 1921–2003.
[23]Al'perovich, 1999, 2003.
[24]Semenov, 2012.
[25]Al'perovich, 2003, pp. 30, 31.
[26]Kostyrchenko, 1995, 2005; Courtois et al., 1999; Berdichevsky, 2005, pp. 109–110; Gruntman, 2007, pp. 44–47; Gruntman, 2010, pp. 47–51.

The appointment of two chief designers for the development of a weapon system, especially one as important and complex as Berkut, was very unusual for the Soviet defense establishment. Al'perovich explained that,

> [t]his appointment of *two* Chief Designers [Kuksenko and Sergo Beria] and *one* Deputy Chief Designer [Raspletin] for *one* program was without precedent and it had definite sense: this was a way to provide the next step in the personal career growth of S. [Sergo] Beria. His career had been carefully nurtured and built up. Here Kuksenko played a special role. His choice was not accidental: in the beginning of the 1930s he was arrested by the OGPU [the predecessor of the KGB] and since those days he worked in the Central Radiolaboratory of the MVD [Ministry of Internal Affairs, one of the incarnations of the KGB] controlled from the prewar [World War II] years by L.P. Beria.[27]

The accelerated career path for Sergo Beria also included his defense of his PhD thesis in 1948. It took place only one year after his graduation with a master's degree, instead of the more common three years or longer. He then quickly receive his second advanced science degree, D.Sc., in 1952.

A recently published celebratory book on the occasion of the 100th anniversary of Alexander Raspletin's birth offers an unflattering characterization of Sergo Beria as "self-confident" (with an arrogant flavor), "unpleasant," and "unpredictable." The publication also noted that contradicting and arguing with him on scientific and technical issues was "useless, impossible, and not safe for life."[28] One has to bear in mind, however, that denigrating and destroying fallen leaders who had been only recently publicly praised represented an unalienable feature of state-controlled life in the Soviet Union and remains deeply ingrained in the culture.

A NEW "EMPIRE" EMERGES

Development of Berkut to defend the capital against a massive raid of bombers from any direction became a huge undertaking. One enemy aircraft breaking through with an atomic bomb onboard was too many. The unprecedented challenge called for significant advances in many areas of science, engineering, and industrial production.

The KB-1 design bureau grew rapidly. It took over desired specialists from other institutes, bureaus, and plants, often without the consent of those organizations or the individuals involved. It also cherry-picked young graduating engineers and scientists. Soon, an imposing building (Fig. 3.4) grew up on its territory, facing a major Moscow thoroughfare, Leningrad Highway (Fig. 3.5).

---

[27]Al'perovich, 2003, p. 12; italics by Al'perovich.
[28]Semenov, 2008, pp. 98, 111, 130.

Fig. 3.4 The imposing building of KB-1 (across the street, right), erected in the early 1950s near the subway station Sokol in Moscow. Photograph from the website of the successor organization of KB-1, Almaz-Antei Corporation, http://www.raspletin.ru/company/history/start/; accessed 28 Nov. 2010.

Kuksenko, S. Beria, and the TGU's Shchukin personally compiled a list of 30 top specialists for transfer to KB-1 by a special order of the Central Committee of the Communist Party.[29] These scientists and engineers would prominently contribute to the development of Berkut and follow-on programs. Those transferred from the Military Academy of Communications in Leningrad included Grigorii V. Kisun'ko and Nakhim A. Livshits.[30] They would later play important roles in the development of the first missile defenses.

The Ministry of Internal Affairs continued to supply German specialists for work on Berkut. Here again, KB-1 raided other organizations. More than 100 Germans worked on ballistic missile projects in a branch of NII-88 at Lake Seliger 200 miles (320 km) west-northwest of Moscow.[31] That group included Johannes Hoch,[32] a leading specialist in control. Chertok described the loss of Hoch from the ballistic missile program in his NII-88:

> Unfortunately, very fruitful activity of Dr. Hoch [in ballistic missiles at NII-88] was not long. His accomplishments became known outside our NII and his fame reached the organization working on control systems for air defense guided missiles. Chief designer there was Sergei Beria, the

---

[29]Kisun'ko, 1996, pp. 197, 213; Al'perovich, 2003, p. 14.
[30]Nakhim Aronovich Livshits.
[31]Groettrup, 1959; Albring and Vinke, 1991; Magnus, 1993; Chertok, 1994.
[32]Johannes Hoch, 1913–1955.

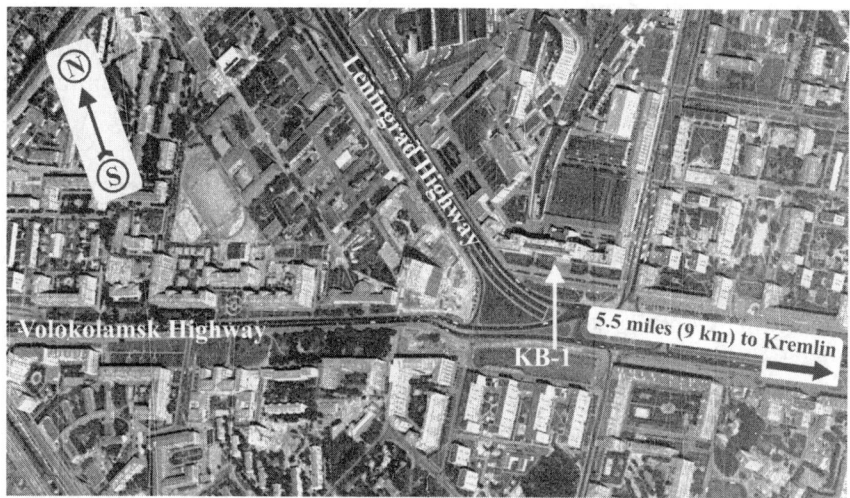

Fig. 3.5   The arrow points at the KB-1 building (Fig. 3.4) near the road fork where Leningrad Avenue branches into Leningrad Highway and Volokolamsk Highway. Leningrad Avenue radiated from the Kremlin and headed northwest; after a turn around the KB-1 building, it became the Leningrad Highway. The Volokolamsk Highway headed west-northwest (to the left in the photo). Original satellite reconnaissance photograph (KH-7; Mission 4030; 16 July 1966) from the U.S. Geological Survey; photograph identification, interpretation, processing, and augmentation by Mike Gruntman.

son of all-powerful Lavrentii Pavlovich [Beria]. The leaders of this organization could transfer to them, without any agreement [or permission], any specialist from anywhere. Doctor Hoch was transferred to the young Beria. According to rumors that reached us, he fit very well [into the new organization], achieved big successes in his work there, and requested to become a Soviet citizen. He suddenly ended up in a hospital where he died after a surgery for an abscessed appendix.[33]

Hoch introduced electromechanical simulators into the development of antiaircraft missiles for Berkut. He also suggested adding some measurements of linear accelerations. Prior to that, "no known ... [to KB-1] autopilot systems of aircraft or missiles (V-1, V-2) had used linear acceleration control signals."[34] In addition, Hoch designed, together with another accomplished captive scientist Kurt Magnus,[35] an integrating gyroscope.

Many Russian publications diminish the importance of the work of German specialists in Soviet post–World War II weapons programs. In contrast, the project engineer for Berkut in TGU, Nikolai K. Ostapenko,[36] and deputy chief designer (from 1953) of the system Al'perovich acknowledged important

---

[33]Chertok, 1994, p. 215; also, Chertok, 2005, p. 62.
[34]Falichev and Sukharev, 2012, p. 3.
[35]Kurt Magnus, 1912–2003.
[36]Nikolai Kuz'mich Ostapenko, 1921–2008.

contributions by the Germans.[37] Ostapenko noted that they provided a "significant scientific and engineering input to research and development of the Berkut system." He particularly credited their work on inflight stabilization, guidance, and control and on detonation of the antiaircraft missile warhead and development of the radio link for missile commands. Ostapenko specifically attributed the success of the antiaircraft missile control system in its first test launch to simulators built by Hoch. Al'perovich singled out Hoch's suggestion for using differences between directions to the target and the missile rather than their absolute coordinates for interceptor guidance. This method took advantage of the adopted design relying on the same radar for tracking both targets and interceptors. In addition, it allowed the design of control units based on electronic circuitry only, without electromechanical devices.

The official publication of the successor organization of KB-1 quoted the former head of the theoretical division of the design bureau that Johannes Hoch had provided "the biggest practical contribution" to "solution of particular problems of control systems in KB-1."[38] The same article also added that "not everybody likes the fact of use [of contributions] of German specialists, but this is the history of our organization."

Ostapenko noted that one Berkut chief designer, Kuksenko, "considered results obtained by Germans critically, subjecting them to rigorous [validation by] scientific analysis. [In contrast, the other chief designer] S. L. Beria listened to reports of deputy chief designer A. A. Raspletin on suggestions of the German specialists in KB-1 and, commonly, instructed 'Do as the Germans suggest.'" Al'perovich adds that Sergo Beria "did not have experience and relied on faith rather than on understanding [of technical issues]. He believed not us [the Soviet engineers involved in the work], but the Germans."[39]

Based on debriefing of German specialists after they had returned to Germany in the 1950s, the U.S. intelligence summarized their role in the development of missile guidance in the USSR: "The foundations for current [in 1957] Soviet capabilities in missile guidance is largely postwar exploitation of German personnel, facilities, equipment, and documents. ... Beginning about 1948, the USSR apparently reached the point where it could largely dispense with German assistance, except in the missile guidance field."[40]

In addition to Germans, the Ministry of Internal Affairs also provided imprisoned Soviet scientists and engineers to Berkut's development. The government eventually released certain prisoners, and some of them continued their work as regular employees. Not only did the KGB officers oversee the administrative units staffed by the captive Germans and Soviet prisoners, but they also headed many science and engineering departments of KB-1.

---

[37]Al'perovich, 2003, p. 58; Belous et al., 2009, pp. 61–63.
[38]Falichev and Sukharev, 2012, p. 3.
[39]Al'perovich, 2003, p. 60; Belous et al., 2009, p. 63.
[40]Central Intelligence Agency, 1957, p. 9.

As Sergei N. Khrushchev,[41] the son of Soviet leader Nikita S. Khrushchev, described it many years later, the presence of these officers "created the atmosphere of ominous uncertainty."[42] Only the death of Stalin in 1953 would lessen the KGB's grip on the organization.

Al'perovich described an unusual setup of work on Berkut:

> [I]n contrast to the established by centuries arrangements, the military was not the procuring organization of Berkut. The development was conducted under conditions of strict secrecy, including—which is hard to imagine today—[in secrecy] from the highest leaders of the Ministry of Defense. Certainly, the fact of that work on a new enormous system of air defense was not concealed from them. It could not be kept secret. But the substance of work on Berkut was kept secret. The government had assigned [to KB-1] a task of creating the system of air defense for Moscow. Then the lead developer of the system, KB-1, acted as both the procuring customer and the organization determining implementation of the system. [In addition,] specialists of the overseeing [Third Chief Directorate] TGU and the developing organization (KB-1) tightly controlled the tasks given to the military: oversight of compliance of articles manufactured by series-production plants to the documentation of chief designers; establishment of a test site for evaluation of the system; creation of a training military organization for preparing military units to operate the system; formation of the First Special Army of Air Defense.[43]

Only with the reorganization of the defense establishment after Stalin's death in 1953 did the military assume an active role in directing Berkut. Then, in August 1954, the government activated the Fourth Directorate of the Ministry of Defense—military unit v/ch 77969[44]—to procure the new weapon and to prepare troops for its deployment. The directorate evolved with time into the main overseeing and procuring agency of Soviet air defense and missile defense systems.

The concept of Berkut included A-100 early warning radars located 125–155 miles (200–250 km) from Moscow. Then, two rings of air defense sites surrounded the city at distances of about 30 miles (45 km) and 55 miles (90 km) from the city center. The system also included interceptor aircraft patrolling the airspace monitored by A-100 radars. The designers later eliminated the aircraft component.

Each air defense site was responsible for one sector-direction. It operated its own tracking and fire control radar, B-200, and had launch pads for 60 antiaircraft guided missiles. The site could engage up to 20 targets simultaneously in the assigned 6–10 mile (10–16 km) wide sector. The central command

---

[41] Sergei Nikitich Khrushchev, b. 1935.
[42] S. Khrushchev, 2000, p. 100.
[43] Al'perovich, 2003, p. 18.
[44] The Ministry of Defense assigned identifying numbers to military units that were described by the letters v/ch, standing for voinskaya chast', or a military unit.

center controlled the entire air defense system, with four subcenters coordinating clusters of sites responsible for four 90-deg-wide directions.

The design bureau KB-1 led the design and integration of the Berkut system. In addition, it developed the innovative tracking and fire control radar B-200. Several other organizations played important roles in building Berkut. Leonid V. Leonov[45] designed the early warning A-100 radar in the Scientific-Research Institute NII-20.[46] This NII-20 belonged to the Ministry of Means of Communications and was not related to the organization of the same name mentioned earlier, the prominent radar institute NII-20 (NIEMI) of Ustinov's Ministry of Armaments.

Leonov's NII-20 began as a research and development organization in Leningrad in the early 1920s, relocated to Moscow in the 1930s, and took an active part in the pioneering work on the first Soviet radars. It changed its name a few times throughout the years and was known for some time during Berkut's development as the Scientific-Research Institute No. 244 (NII-244). The institute has continued building various radar systems to the present day. Now, it is the All-Russia Scientific-Research Institute of Radiotechnology, or Vserossiiskii Nauchno-Tekhnicheskii Institut Radiotekhniki (VNIIRT), and it is located in northeastern Moscow 3 miles from the city center.

A leading Soviet specialist in radio technology, Alexander L. Mints, headed another important Berkut contractor, the Radiotechnical Laboratory of the USSR Academy of Sciences, or Radiotekhnicheskaya Laboratoriya Akademii Nauk (RALAN). The laboratory also changed its name a few times through its history. In 1957, it became the Radiotechnical Institute (RTI) of the USSR Academy of Sciences. The organization later excelled in high-energy charged particle accelerators, plasma studies, and beam and laser weapons. It has been continuously engaged in Soviet missile defense programs to this day and built numerous radar systems for early warning of ballistic missile attack, missile defense fire control, and monitoring space objects in orbit, including Dnestr, Dnepr, Daryal, and Don. Today, it is the A. L. Mints Radiotechnical Institute.

As a principal Berkut contractor, Mints's laboratory designed and produced powerful transmitters for the radars, under the direction of Nikolai I. Oganov.[47] Before joining RALAN, Oganov had worked in the TsNII-108 radar institute on the first radar-jamming systems.

Although formally part of the USSR Academy of Sciences, RALAN directly reported to the Third Chief Directorate (TGU). In fact, the same government decree that had activated TGU in February 1951 also formed RALAN. In addition to transmitters, the laboratory oversaw the civil engineering design of Berkut's supporting infrastructure, including roads, bridges, storage depots, blockhouses, barracks for soldiers and living quarters for

---

[45]Leonid Vasil'evich Leonov.
[46]Pervov, 2001, p. 64.
[47]Nikolai Ivanovich Oganov, 1902–1966.

officers and their families, electric power generators and lines, and various other facilities. RALAN expanded throughout the 1950s, and eventually relocated to new premises 2 miles away from KB-1.

A famed aircraft designer, Semen A. Lavochkin,[48] headed another key contributor to the program, Special Design Bureau No. 301 (Osoboe Konstruktorskoe Byuro No. 301, or OKB-301). He designed and built the V-300 antiaircraft guided missiles for Berkut.

A thorough evaluation of the new air defense system required a specialized proving ground. Therefore, the military expanded the facilities at Kapustin Yar, initially occupied with testing the first ballistic missiles. On 28 May 1951, the Ministry of Defense activated a new military unit, v/ch 29139 or the State Scientific-Research Test Range No. 8 (Gosudarstvennyi Nauchno-Issledovatel'skii Ispytatel'nyi Poligon No. 8, or GNIIP-8), to specifically engage in the development of Berkut. Lt. Gen. Sergei F. Nilovsky[49] became the first head of GNIIP-8. (In 1957, Nilovsky assumed command of the leading research institute of the air defense branch of the Soviet Army.[50]) The air defense proving ground rapidly grew, with tests of follow-on systems later expanding to a new and bigger site at Saryshagan in Kazakhstan.

TRACK-WHILE-SCAN RADAR

A common approach to aircraft missile intercepts in the late 1940s relied on two narrow "pencil-beam" radar units. One radar continuously tracked an aerial target and determined its elevation, azimuth, and range (distance). The other radar tracked and guided the antiaircraft missile, and an analog computing device solved the intercept problem. The concept had originated from World War II work on the first air defense systems in Germany.[51]

Following such an approach, the defense of Moscow against a 1000-bomber raid from all directions would have required 1000 antiaircraft missile sites, with two radar units at each. Such an enormous system would have been prohibitively cumbersome, expensive, and difficult to control. In short, the traditional concept looked impractical, and Berkut called for radically new solutions. The emergence of digital computing with rapidly improving capabilities provided essentially new possibilities.

Alexander Raspletin devised such an alternative innovative concept that tracked multiple aircraft targets as well as interceptor missiles. In his design, two separate transmitting antennas emitted two narrow fan beams, independently scanning the wide defended sector in elevation and azimuth, respectively. One radar beam determined only the vertical (elevation) angular

---

[48]Semen Alexeevich Lavochkin, 1900–1960.
[49]Sergei Fedorovich Nilovsky, 1906–1973.
[50]Belous et al., 2009, p. 69.
[51]e.g., Schirrmacher, 1957; von Zborowski, 1957.

coordinate of a target and the other only the horizontal (azimuth) coordinate. Then, a control center combined the information and stored the target position in memory.

In essence, the concept relied on "remembering" target positions in a computing device while periodically scanning a wide sector and updating their coordinates. In contrast, a pencil beam radar system continuously tracked only one selected target. Raspletin's new "track-while-scan" radar could thus simultaneously follow multiple airplanes as well as intercepting antiaircraft missiles in the periodically scanned sector. Importantly, the same radar determined coordinates of both a target and an antiaircraft missile, which resulted in higher precision of intercepts.

An antenna with a narrow rectangular aperture formed a narrow fan beam of radio waves perpendicular to the orientation of the antenna aperture (Fig. 3.6). The beam of the Raspletin radar extended only 1 deg in one direction but spread 60 deg wide in the perpendicular direction. Soviet engineers nicknamed such fan beams "spades."

Two such antennas mechanically scanned the sector in two perpendicular directions, measuring target elevation and azimuth, respectively. The technique provided angular accuracy in the scanning direction significantly better than the 1-deg beamwidth by determining the "center of gravity" of the directional dependence of intensity of returned, "echo" radar pulses. The new radar tracked and produced targeting information on up to 20 airplanes. Consequently, only

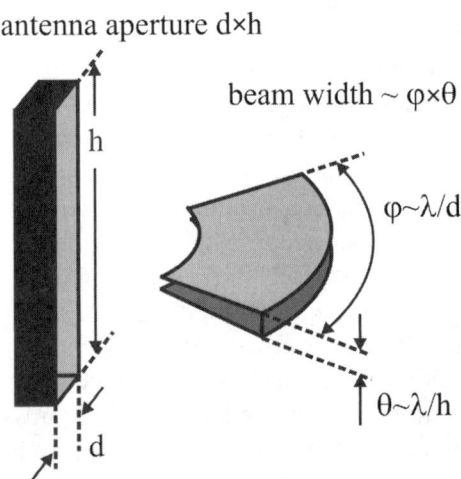

Fig. 3.6 An antenna with a physical aperture $d \times h$ (left) would produce a beam of electromagnetic radiation (radio waves) with angular dimensions $\varphi \times \theta$ (right), where $\varphi \approx \lambda/d$ and $\theta \approx \lambda/h$ are in radians and λ is the wavelength of the electromagnetic radiation; 1 rad $\approx$ 57 deg. An antenna with a narrow vertically oriented aperture, as shown in the figure, produces a fan beam of radio waves narrow in the vertical direction and widely spread horizontally.

50 or 60 such track-while-scan radar units could cover all directions around the capital, thus enabling a realistic air defense system.

Raspletin worried that his bosses, S. Beria and Kuksenko, might reject the new concept for a number of technical and personal reasons, so he cautiously introduced new ideas in increments.

Al'perovich noted that Raspletin's concept rejected traditional narrow pencil-beam radars and achieved

> the [required] functionality by [novel] sector radars. It was impossible to predict how S. Beria (whose word was final) would react to this part of Raspletin's concept, breaking the accepted design of Berkut. Therefore, Raspletin acted with maximum caution. He followed the known rule that if one wanted realization of an idea, then one should tell it to the boss and "forget" about it. The concept, as originating from the boss, would then come back to you some time later for implementation. Raspletin "injected" his entire concept into Chief Designers [S. Beria and Kuksenko] in the middle of November [1950].[52]

Radar B-200 (Yo-Yo)

In the early 1950s, U.S. and British intelligence began to search for information on surface-to-air missiles being deployed around Moscow. In September 1954, two U.S. military attachés left Moscow on a plane to Leningrad. Through the window, one attaché "noticed an unusual installation on the ground. ... His eye caught the motion of two large wheels ... each wheel ... was like a thin yo-yo, with twin flat disks spinning at an angle to the horizontal ... at about 60 rpm." Another attaché, seated at the other side of the plane, "spotted one of the herringbone [road pattern] sites" at the same time (Ahern, 1961, pp. 12–14). The attachés' report gave the radar its nickname, "Yo-Yo."

By October 1955, the CIA had successfully correlated bits of information from various sources to reconstruct the concept of the air defense missile guidance system based on the Yo-Yo radar. The Agency's analysts called the implications of the new approach "startling." They concluded that

> the Soviets had not continued in the direction taken by the original German wartime development of surface-to-air missile guidance nor in that of postwar Western efforts, which were based on extensions of the German work. Instead, making a clean break with precedent, they had

---

[52]Al'perovich, 2003, pp. 28

arrived at a design that was inherently capable of dealing with multiple targets simultaneously (Ahern, 1961, pp. 19–20).

Soon a U-2 overflight of Moscow in July 1956 confirmed the conclusions of the provisional CIA report (6 October 1955) on the Yo-Yo radar. Then, debriefing of German specialist Christian Sorge, returning from the USSR through the Dragon Returnee Program, provided other details, including the Soviet name for the radar, "B-200" (Ahern, 1961). The U.S. government then built and tested (in early 1959) a prototype, or mockup, of the radar, in order to evaluate its capabilities. The CIA summarized the system performance

> The results of the test program showed the Soviet B-200 to constitute a major technological advance in radar tracking system. An additional surprise was that it performed much better than expected when tested against electronic countermeasures, jamming; but the technique dropping chaff was effective against it if properly employed. The B-200 was found to have an angle accuracy as great as 0.05° on strong targets and a range accuracy of 25 yards; this meant that missiles in the range 20 to 25 miles would not need a homing radar of their own. ...
>
> The ability of the system to cope with multiple targets was confirmed; the ability of one installation to direct as many as 20 or 25 simultaneous target-missile interceptions, as claimed by the [debriefed] Germans, seems to depend only on whether the Soviets choose to provide the necessary computer for each interception (Ahern, 1961, pp. 22, 23).

AMERICAN MILITARY ATTACHÉS EXPELLED

Intelligence gathering in tightly controlled totalitarian societies presented special challenges. U.S. Ambassador to the Soviet Union Charles (Chip) E. Bohlen (1904–1974) described that (Bohlen, 1973, p. 343)

> [w]ith trade and tourism operating at only a trickle, the principal tasks of the [U.S.] Embassy staff were intelligence-gathering, political reporting, and negotiation of differences. Our [military] attachés had a difficult time operating [in Moscow in the 1950s]. The Soviets considered normal contacts and exchanges of information as espionage. As a result, seven American military attachés, three on a single occasion, were ordered to leave Moscow during my tenure [as the U.S. Ambassador] of more than four years [from 1953–1957].

The chief designers approved the new approach in January 1951. The proposed radar would become known as B-200. A few years later, U.S. specialists praised this innovative solution [see the box "Radar B-200 (Yo-Yo)"].

The designed radar antenna consisted of two large parallel triangular structures, turned at a 60-deg angle to each other (Figs. 3.7 and 3.8). Each side of each triangle served as an antenna aperture, producing a 1-deg-by-60-deg beam that was narrow in the triangle plane and wide in the direction perpendicular to

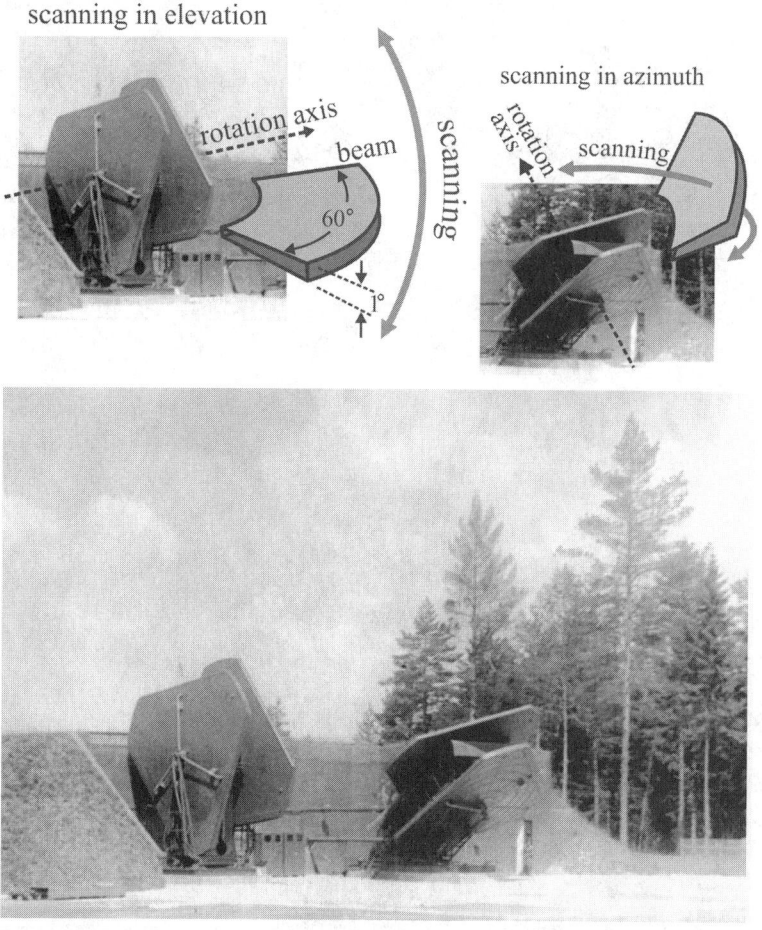

Fig. 3.7 Track-while-scan radar B-200 (bottom). The left radar (top left) rotated about the horizontal axis, normal to the plane of the triangular structures, and thus scanned in elevation (up–down) with a 1-deg × 60-deg beam (1-deg wide in the vertical direction and 60-deg wide in the horizontal direction). The plane of the radar triangles was aligned away from the center of Moscow. The other radar (top right) spun about the axis tilted by 30 deg from the vertical and toward Moscow. It scanned the sector in azimuth (left–right) with a vertically oriented 1-deg × 60-deg beam. The heights of the elevation and azimuth antennas were 30 ft (9 m) and 26 ft (8 m), respectively. Original photograph (bottom) from *Sistema-25*, p. 8; photograph augmentation (top) by Mike Gruntman.

Fig. 3.8 B-200 radar, nicknamed by U.S. intelligence analysts Yo-Yo, scanning in the azimuthal direction (top) and in elevation (bottom). Photographs from *Sistema-25*, pp. 9, 10.

that plane. The structure rotated about the axis normal to the plane of the triangles, and the transmitter commutated sequentially to the emitting antennas. Therefore, the 1-deg-wide fan beam swept through the 60-deg sector six times during each full rotation of the triangle pairs. Use of six physically different antennas posed challenges because their beam patterns, gains, and side lobes were not identical, and the antenna main lobes did not point at exactly 60 deg away from each other. All of these imperfections affected the precision of interceptions.

The plane of the antenna triangles of the elevation angle radar was oriented vertically and aligned with the radial direction away from Moscow's center. Its rotation axis correspondingly pointed horizontally and perpendicular to the direction of Moscow. It thus scanned in elevation (up–down) the 60-deg-wide azimuthal (horizontally spread) sector centered on the city (Figs. 3.7 and 3.8, bottom).

The other, azimuthal angle radar had the plane of the triangles tilted by 30 deg from the ground with the spin axis pointing toward Moscow (Fig. 3.7). The antenna thus rotated about the tilted vertical axis and scanned in azimuth (left–right) the 30- to 40-deg-wide sector centered on the capital. The radars operated at the 10-cm wavelength (3-GHz frequency). The antennas rotated at the rate of 50 rpm, and the radar beams thus swept the defended sector five times each second.

ANTIAIRCRAFT MISSILE V-300

The government assigned the task of developing antiaircraft guided missiles for Berkut to special design bureau OKB-301 in September 1950. An accomplished aircraft designer, Semen Lavochkin (Fig. 3.9), directed the bureau in the town of Khimki at the outskirts of Moscow. The buildings of OKB-301 faced the Leningrad Highway, a main road leading to today's Moscow international airport, Sheremetyevo.

Semen Lavochkin had earned fame for his family of fighter planes, such as La-5 and La-7, during World War II, with more than 22,000 aircraft produced. After the war he pursued the development of jet aircraft and built the first Soviet plane that reached the speed of sound. In addition, Lavochkin's OKB-301 expanded into the development of antiaircraft weapons and air defense systems, beginning with work on reproduction of the German Wasserfall missile.[53]

The new antiaircraft guided missile designed for Berkut, the V-300, represented a subsonic, single-stage liquid-propellant rocket, carrying a 520-lb (235-kg) high-explosive warhead. In 1959, the CIA's National Intelligence Estimate erroneously listed the missile as V-301,[54] either because of a typo or possibly assuming its number was derived from the designation number of the

---

[53]Gruntman, 2004, p. 279.
[54]Central Intelligence Agency, 1959, p. 7.

Fig. 3.9 Aircraft designer Semen A. Lavochkin directed development of the surface-to-air missile V-300 (SA-1) for Berkut in his Special Design Bureau OKB-301 in Khimki near Moscow. Photograph (1947) courtesy of ITAR-TASS.

developing organization (Plant No. 301 and OKB-301). The numbers in missile nomenclature (e.g., V-300, V-750, or V-1000) corresponded to average missile velocities (300, 750, or 1000 m/s, respectively).[55] The V-300 missile and the associated air defense system would become known as SA-1 or Guild in the West.

The Lavochkin rocket launched vertically from a simple pad. Internally, OKB-301 designated it the "antiaircraft missile 205" or ZUR-205 (Fig. 3.10). (ZUR stands for zenitnaya upravlyaemaya raketa, or an antiaircraft guided missile.) Later, OKB-301 introduced more capable modifications of the missile. The variants 207, 207A, and 215 went into series production and deployed in air defense units. The 215 model became the first Soviet antiaircraft missile carrying a nuclear warhead.[56]

The 205 missile was almost 38 ft (12 m) tall with a body diameter of 25.6 in. (0.65 m); its fully fueled mass was close to 7950 lb (3600 kg). Alexei M. Isaev[57] (Fig. 3.11) designed the engine for the missile at NII-88 in Podlipki. Another group in the same institute, headed by Dominik D. Sevruk, had been

---
[55]Svetlov, 2010
[56]Pervov, 2001, pp. 102–105.
[57]Alexei Mikhailovich Isaev, 1908–1971.

Fig. 3.10 Early variant ZUR-205 of the antiaircraft guided missile V-300 developed by Semen Lavochkin's design bureau OKB-301 in Khimki near Moscow. Photo from the website of the S.A. Lavochkin Scientific-Industrial Association, http://www.laspace.ru/rus/zur205.html; accessed 28 November 2010.

Fig. 3.11 Alexei M. Isaev (1908–1971), designer of liquid-propellant rocket engines for Berkut's V-300 antiaircraft missile and the V-1000 missile that achieved the first ballistic missile intercept. Photo (1971) courtesy ITAR-TASS.

working on a competing engine for Lavochkin's missiles. Sevruk's designs had not reached series production. His later attempts to develop engines for the main stage of the interceptor for missile defense, the V-1000, also ran into difficulties. Consequently, his team disbanded and dispersed among other propulsion organizations in the 1960s.

Isaev's bipropellant liquid rocket engine produced 9 t of thrust (88 kN or 20 klbf). The 205 antiaircraft missile carried 880 lb (400 kg) of fuel and 3840 lb (1740 kg) of oxidizer. The engine used a hypergolic combination of toxic propellants: TG-02 fuel and concentrated nitric acid oxidizer. In a hypergolic propulsion system, a fuel and an oxidizer ignite on contact.

The TG-02 represented the Soviet version of a rocket fuel known as Tonka that was developed by Germany during World War II. Tonka consisted of a main component from the aniline family with aliphatic and aromatic amino additives such as triethylamine and xylidines.[58] Combustion instabilities caused major problems in the engine, so Isaev first achieved the required thrust by using four smaller chambers. He succeeded in building an improved single-chamber engine for the follow-on missile variants.

The ramping up work in OKB-301 on antiaircraft guided missiles for Berkut coincided with the termination of research and development of air defense systems in NII-88 in Podlipki. The institute disbanded groups engaged in reproduction of German World War II–era antiaircraft missiles, and some specialists transferred to Lavochkin's design bureau. The government also used this reorganization to advance its anti-Semitic campaign and purge NII-88 of some leading specialists who were Jews.[59] All organizations in the Soviet Union belonged to the same government and the state controlled many aspects of people's lives, so finding new jobs was exceptionally difficult.[60]

Building rocket engines for Lavochkin's V-300 antiaircraft missiles protected the department of Alexei Isaev in NII-88 from the reorganization. Isaev had been successfully growing his programs and eventually split off from the parent institute in 1959, forming an independent design bureau. (Earlier, in 1956, Sergei Korolev had separated from NII-88 and formed OKB-1, now NPO Energia, also in Podlipki.) Isaev's new organization, located next to NII-88, became known as the Design Bureau of Chemical Machine Building (KB Khimmash).[61] It later built submarine-launched liquid propellant ballistic missiles.

Since its inception in 1946, NII-88 remained primarily focused on ballistic missiles, and then expanded into space technology. The government renamed it the Central Scientific-Research Institute of Machine Building, or Tsentral'nyi Nauchno-Issledovatel'skii Institut Mashinostroeniya

---

[58]Lutz, 1957; von Zborowski, 1957; Clark, 1972, pp. 15, 116; Sutton, 2006, p. 678.
[59]Evtif'ev, 2000, p. 48.
[60]e.g., Berdichevsky, 2005, pp. 109, 110.
[61]Gruntman, 2004, p. 282.

Fig. 3.12 Vladimir P. Barmin designed the ground support and launch equipment for Berkut's V-300 missiles. Later he built launch complexes for numerous strategic ballistic missiles and space launch vehicles. Barmin's tombstone (left; magnified bottom section on the right) in Moscow highlights his contribution to Soviet rocketry. Photo (2006) courtesy of Mike Gruntman.

(TsNIIMash). With time, the institute has become the leading national certification and system engineering organization in rocket and space technologies. It is also the home of the Russian, formerly Soviet, manned spaceflight control center (Tsentr Upravleniya Poletami, or TsUP).[62]

Another design bureau headed by Vladimir P. Barmin[63] (Fig. 3.12) in Moscow built the ground support equipment for Lavochkin's V-300 missiles. It included fueling, maintenance, and transporting mechanisms for preparation, servicing, and launch of the rockets. Barmin, a specialist in refrigerators, got involved with rocketry during World War II. Then, he organized mass manufacturing of mobile launch stands for the solid-propellant barrage missiles M-8 and M-13, the latter popularly known in the USSR as Katyusha.[64]

In the mid-1950s, Barmin's design bureau built launch complexes for the first Soviet ICBM R-7, or SS-6 (Sapwood), and the first space launchers. The

---

[62]Gruntman, 2004, p. 280.
[63]Vladimir Pavlovich Barmin, 1909–1993.
[64]Gruntman, 2004, pp. 271, 273.

organization, known first as the Design Bureau of Special Machine Building, or KB Spetsmash, and then Design Bureau of General Machine Building (KB Obshchego Mashinostroeniya, or KB OM), produced ground launch systems for numerous silo-based and mobile ballistic missiles, including SS-4, SS-5, SS-8, and SS-11, as well as the space launchers Proton and Energia-Buran.[65] In the 1970s, KB OM designed and built an engineering development mockup of key components of a planned permanent base on the moon for a projected crew of one dozen cosmonaut settlers.

Engineers launched the first V-300 missile in July 1951. Numerous trials followed at the proving ground in Kapustin Yar. These first successes in the summer of 1951 led to authorization by the Council of Ministers in December of the same year for construction of operational air defense sites around Moscow.

By the mid-1950s, Semen Lavochkin had expanded the work of OKB-301 beyond Berkut into two other major weapon systems. First, Lavochkin began development of a new ambitious antiaircraft complex Dal' (meaning "great distance" in Russian) for defense against planes approaching from any direction. In contrast, each individual site of Berkut protected only a limited and predetermined stationary 30- to 40-deg sector. The other new system, the supersonic intercontinental missile Burya (Tempest), represented an analog of the American Navaho, being developed in the 1950s. Both the United States and the Soviet Union considered such missiles as a possible alternative to intercontinental ballistic missiles (ICBMs).[66] Lavochkin's design bureau was thus evolving into a competitor in air defense to KB-1, which was also actively exploring its own successor systems to Berkut.

Simultaneously with contracting the V-300 for Berkut to the Lavochkin design bureau, Kuksenko and Beria initiated an independent study of an alternative antiaircraft missile, the 32-B, internally in KB-1. The tendency to expand the scope of in-house work and capabilities, "empire building," constituted an inalienable feature of the state-owned economy of the Soviet Union, leading to costly duplication of effort. Al'perovich recalled that

> [o]fficially this [new] missile was not included in Berkut but its development was conducted with the perspective of using it in Berkut [in the future]. Dmitrii Lyudvigovich Tomashevich led development of the missile called 32-B (B stood for Beria, and 32 stood for the [administrative] number of a separate unit in KB-1 where the rocket was being developed).[67]

Whereas the V-300 took off vertically, the new competing 32-B launched from an inclined ramp.

---

[65]Gruntman, 2004, pp. 317–319.
[66]Gruntman, 2004, pp. 221, 311.
[67]Al'perovich, 2003, pp. 96, 97.

Fig. 3.13 Commander of the Kapustin Yar test range from 1946 to 1973, Vasilii I. Voznyuk (left), with Sergei I. Vetoshkin (center) and Sergei P. Korolev (right) in 1948. The range's rocket units launched intermediate-range ballistic missile (IRBM) targets towards Saryshagan in missile defense tests of the early 1960s. Most trials of Berkut took place at the nearby split-off proving ground GNIIP-8. Voznyuk also headed a government survey group that chose the location for the new long-range missile test range and future space port at Tyuratam (Baikonur). Vetoshkin served from 1951 to 1953 as the first deputy head of the Third Chief Directorate (TGU) overseeing development of Berkut. He became first deputy chairman (1958–1966) of the exceptionally powerful Military-Industrial Commission (known under its Russian abbreviation VPK) of the USSR Council of Ministers. From its creation, the commission directed and coordinated weapons development in the Soviet Union until the country's disintegration in the 1990s. Korolev led the development of the first ICBM and the first artificial satellite, Sputnik; launched the first man, Yuri Gagarin, into space; and achieved numerous other firsts in Soviet ballistic missile and space programs. Collection of the Russian State Archives of Scientific-Technical Documentation (RGANTD), f. 33, op. 2, d. 189. Photograph courtesy of RGANTD.

Engineers conducted the first tests of the new 32-B missile in late 1952. In order to accelerate its development, the government transferred Plant No. 293 to KB-1. The plant, also located in Khimki as OKB-301, had some prior experience in building missiles. This addition of manufacturing facilities to KB-1, along with the growing interests of Semen Lavochkin in the competing Dal', would have important consequences for the organization of work on the future first ballistic missile intercept. (The government terminated Dal' after the sudden death of Lavochkin in 1960 and ultimately reoriented OKB-301 toward spacecraft design. The design bureau built numerous Soviet lunar and planetary probes as well as satellites carrying sensors for detection of ballistic missile launches.)

Spring 1953 marked the first tests of Berkut's V-300 missiles guided by the B-200 radar against real aircraft. It was a success: the system shot down all five target Tu-4 bombers at Kapustin Yar (Fig. 3.13).

The Death of Stalin

The feared and revered communist dictator Joseph Stalin died on 5 March 1953. Soon the power struggle in the Kremlin led to the arrest (on 26 June 1953) and execution (on 23 December 1953) of Lavrentii P. Beria. The secret services also arrested his son Sergo, chief designer of Berkut, and sent him into exile in the town of Sverdlovsk, a regional capital in the Ural mountains. (Sverdlovsk today has reverted to its original pre-Bolshevik-revolution name, Yekaterinburg.) The other chief designer of Berkut, Pavel Kuksenko, lost his powerful position and became a low-level deputy to the chief engineer of KB-1 responsible for science.

Major changes took place in KB-1. The air defense system Berkut received a new name, Sistema 25 (System 25), or S-25. Many associated the old name Berkut with a combination of the names of BERia and KUksenko. A time-honored tradition of the communist state required the obliteration of any traces of leaders who had fallen out of power.

Alexander Raspletin became chief designer of the S-25, with three principal deputies: Al'perovich (signal channel), Anatolii V. Pivovarov (high-frequency units), and Vladimir I. Markov (building air defense sites around Moscow).[68] The government promoted Markov to Deputy Minister of Radio Industry responsible for missile defense in 1967. Then, in 1970, he assumed the directorship of the specially established scientific-production association, which consolidated national missile defense work under one organizational roof. Markov would also play a prominent role in the struggle for control and development direction of over-the-horizon radars for ballistic missile early warning.[69]

KGB officers in the senior positions of department heads in KB-1 departed to other assignments. Captive German specialists also soon disappeared from the bureau. They joined many other forcibly "imported" Germans in research groups in the Sukhumi area[70] on the Black Sea shore in Abkhaziya (today a

---

[68]Anatolii Vasil'evich Pivovarov, b. 1916; Vladimir Ivanovich Markov, b. 1921.
[69]Babakin, 2008.
[70]Lowenhaupt, 1967.

The "Second Life" of Sergo L. Beria

After the arrest and execution in 1953 of his feared and powerful father, Lavrentii P. Beria, the secret services banished the chief designer of Berkut, Sergo Beria, to Sverdlovsk (now Yekaterinburg) in the Ural mountains. The authorities also sent with him his mother Nina, the wife of the executed Beria.

In Sverdlovsk, Sergo worked as an engineer. He lived under the maiden surname of his mother as Sergei Alexeevich Gegechkori, with the government-ordered fake patronymic Alexeevich (instead of the proper Lavrentievich). Later, Soviet leader Nikita Khrushchev allowed them to relocate to Kiev, Ukraine. There, Sergo continued his work in defense-related research and development institutions. He also published a book (Beria, 1994) in which he attempted to whitewash his father's record. Sergo Beria died in 2000 (Fig. 3.14).

Fig. 3.14 Sergo Beria in 2000. Photograph from http://www.gordon.com.ua/books/heroes/beriya/[Ins]; accessed 6 September 2011, courtesy of Dmitrii Gordon, Kiev, Ukraine.

---

contested Russia-controlled region in Georgia). Another large group of German engineers engaged in ballistic missile work assembled at an island on Lake Seliger.[71] Eventually all Germans, after a "cooling off" period, repatriated to the Soviet-occupied East Germany.

The government again reorganized KB-1 in September 1953. The bureau now had three chief designers responsible for three major directions of work in the bureau. Alexander Raspletin became chief designer in charge of air

---

[71] Albring and Vinke, 1991; Riehl and Seitz, 1996; Gruntman, 2004.

defense systems. Two other chief designers oversaw development of air-to-sea and air-to-air guided missile systems, building on the successful Kometa. The bureau also included two lead, or main, research departments and four smaller specialized units. Grigorii Kisun'ko headed one such lead department, focused on air-defense systems. It employed more than 1000 people. The other lead department concentrated on guidance of air-launched missiles. In December 1953 the Ministry of Defense introduced in KB-1 the conventional system of military representatives for quality control, common in the development and procurement of weapons.

Testing of the radars and missiles of the S-25 system steadily progressed at Kapustin Yar. The time had come to plan for the future. Raspletin and TGU science chief Shchukin decided to focus KB-1 on two areas, improving air defense efficiency against countermeasures and transportability. The stationary S-25 could defend only a unique object such as an important city. The technology for a mobile system capable of defending against multiple aircraft, similarly to the S-25 but at the same time transportable and quickly deployable in a desired location, was not yet attainable.

Therefore, Raspletin decided first to concentrate on a transportable system with 360-deg coverage but engaging a single target aircraft. He assigned Boris Bunkin to lead preliminary studies of the new weapon. One month later, on 20 November 1953, the USSR Council of Ministers issued a decree, No. 2838-1201, authorizing development of this new air defense system, with KB-1 as the lead, and designated it the S-75.

At this time KB-1 also stepped up its effort to produce an antiaircraft missile alternative to Lavochkin's V-300. Initiated earlier, the in-house development of the 32-B missile provided an important foundation for this effort. Al'perovich, who by that time had become deputy chief designer of the S-25 under Raspletin, explained that,

> For our [KB-1] rocketeers working on the 32-B missile and operating inside KB-1, the task [of building the missile for the S-25] was beyond their capabilities. At the same time, Petr Dmitrievich Grushin,[72] working since 1951 in Lavochkin's OKB as the first deputy of the chief designer, was striving for independent work. As a result, the new separate rocket design bureau OKB-2 headed by Grushin ... was formed in the administrative organization of Glavspetsmash [Directorate of Special Machine Building formed from TGU] in late 1953 on the foundation of the group of [Dmitrii L.] Tomashevich [in KB-1]. It was located on the territory of the same plant, No. 293, transferred in the beginning of the year to KB-1 to support work on the 32-B missile. ...
>
> Excellent designer Grushin ... was not a simple man. He valued specialists and built up a great creative collective. At the same time not all

---

[72]Petr Dmitrievich Grushin, 1906–1993.

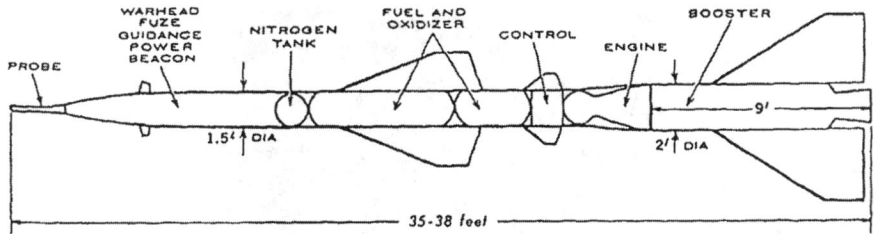

Fig. 3.15 Drawing of an early operational surface-to-air S-75 (SA-2, Guideline) missile observed at a military parade in Moscow on 7 November 1957. Figure 3 (following page 7) from Central Intelligence Agency, 1959.

talented people could work smoothly with him. Tomashevich was among those. He returned back to KB-1 after spending a little more than two years in the new OKB of Grushin.[73]

The deployed S-25 air defense system relied on Lavochkin's missiles. Raspletin's next antiaircraft system, the S-75 (Fig. 3.15), would use missiles designed and built by the new OKB headed by Petr D. Grushin (Fig. 3.16). This organization would become known as the special design bureau OKB Fakel. (Fakel means a plume, as in a rocket plume.)

The first variant of the new S-75 entered service in 1957. It was this missile that shot down the U.S. U-2 reconnaissance aircraft piloted by Francis Gary Powers[74] near Sverdlovsk in 1960. Bunkin and Grushin later earned the highest awards of the Heroes of Socialist Labor for development of the S-75. The antiaircraft missile and the associated air defense system became known in the West as SA-2 and Guideline. In addition to wide deployment in the Soviet Union, more than 800 S-75 units were delivered to two dozen foreign countries. Some of the missiles remain operational to this day.

Raspletin's KB-1 and its successor design bureau Almaz (part of the present day Almaz-Antei Corporation) built on the S-25 and S-75 and developed many advanced air defense systems, including S-125 (SA-3), S-200 (SA-5), S-300 and its variants (SA-10, SA-20), and S-400 (SA-21). Grushin, a man of rough character and a forceful, uncompromising personality,[75] designed missiles for numerous air defense systems for Almaz (S-125, S-200, S-300) as well as for other organizations, such as Osa (SA-8) and Tor (SA-15). Important for this story, he built the V-1000 missile for the first intercept of a ballistic missile warhead in 1961.

In the midst of the reorganizations of KB-1, testing of the maturing first Soviet antiaircraft guided missile system (S-25) continued at Kapustin Yar. By fall 1954, the military had fired almost 65 missiles against parachute-

---

[73]Al'perovich, 2003, pp. 120, 121.
[74]Francis Gary Powers, 1929–1977.
[75]Semenov, 2008, pp. 407–408.

Fig. 3.16 Petr D. Grushin (1906–1993) developed numerous antiaircraft missiles as well as the V-1000 that first intercepted a ballistic missile warhead. Photograph courtesy ITAR-TASS.

deployed and then airplane targets. The tests culminated with engaging 20 targets by 20 missiles.

The time had come for the series production and deployment of the operational air defense system of Moscow.

Chapter 4

AIR DEFENSE SYSTEM OF MOSCOW

EARLY WARNING RADARS

The S-25 (Berkut) air defense system of Moscow included a network of early warning radars; antiaircraft guided missile launch sites with fire control radars; depots for storage, preparation, and maintenance of missiles; and command and control centers. The military built extensive supporting infrastructure of roads, power-generating and distributing facilities, communication circuits, military bases, barracks, and living quarters.

The series production of radars, missiles, and ground equipment evolved into a huge undertaking. The Soviet government assigned numerous industrial facilities to the effort. Many mobilized organizations had expertise in entirely different manufacturing areas, which required retraining engineers and workers. Fifty industrial plants in Moscow, Leningrad, Gorky, Zagorsk, Tashkent, and many other towns across the USSR built antennas, mechanisms, vacuum tubes, electronic components, radio command systems, and computing devices.

The Soviet state did not stint in providing resources for weapon programs. As a former high-ranking Soviet intelligence officer, Alexander Orlov, described such policies a few years later in his testimony to the U.S. Senate, "... stress on heavy industry for war armaments and nothing for the consumer, no consumer goods, very little food, and the shortages of food and goods and the hardship of the Russian people continue."[1] Occasionally, even consumers enjoyed unexpected benefits from the gigantic military effort. For example, when the first television sets appeared in the country, coaxial cables were in short supply. Ramping up cable production for air defense resolved this issue.

For early warning, the new Moscow air defense system relied on a ring of A-100 radars operating at the 10-cm wavelength (frequency 3 GHz) and

---
[1]Orlov, 1957, p. 3469.

deployed at distances of 125–155 miles (200–250 km) from the city. A typical A-100 site included two Kama radar units (Fig. 4.1), measuring distances to aerial targets and their altitudes, respectively.

By the early 1960s, U.S. intelligence analysts had identified 10 early warning radar sites around Moscow (Fig. 4.2). A U.S. U-2 reconnaissance aircraft

Fig. 4.1 Operations bunker built to withstand significant overpressure of an early warning A-100 radar site of the Moscow air defense system S-25 (bottom). The installation included Kama radars for measuring distances to targets (top left) and determining their altitudes (middle right). Photographs courtesy of Mikhail Khodarenok.

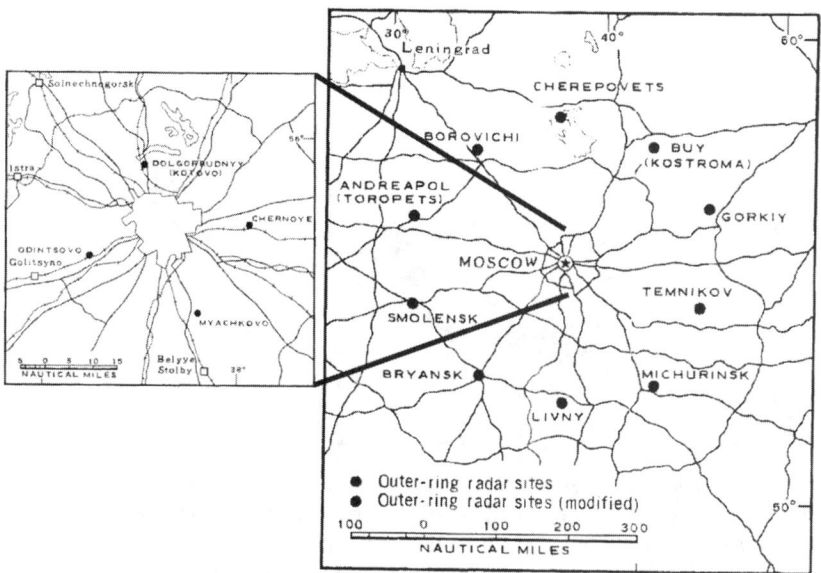

Fig. 4.2 U.S. intelligence report maps showing the locations of 10 early warning A-100 radar sites around Moscow (right) and four similar sites in the immediate vicinity of the city (left). The radar at Dolgoprudny (Dolgoprudnyi) north of Moscow served as a pilot site for A-100 development. Figures from National Photographic Information Center, 1964b, p. 2 (left map) and p. 4 (right).

reached Moscow for the first and only time on 5 July 1956.[2] On its way, it flew over a number of industrial and military installations, including S-25 air defense sites. The U-2 plane photographed the A-100 radar near Smolensk. Then, another deep-penetration overflight on 5 February 1960, reached Temnikov (near the nuclear weapon development center Arzamas-16) and Michurinsk southeast of Moscow.

A contemporary U.S. report observed that these three A-100 sites at Michurinsk, Temnikov, and Smolensk

> ... are similar in design and contain a secured (double-fenced) operational area and a nearby fenced support area. Each operations area includes up to seven radars, an operations bunker, operational support structures, and probable transmitting and receiving antennas oriented toward Lozhki, 26 n.m. [48 km] northwest of Moscow. The support area contains single-family-type housing units, barracks, and recreational facilities.[3]

Figure 4.3 shows the secured area of an A-100 base near Michurinsk with an operations bunker designed to withstand significant overpressure, radar

---
[2]Pedlow and Welzenbach, 1998, p. 105.
[3]Central Intelligence Agency, Photographic Intelligence Center, May 1960, pp. 5, 8.

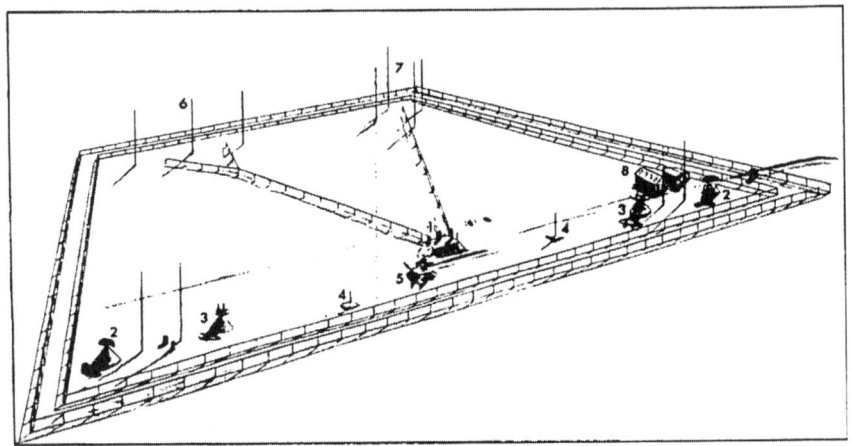

Fig. 4.3 Perspective sketch of an A-100 early warning radar site 5.5 nautical miles (10 km) from Michurinsk (Fig. 4.2) southeast of Moscow, overflown by a U-2 aircraft (Mission 8009) on 5 February 1960. 1: operations bunker; 2, 3: two types of radar on mounds; 4, 5: possible (not identified) radar; 6: two communications antennas; 7: diamond-shaped antenna; 8: two gable-roofed support structures. From Central Intelligence Agency, Photographic Intelligence Center, 25 Feb. 1960, p. 3.

units on mounds, communications antennas, and support structures. Another intelligence report described that "[t]he operations area is double fenced, has the shape of a trapezoid, and measures approximately 3,000 by 2,400 feet [910 by 730 m]. ... The support area contains bachelor quarters, family quarters, support structures, and utility facilities."[4]

From August 1960, U.S. Corona satellites began to routinely provide space reconnaissance photographs of the Soviet Union, with gradually improving resolution. In addition to the external ring of 10 A-100 sites, satellite photographs revealed 4 other A-100 radars in close proximity to Moscow, at distances of 14 nautical miles (26 km) from the city center (Fig. 4.2, left). One of these, at Dolgoprudny north of Moscow, served as a pilot site for A-100 development.

A U.S. intelligence report noted in 1964 that

> [t]he greatest apparent difference between an inner-ring and outer-ring [A-100] site is in the configuration of the fenced operations area—rectangular for the inner and trapezoidal for the outer. This difference occurs because the outer ring site fence is extended to enclose high-frequency communications antennas situated to the rear of the radar positions, antennas which are apparently not needed at the inner-ring sites in view of the proximity to Moscow.[5]

---

[4]Central Intelligence Agency, Photographic Intelligence Center, 26 Feb. 1960.
[5]National Photographic Interpretation Center, 1964b, p. 1.

Two Rings of Fire

Two rings containing a total of 56 air defense sites, each with a B-200 radar and antiaircraft missile launch pads, surrounded Moscow. Al'perovich described that

> [t]he volume of construction required for deployment of the Moscow [S-25] air defense system was enormous. It was necessary to lay two automobile ring roads (for transportation of antiaircraft missiles from their storage depots to air defense sites) at distances 50 and 90 km from Moscow with overpasses and bridges at the intersections with highways, railroads, and waterways; build high-capacity [high-voltage] lines for supplying electric power [and manufacture and install] backup diesel power generators; construct bases for missile storage and preparing them for operational deployment; and build at each of the 56 air defense sites command centers with concrete bunkers for the Central Guidance Radar [B-200], missile positions with 60 launch pads [at each site], and a network of access roads as well as small towns as living quarters for officers [and their families] and barracks for soldiers.
>
> The [Ministry of Internal Affairs] MVD conducted construction [of the air defense sites] using prisoners. ...[6]

Two new automobile roads encircling Moscow had to be built first. People later nicknamed them betonkas, which literally meant roads made of concrete, or beton in Russian. These ring roads naturally fit into the centralized transportation infrastructure of the region—highways radiated in all directions, as spokes in a wheel, from the capital Moscow. A third, smaller ring road, about 20 miles (32 km) in diameter, would mark the city's administrative boundaries for many years. In total, the government built 310 miles (500 km) of concrete roads, which prominently stand out in maps of the region to this day (see also Fig. 9.6, Chapter 9).

Betonka Ring Roads

Initially, sections of the two new betonka ring roads around Moscow were kept closed to the public. Then, authorities opened them for general traffic. The roads also provided convenient access to forested areas favored by many for highly popular mushroom hunting.

Amusingly, many Soviet maps did not show betonka roads in the 1960s and 1970s, ostensibly as secret military assets. Obviously, space reconnaissance had laid bare road networks by that time. The ways of

---

[6]Al'perovich, 2003; p. 128.

> bureaucracies are often incomprehensible and irrational, especially in all-powerful totalitarian states. With time, betonka roads became integral parts of regional transportation networks.

Fifty-six air defense sites, 24 in the inner ring and 32 in the outer ring, formed the S-25 system (Fig. 4.4). Each site housed an air defense regiment. Prisoners usually first constructed the barracks for soldiers of the unit. Then they lived in these barracks while working on other installations of the base and supporting infrastructure.[7] Prisoners had no idea what they were building. Officers and guards described to them radar blockhouses as vegetable storage buildings and referred to missile launch fields as pasture for cattle.[8]

Each of the 56 sites included the Radiotechnical Center (Radiotekhnicheskii Tsentr) operating the Central Guidance Radar (Tsentral'nyi Radiolokator Navedeniya) B-200; the missile launch area; technical positions for missile storage, preparation, and maintenance; barracks for enlisted men; a settlement for officers and their families; and an electric power distribution station. In addition, there were regimental headquarters, sports fields, and other supporting facilities typical for an isolated military post.

THE S-25 SITE

With a few exceptions, the military built S-25 bases in forests, which were common for the area, and away from the existing towns and villages. Figure 4.5 shows an aerial photograph of one such site of the inner ring near Kubinka (Fig. 4.4) that a U-2 reconnaissance aircraft passed over on 5 July 1956, approaching Moscow from the west-southwest.

At each site, its air defense regiment consisted of the headquarters, a unit operating the guidance radar, a missile launch divizion (a battalion),[9] a unit operating technical facilities for missile preparation and maintenance (Missile Preparation Position, or Punkt Podgotovki Raket, PPR), and support units such as guard, automobile, and logistics platoons.[10]

The missile launch divizion included two batteries, with five missile platoons each. A platoon consisted of two squads, each with a sergeant in charge of two three-man launch crews. During mobilization, platoons expanded to three launch crews each. After deployment of the new variant 215 (antiaircraft

---

[7]Veideman, 2007, Part 2.
    The author of this manuscript, Oleg Veideman, served as a conscript soldier from 1971 to 1973 at an S-25 air defense site near Moscow. A keen observer, he wrote his extensive recollections of those days with minute details of routine operations and life, seen through a soldier's eyes. His curiosity also captured some history of system development and construction from veteran officers of his regiment.
[8]Semenov, 2008, p. 447.
[9]A divizion is a unit equivalent to a battalion and was common in Soviet artillery and missile troops.
[10]Veideman, 2007, Part 2.

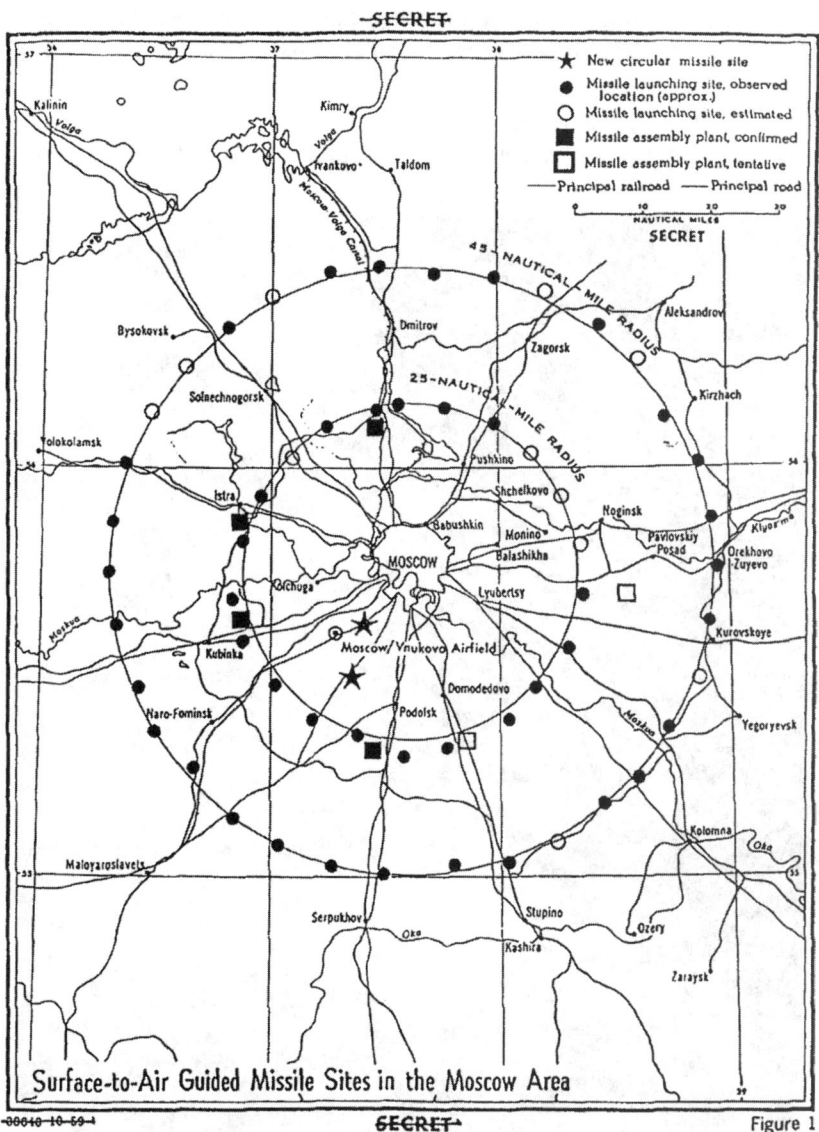

Fig. 4.4 Locations of 56 air defense sites, each housing a regiment, in two rings around Moscow. The CIA had compiled this map in November 1959 before the advent of space reconnaissance that would reveal many of the country's secrets. It shows 22 and 34 sites in the inner and outer rings, respectively, instead of the correct 24 and 32 sites. From Central Intelligence Agency, 1959, figure 1, following p. 6.

missiles that carried nuclear warheads), one platoon underwent special training to operate them. In addition, the launch battalion included a separate detachment conducting final checks of the fully fueled missiles delivered to launch pads. This unit also attached radio-controlled fuses to warheads.

When viewed from above, the launch area roads resembled a herringbone pattern (Figs. 4.5 and 4.6). The central road, the "backbone," was aligned along the radial direction away from Moscow and determined the orientation of the defended sector. Ten side roads on both sides constituted two launch areas assigned to two batteries, respectively, of the launch battalion. Each area provided

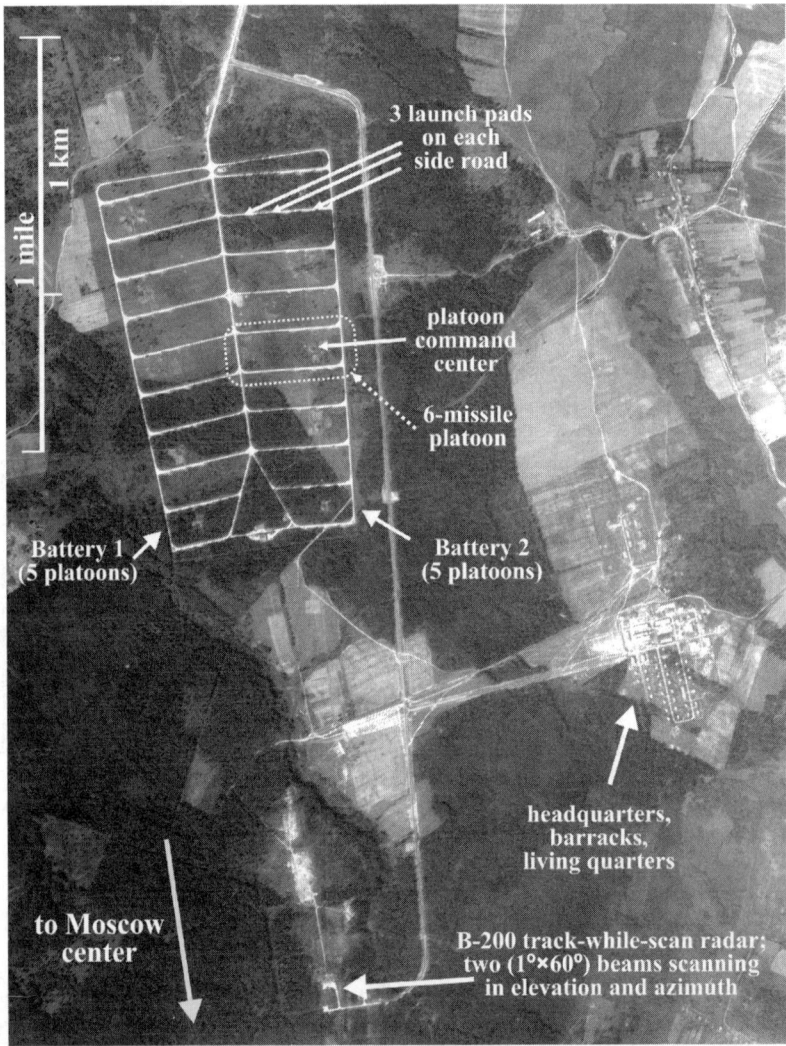

Fig. 4.5 Aerial photograph of a typical air defense site of the S-25 system near Kubinka. The vertical camera obtained this photograph from a U-2 reconnaissance aircraft approaching Moscow from the west-southwest on 5 July 1956. The characteristic herringbone road pattern stands out, identifying the S-25, or SA-1, antiaircraft system. Original photograph (U-2 Mission 2014; 5 July 1956) from the U.S. National Archives and Records Administration; photograph identification, interpretation, processing, and augmentation by Mike Gruntman.

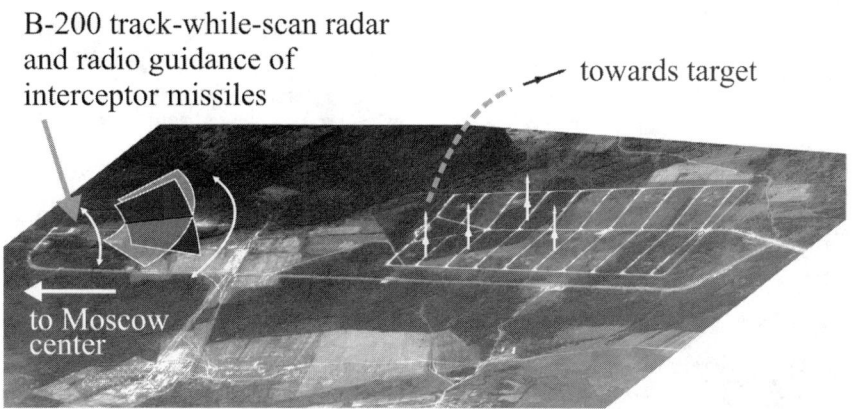

Fig. 4.6 Schematic of an S-25 air defense site (Fig. 4.5). The B-200 radar (left) scans the defended sector with two independent fan beams in elevation and azimuth. Antiaircraft missiles are launched from the pads (on the roads forming the herringbone pattern) and guided to their targets by the radio control system of the B-200 radar. (Drawn missiles are not to scale.) Original photograph (U-2 Mission 2014; 5 July 1956) from the U.S. National Archives and Records Administration; photograph identification, interpretation, processing, and augmentation by Mike Gruntman.

launch positions for 30 missiles, with 6 missiles per platoon, for a total of 60 launch pads for the regiment. Each launch position had mechanisms for erecting the missile, the launch table, and the control box (Fig. 4.7).

After 1972, camouflage nets periodically covered many air defense sites during overflights by U.S. reconnaissance satellites. As former serviceman Veideman recalled,

> In January 1972 ... after 9 o'clock in the morning there was a telephone call [to the units] from an operations duty officer who passed on the so-called signal "Omega" for the day. The message listed time intervals of [expected] overflights by American photoreconnaissance satellites. ... During these intervals it was forbidden to conduct any work [outdoors] and to move on roads in unit formations; all technical positions and missiles had to be with their [camouflage] covers put on. At first the list had only a few 5–15 minute intervals. By the end of the year, the daily list included a couple dozen intervals, with the duration of some [intervals] reaching 45 minutes. ...
>
> In July, the new signal "Omega-2" was added ... providing time intervals of overflights by electronic intelligence satellites.[11]

The satellite overflight warning system originated from the Soviet missile defense and antisatellite weapon efforts of the 1960s. These early programs led to the establishment of dedicated radar and optical facilities built by

---

[11] Veideman, 2007, Part 6.

Fig. 4.7   Top: V-300 antiaircraft guided missile of the S-25 system being erected on a launch pad; bottom: V-300 missiles ready for launch. Photographs from *Sistema-25* ... , pp. 39, 41.

leading design bureaus. In January 1970, the Minister of Defense activated the Center for Control of Space (Tsentr Kontrolya Kosmicheskogo Prostranstva, or TsKKP),[12] which was charged with observing, cataloging, and characterizing orbital objects. Later in the year, TsKKP began issuing warnings to military units with predictions of overflights by reconnaissance satellites.[13]

Each S-25 air defense regiment represented an autonomous unit, with all the infrastructure and living quarters for its officers, their families, and enlisted men. It included sports facilities, medical units, and even a small guardhouse for punished soldiers. A civilian organization, Mosenergo, supplied electric power to the sites as well as to everybody else in the Moscow region. High-voltage lines carried power to distribution stations with transformers near the locations of the regiments. The stations provided power to technical positions and living quarters. Each site also operated backup diesel generators, which started automatically, for supplying emergency power in case of disruption. In addition, soldiers always turned on one such generator manually when units went on combat alert, so it stayed in "hot" reserve.[14]

S-25 OPERATIONAL

Military and industrial teams had completed the installation, assembly, and tests of all 56 air defense sites around Moscow by early 1955. The Central Committee of the CPSU and the USSR Council of Ministers formally accepted the system as operational by special ordinances on 14 April 1955 (No. 720-435) and 7 May 1955 (No. 893-533), respectively. The new First Special Army of Air Defense took control of the sites. The Army consisted of four air defense corps responsible for four directions. Each corps included 14 regiments, with 8 regiments and their bases in the outer ring and 6 in the inner ring.

The completed S-25 system was to simultaneously engage up to 1120 aerial targets with velocities up to 310 mph (500 kph). The missiles could shoot down aircraft at distances of 16–28 miles (35–45 km) and altitudes of 2–16 miles (3–25 km or 10,000–80,000 ft). Operational missiles demonstrated miss distances better than 165 ft (50 m) in intercepts. Initially, antiaircraft missiles carried high-explosive warheads with a mass of 520 lb (235 kg). As rockets and engines improved with time, the warhead mass also increased to almost 900 lb (400 kg). It took about five minutes for the radar to reach combat readiness after receiving the activation order.

The same ordinance of the Council of Ministers that accepted the S-25 as operational and the following order of the Minister of Defense, No. 00112 from 21 May 1955, reorganized the Fourth Directorate into the Fourth Chief

---

[12]*Control of space* is a Russian expression meaning *space situational awareness* rather than its literal translation, *space control*; the latter has a substantially different meaning.
[13]Pervov, 2003, p. 164.
[14]Veideman, 2007, Part 3.3.

Directorate of the Ministry of Defense. The new Chief Directorate reported to the Commander of Air Defense of the Soviet Army. Lieutenant General Pavel N. Kuleshov became head of the Fourth Chief Directorate with Lieutenant General Georgii F. Baidukov[15] as his deputy.

At first, the Fourth Chief Directorate oversaw technical support for the S-25. With time, it evolved into the main procuring organization of all major air defense systems, on behalf of the Ministry of Defense and the powerful Military-Industrial Commission of the Council of Ministers. (The latter government body, Voenno-Promyshlennaya Komissiya, or VPK, emerged in its final and lasting organizational shape in December 1957.) The Fourth Chief Directorate later would expand to procure missile defense systems.

The Soviet state showered developers of the "practically impenetrable air defense system" S-25 with awards. Seven men—Raspletin, Kisun'ko, Shchukin, Vetoshkin, Lavochkin, Isaev, and Mints—received the highest nonmilitary state decorations of the USSR, the titles and gold medals of the Heroes of Socialist Labor.[16] For Lavochkin, who had received the Hero award during World War II for his fighter planes, this was the second gold medal.

Two of the awarded men (Raspletin and Kisun'ko) were the leading specialists of KB-1, two (Shchukin and Vetoshkin) represented the overseeing government organizations, and three others headed main contractors (Lavochkin, the antiaircraft guided missile; Isaev, the engine for the missile; and Mints, powerful transmitters for the radar as well as civil engineering design of air defense sites and supporting infrastructure). The government also decorated a large number of managers, engineers, military officers, and Communist Party and state officials with various lesser orders and medals.

The S-25 air defense system was continuously modernized and remained operational until 1984. The missile variant 215 introduced a major modification—an atomic warhead intended to intercept a group of approaching airplanes in a tight formation. The military conducted its live test at Kapustin Yar on 19 January 1957. A 10-kton warhead successfully detonated at an altitude of 34,000 ft (10.4 km) and destroyed two IL-28 bombers separated by 2600 ft (800 m).[17] Later, more powerful radar transmitters improved the S-25's sensitivity to smaller targets, and faster interceptor missiles enabled engagement of aircraft with higher velocities.

In 1956, KB-1 began work on another, stationary air defense system around Leningrad, similar to S-25 and codenamed S-50. The S-50 plan called for a ring of 35 air defense sites with the upgraded B-200 radar and V-300 missiles

---

[15]Pavel Nikolaevich Kuleshov, 1908–2000; Georgii Filippovich Baidukov, 1907–1994.

[16]The Order of Lenin was the highest state decoration in the USSR. The title of the Hero of the Soviet Union (for exceptional military exploits) and the Hero of Socialist Labor (for exceptional nonmilitary accomplishments) included a special gold star medal in addition to the Order of Lenin. The gold stars of the Hero of the Soviet Union and the Hero of Socialist Labor had slightly different and easily recognizable shapes.

[17]Pervov, 2001, pp. 121, 122; Greshilov et al., 2008, p. 267.

around the city at distances of about 30 miles (50 km). As development and deployment of the new mobile S-75 (SA-2) antiaircraft system rapidly progressed, the government terminated work on the S-50 in 1958.

In the early 1960s, the military collocated new S-125 (SA-3, Goa) air defense batteries, also developed by KB-1, at some S-25 sites in the external ring around Moscow.[18] The S-125 antiaircraft missiles provided complementary capabilities against low-altitude targets. In addition, several external ring sites would later expand to accommodate the firing complexes of the A-35 operational missile defense system of Moscow.

U-2 AIRCRAFT OVER MOSCOW

The first unexpected test of the S-25 came one year after declaring the air defense system operational and decorating its developers. A U.S. U-2 reconnaissance plane flew over Moscow on 5 July 1956, and the S-25 failed to stop it. It was the only U-2 passage over the Soviet capital. This flight, Mission 2014, and another, Mission 2013, over Leningrad on the previous day piloted by Carmine Vito and Harvey Stockman,[19] respectively, began a program of deep-penetration reconnaissance overflights of the USSR that lasted until 1960. The United States implemented a policy of peacetime overhead reconnaissance of the denied areas to avoid a fatal miscalculation during the Cold War confrontation and thus reduce the risk of war.

U.S. President Dwight D. Eisenhower considered aircraft overflights of the USSR as a temporary measure of last resort and kept the number of intrusions down to the absolute minimum. He stopped the program after an S-75 missile shot down an airplane piloted by Francis Gary Powers over the Ural mountains on 1 May 1960. This was the last, 24th deep-penetration U-2 flight over the Soviet Union. Less than four months later, the first successful Corona space mission photographed the Soviet Union and opened a new era in monitoring denied areas. Satellite reconnaissance would eventually make possible verifiable arms control and limitation agreements between the USSR and the United States.[20]

FIRST DEEP-PENETRATION OVERFLIGHTS OF THE USSR

Reports on the Soviet radar coverage of the first two overflights of the Soviet Union [on 4 and 5 July 1956 that overflew Leningrad and Moscow, respectively] became available on 6 July. These reports showed

---

[18]e.g., National Photographic Interpretation Center, 1964a.
[19]Reed, 2004, p. 52.
[20]e.g., Ruffner, 1995; Pedlow and Welzenbach, 1998; Gruntman, 2004, Chapter 16.

that, although the Soviets did detect the aircraft and made several very unsuccessful attempts at interception, they could not track U-2s consistently. Interestingly, the Soviet radar coverage was weakest around the most important targets, Moscow and Leningrad, and the Soviets did not realize that U-2s had overflown these two cities (Pedlow and Welzenbach, 1998, p. 106).

The 4 and 5 July overflights brought strong protest from the Soviet Union on 10 July in the form of a note handed to the US Embassy in Moscow. The note said that the overflights had been made by a "twin-engine medium bomber of the United States Air Force" and gave details of the routes flown by the first two missions. The note did not mention Moscow or Leningrad, however, because the Soviets had not been able to track these portions of the overflights (Pedlow and Welzenbach, 1998, p. 109).

Soviet Diplomatic Note of 10 July 1956

The Embassy of the Union of Soviet Socialist Republics presents its compliments to the Department of State of the United States of America and, on instructions from the Soviet Government, has the honor to state the following.

According to precisely established data, on July 4 of this year at 8:18 a.m. Moscow time, a twin-engine medium bomber of the United States Air Force appeared from the direction of the Zone of Occupation in West Germany, crossed over the territory of the German Democratic Republic, and invaded the airspace of the Soviet Union from the direction of the Polish People's Republic at 9:35 a.m. The aircraft which violated the airspace of the Soviet Union flew along the Minsk–Vilnius–Kaunas–Kaliningrad route, penetrating the territory of the Soviet Union in the Grodno area to a depth of 320 km and remaining over the said territory for one hour and 32 minutes.

On July 5 of this year at 7:41 a.m. Moscow time, a twin-engine medium bomber of the United States Air Force appeared from the direction of the American Zone of Occupation in West Germany ... and at 8:54 a.m. invaded the airspace of the Soviet Union in the area of Brest. ... The aircraft violating the air boundary of the Soviet Union flew along the Brest–Pinsk–Baranovichi–Kaunas–Kaliningrad route, penetrating Soviet territory to a depth of 150 kilometers and remaining one hour and 20 minutes over the said territory.... The same day another twin-engine

> bomber of the United States Air Force invaded the airspace of the Soviet Union and penetrated Soviet territory to a considerable depth.
>
> (U.S. Department of State, 1956, pp. 191–192)

On 5 July 1956, a U-2 plane took off from an airfield in West Germany, crossed East Germany and Poland, and flew over Baranovichi, Minsk, and Smolensk toward the Soviet capital. On that day heavy clouds covered the center of Moscow and the areas to the north and east of the city (Fig. 4.8). If the cloud edge were only 1.3 miles (2 km) farther north, then the U-2 would have photographed the main building and other facilities of the S-25 developer, KB-1 (Fig. 4.9).

Heavy clouds also prevented obtaining photographs of the Kremlin and such primary targets of the mission as a rocket engine design bureau and plant in Khimki, headed by Valentin P. Glushko, and a ballistic missile development center headed by Sergei P. Korolev in Podlipki.[21] The mission did return, however, a photograph of the Fili airframe plant (Fig. 4.9) in Moscow that built the strategic Bison (Myasishchev-4) bombers. The photographs also showed the southwestern part of the city with the landmark Luzhniki sports stadium and newly erected buildings of Moscow State University (Figs. 4.8 and 4.10). The U-2 aircraft ventured farther east for about 125 miles (200 km) and then returned back to a home airfield in West Germany on a path to the north.

In an unusual twist, the overflight helped to improve the conditions for U.S. officials stationed in the Soviet capital. Until the 1960s, the Soviet security establishment had prevented the publication of detailed street maps of Moscow, so essential for the everyday life of city inhabitants and for numerous visitors. This restriction was not unlike the omission of betonka roads in the maps of the region. Consequently, the Central Intelligence Agency (CIA) initiated a special program in the 1950s to make accurate city maps to be used by the personnel of the U.S. Embassy in Moscow.[22] Later, they were also provided to diplomats of friendly countries. Aerial and later space reconnaissance photographs significantly contributed to map production.

Several days after the U-2 had reached Moscow, two other aircraft flew over Lithuania, Belarus, and the capital of Ukraine, Kiev. Then, another airplane photographed southern parts of Ukraine, concluding the first set of missions during a 10-day window authorized by President Eisenhower.

The leading designers of the most advanced Soviet operational air defense system did not know—incredibly—about this overflight of Moscow. Al'perovich, who had been deputy chief designer of the S-25, even explicitly

---

[21] Pedlow and Welzenbach, 1998, p. 105; Gruntman, 2004, p. 279.
[22] Baclawski, 1997.

Fig. 4.8 Oblique aerial photograph of the western part of Moscow, about 7 × 12 miles (10 × 19 km), obtained by the left camera of the U-2 aircraft (Mission 2014) on 5 July 1956. The black box area is shown in Fig. 4.9. One can see an easily recognizable Luzhniki stadium in the southeast corner of the photograph. Original photograph from the U.S. National Archives and Records Administration; photograph identification, interpretation, processing, and augmentation by Mike Gruntman.

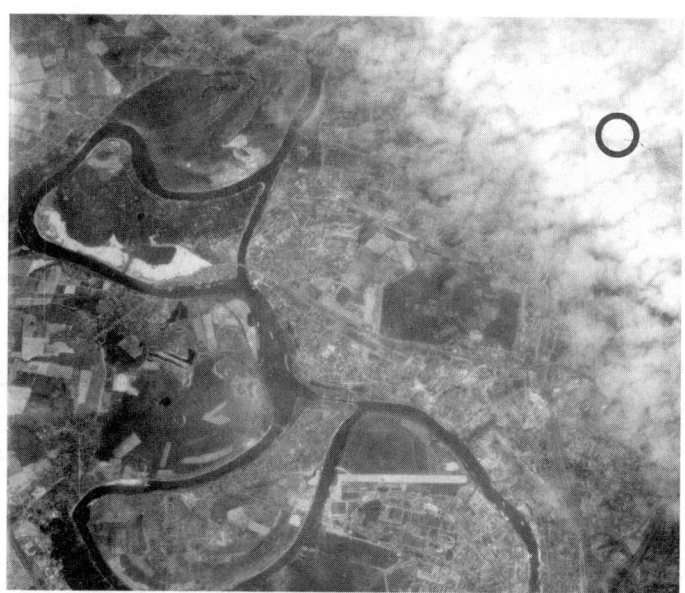

Fig. 4.9 Aerial photograph (black box in Fig. 4.8) showing characteristic bends of the Moscow river in western Moscow. The oblique (left) camera of the U-2 aircraft (Mission 2014) obtained this photograph on 5 July 1956.

The circle shows the location of KB-1, the developer of the S-25 (Berkut) Moscow air defense system designed to prevent such an intrusion. (The U-2 film is slightly damaged in this area of the photograph, with the transparent tape used to repair it visible.) The main building of KB-1 and its surrounding facilities would have been clearly seen if the cloud edge were only 1.3 miles (2 km) farther north. The aircraft is roughly 8 miles (13 km) to the south of KB-1. Heavy clouds also covered clusters of missile and future space development centers and design bureaus in Khimki and Podlipki to the north, which were among priority targets of this U-2 mission.

A white horizontal strip, limited by the Moscow river on both sides, in the lower part of the photograph, is the runway at the Fili airframe plant that built the strategic heavy bomber Bison (Myasishchev-4). Such photographs of design bureaus and manufacturing plants as well as those of operational strategic aviation airfields helped to resolve disagreements over the possible "bomber gap."

Original photograph from the National Archives and Records Administration; photograph identification, interpretation, and processing by Mike Gruntman.

stated in the print edition of his extensive memoirs, published in 1999, that the U-2 had not overflown Moscow:

> Near Kiev, air defense artillery opened fire [at a U-2 aircraft]. However, the U-2 was at an altitude higher than 20 km. Only guided antiaircraft missiles could have destroyed the aircraft at such altitudes. Guided missile complexes [S-25] did exist only near Moscow that the U-2 had not approached. ...[23]

---

[23]Al'perovich, 1999, p. 76 (printed version of the book).

Fig. 4.10 Aerial reconnaissance photograph of southwestern Moscow showing a bend of the Moscow river around the Luzhniki sports arena (top) and the recently constructed main building of the Lomonosov Moscow State University (bottom) on the Lenin (Vorob'evy) Hills across the river. This area is at the bottom-right (southeast) corner of Fig. 4.8. The vertical camera of the U-2 aircraft (Mission 2014) obtained this photograph at 10:08 am Moscow time on 5 July 1956.

The U-2 planes made their first five deep-penetration flights over the Soviet Union in early July 1956. Mission 2014 over Moscow followed Mission 2013, which had flown over Leningrad on the previous day. Then Missions 2020 (9 July), 2021 (9 July), and 2023 (10 July) covered Lithuania; Belarus; and Kiev, Odessa, and the Crimea in Ukraine. The overflights ended the "bomber gap" disagreements. Many in the United States at that time believed that the Soviet Union had been rapidly building up its fleet of Bison strategic heavy bombers. They argued for deployment of additional U.S. B-52 bombers in response. The U-2 photographs did not confirm these alarming estimates and allowed President Eisenhower to rely on accurate intelligence assessments in weapons procurement. Similarly, space reconnaissance Corona satellites helped later to resolve the uncertainty of the "missile gap."

Original photograph from the National Archives and Records Administration; photograph identification, interpretation, and processing by Mike Gruntman.

Later, Al'perovich corrected the text in the electronic posting (PDF file) of the book on the website of the Almaz-Antei Corporation (the successor organization to KB-1): "On July 4, the Day of Independence of the USA, a U-2 airplane was sent to our airspace—the latest achievement of transoceanic

[American] aircraft designers. Then, U-2 aircraft committed several intrusions into our skies, flew over western parts of the country, over Ukraine, and flew over Moscow."[24]

Al'perovich specifically pointed out in a footnote that the earlier print edition of his book had not mentioned—erroneously—the overflight of Moscow, and he corrected the error. He then offered an explanation for why the U-2 had not been intercepted:

> The U-2 flew at altitudes above 20 km and neither fighter airplanes nor air defense artillery could deal with it. It was possible to stop the flight only near Moscow, where the first air defense missile system [S-25] had been deployed one year earlier. But this did not happen. The military did not rush [to put] the new system on operational combat duty, and they conducted training using engineering [not operational combat-ready] versions of antiaircraft missiles. After the flight of the U-2, the combat-ready missiles were deployed at launch sites.[25]

Prior to the first U-2 deep overflights of the Soviet Union in 1956, U.S. intelligence had underestimated the abilities of Soviet air defense radars to detect and track U-2 aircraft.[26] The first flights revealed this misconception:

> Photointerpreters examining the films [brought by U-2s] eventually discovered the tiny images of [fighter aircraft] MiG-15s and MiG-17s beneath the U-2s in various pursuit and attack attitudes: climbing, flipping over, and falling toward Earth. It was even possible to determine their approximate altitudes. The photographs showed that the Soviet air defense system was able to track U-2s well enough to attempt interception, but they also provided proof that the fighter aircraft available to the Soviet Union in 1956 could not bring down a U-2 at operational altitude.[27]

Who Knew What and When about the U-2 in the Soviet Union?

Soviet radar detected, but could not continuously follow, the first deep-penetration U-2 overflights of the Soviet Union in July 1956. The U-2 photography showed Soviet fighters attempting interception of the

---

[24]Al'perovich, 1999, p. 75. (PDF file of the book posted at the website of Almaz-Antei Corporation, http://www.raspletin.ru/press-centre/books/alperovich/s75/files/Alperovich.pdf; accessed 30 July 2010.)
[25]Al'perovich, 1999, p. 76 (PDF file of the book).
[26]Pedlow and Welzenbach, 1998, p. 97.
[27]Pedlow and Welzenbach, 1998, p. 108.

planes. At the same time, many top Soviet government officials, officers, and top designers of advanced air defense systems did not know about the event.

In the late 1950s and early 1960s, the future first commander of the Missile and Space Defense Forces, General Yuri V. Votintsev, was in charge of air defenses in the Turkestan military district in the Central Asian part of the USSR. On his watch, the final series of U-2 missions flew over missile proving grounds and nuclear test facilities in Kazakhstan in 1959 and 1960. Votintsev's radar units had first detected the U-2 piloted by Francis Gary Powers, which was then shot down over the Ural mountains by an S-75 missile (Votintsev, 2003).

Votintsev wrote later that the U-2

> ... was deployed in 1956 about which our [Soviet] intelligence did not know anything. Otherwise Chairman of the State Committee for Aviation Technology P. V. Dementiev and aircraft designer A. I. Mikoyan [the letter "M" in the MiG family of aircraft] would not have stated at a meeting of the Politburo of the CC [Central Committee] of the CPSU in April 1960 that there were no airplanes in the world capable of flying for 6 hours and 48 minutes at an altitude of 20 thousand meters [65,600 ft]. It happened so that they [Americans] could and had done [exactly that] and even completed almost 30 intelligence gathering overflights [of the USSR] by that time (Votintsev, 1999, Part 2, p. 27).

Votintsev certainly knew, but many others apparently did not, about the overflights that had begun in 1956.

In 1961, in order to register the record flight of the first man into space (Yuri Gagarin), the Soviet government deliberately identified his launch site as being located at Baikonur, a small town 200 miles (320 km) away from the real launch base at Tyuratam (Gruntman, 2004, p. 318). Top levels of the Soviet government made the decision to announce fake coordinates (e.g., Kamanin, 1995, p. 47; Beloglazova, 2005, pp. 46, 47) in a clumsy attempt to preserve the secrecy of the location. The official Soviet filing of Gagarin's record flight with the Fédération Aéronautique Internationale listed the launch site as Baikonur with coordinates 47°22'00" N and 65°29'00" E instead of the real coordinates (using the same accuracy of 1') 45°55'00" N and 63°21'00" E.

At the same time, a number of people in the Soviet defense establishment and among Communist Party and government officials should have known, as General Votintsev and others did, about earlier multiple U-2 overflights of the Tyuratam launch site, first in 1957 and then in 1959 and 1960. The country could have avoided an embarrassment of

> unnecessary lying about the precisely known location of Gagarin's launch site. The total control of information in a socialist society inevitably becomes counterproductive and harmful.

Soviet leader Nikita Khrushchev put the U-2 overflights and improving Soviet air defenses into perspective:

> We'd been acquainted with the U-2 for some time.... We did everything we could to intercept the U-2 and shoot it down with our fighters, but they could not reach the altitude the U-2 was flying at.... Fortunately, by that time our surface-to-air missiles had already started rolling off the production line. It looked that they were going to be the answer to our problem.[28]

This new air defense system S-75, or SA-2, alluded to by Khrushchev, had been under development by KB-1 since 1953. The CIA accurately predicted in 1957 that "the USSR will develop and could have in operation in 1959 a surface-to-air system capable of carrying 500–800 pound payload to a maximum of 80,000 feet altitude and 50 n.m. range."[29]

Two deep-penetration U-2 missions over the Soviet Union in February (Mission 8009) and April (Mission 4155) 1960 revealed deployment of the S-75 surface-to-air missile (SAM) batteries. The sites showed a characteristic hexagon pattern of roads connecting individual missiles, described as "hexadic SAM sites" in intelligence reports. Mission 8009 alone identified 18 new sites and 4 support facilities (Fig. 4.11) on its path, "swinging in a wide arc from the Aral Sea to the Crimea."[30]

A number of the observed sites displayed operational capabilities. By 1960, the Soviet Union had deployed 80 regiments armed with the newly capable S-75 missiles (Fig. 4.12). One year later, there were 154 such regiments deployed and the unified national system of air defense in place.[31] The number of operational S-75 sites with six launchers at each reached 870 by the end of 1965.[32]

Finally, an S-75 missile brought down the U-2 aircraft piloted by Powers on 1 May 1960. By this time, the Soviet military had installed an operational radar at a 15,100-ft (4,600-m) altitude in the mountains of the Turkestan military district. It had detected the approaching aircraft even before it crossed the border into Central Asia and alerted air defense units.[33]

---

[28]N.S. Khrushchev, 1974, pp. 443, 444.
[29]Central Intelligence Agency, 1957, p. 15.
[30]Central Intelligence Agency, Feb. 1960.
[31]Semenov, 2008, pp. 179, 180.
[32]Central Intelligence Agency, Nov. 1965, p. 8; Central Intelligence Agency, 1967, p. 9.
[33]Votintsev, 2003, p. 16.

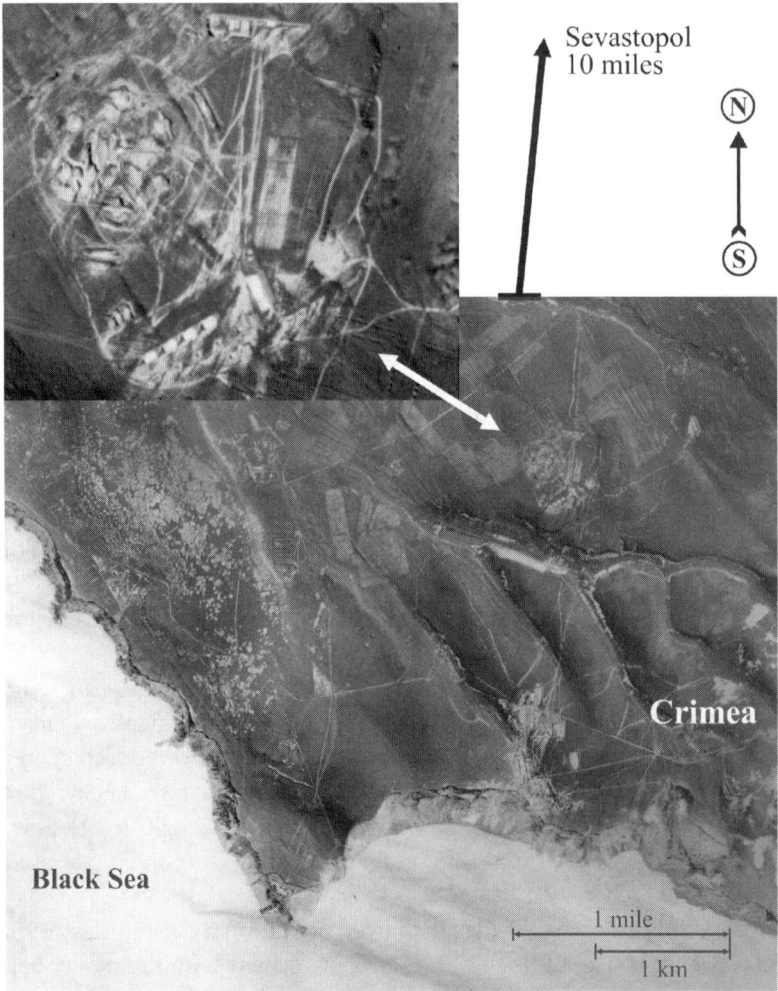

Fig. 4.11 Aerial reconnaissance photograph of new S-75 air defense battery support facilities 10 miles (16 km) south of a major Soviet naval base at Sevastopol in the Crimea, Ukraine. The right camera obtained this photograph from a U-2 reconnaissance aircraft (Mission 8009) heading south, crossing the Black Sea shoreline, and leaving the Soviet Union toward a landing site in Turkey on 5 February 1960. The characteristic "hexadic SAM site" pattern stands out, identifying the S-75, or SA-2, antiaircraft system. Original photograph from the National Archives and Records Administration; photograph identification, interpretation, and processing by Mike Gruntman.

President Eisenhower promptly ordered termination of U-2 deep-penetration flights over the Soviet Union. In 1962, another S-75 missile would shoot down a U-2 reconnaissance aircraft flown by the Republic of China in Taiwan over the communist mainland People's Republic of China (PRC).[34] The Soviet

---

[34]Hua, 2002; Hua, 2005, p. 5.

Fig. 4.12 The author of this book next to a later variant of the S-75 (SA-2) antiaircraft missile on display in Yerevan, Armenia, in 2014. Photograph courtesy of Mike Gruntman.

Union had supplied five S-75 batteries with 62 missiles to the PRC as well as technical documentation for air defense system production[35] before the ideological breakup between these two communist giants terminated the Soviet aid in 1960.[36]

Two Bears in One Lair

The highest state decorations and the title of Hero of Socialist Labor clearly identified Alexander Raspletin and Grigorii Kisun'ko (Fig. 4.13) as the two most powerful leaders in KB-1. The rivalry and tension between them would have a profound effect on the organization of work on missile defense in the Soviet Union. Eventually Kisun'ko would succeed in separating his missile defense program and forming an independent design bureau.

Tensions between Kisun'ko and Raspletin had been building for some time. In February 1953, when Stalin still ruled the country, Raspletin and chief engineer of TGU Valerii D. Kalmykov[37] formally complained in a secret cable to the feared Lavrentii Beria about technical problems with the antennas of the

---

[35] Hua, 2002; Hua, 2005, pp. 4, 5; Semenov, 2008, p. 185.
[36] Hua, 2002; Gruntman, 2004, pp. 440, 441.
[37] Valerii Dmitrievich Kalmykov, 1908–1974.

Fig. 4.13 Grigorii V. Kisun'ko (1918–1998). Photograph courtesy of the Russian Academy of Sciences.

B-200 radar.[38] (Kalmykov would soon rise to prominence as a minister overseeing development of missile defense from the late 1950s until the early 1970s.)

At that time, engineers had been struggling with a nonuniform power output among the six sectors of the rotating antennas of the radar. In their complaint, Raspletin and Kalmykov blamed Kisun'ko, who had the overall responsibility for the antennas, and his close collaborator Mikhail Zakson. (Zakson had been one of the four specialists transferred by Raspletin and Shchukin in 1950 to KB-1 from Berg's TsNII-108.)

Raspletin and Kalmykov's cable triggered a meeting in Beria's office. At the meeting, Kisun'ko took responsibility for the actions of his subordinate, Zakson. Subsequently, Beria ordered Kisun'ko to immediately proceed to the Kapustin Yar test range and fix the problem.

Al'perovich described that when Beria heard the name of Zakson as the chief designer of the antennas, he ominously asked, "Who is that citizen Zakson?"[39] The word "citizen" (rather than a common "comrade") was used in criminal investigations to address the accused. It was also reserved for convicts in prisons and concentration labor camps. Beria's question clearly pointed to this possibility with a predetermined devastating outcome for those involved.

---

[38]Kisun'ko, 1996, pp. 20–24; Belous et al., 2009, p. 58.
[39]Al'perovich, 2003, p. 101.

Importantly, the complaint to Beria by Raspletin and Kalmykov followed a denunciation of Kisun'ko and Zakson one year earlier. Then somebody had written a letter to Stalin, who passed it to Beria for action, accusing them of "wrecking activities."[40] These were the specific words employed to denounce and destroy so many people in the Soviet Union. Since the late 1920s, the country had witnessed a continuous parade of "wreckers" and their "wrecking activities" incessantly uncovered by loyal and "vigilant" common people and trusted communists. The Soviet propaganda demonized these enemies of the state and incited hatred toward them.

Such denunciations became common in totalitarian states where communists controlled all aspects of people's lives. They often led to arrest, torture, execution, imprisonment, and banishment of those involved. In addition to being a tool in political struggle, reporting on somebody also provided a way of settling personal scores among people, removing rivals, and achieving gains. As a result, millions of innocent men, women, and children lost their lives in the Soviet Union alone.[41]

As Kisun'ko recalled, "They [Raspletin and Kalmykov] certainly believed that their letter [about antenna problems to Lavrentii Beria in 1953] would be a good addition to the [earlier] denunciation, as gasoline added on the smoldering coal. It is scary to work with such people."[42] The TGU program engineer for Berkut, Nikolai Ostapenko, who had been present at that memorable meeting in Beria's office, observed that the events had "made an imprint on interpersonal relations" of those involved.[43]

Another major consequential flash point between Raspletin and Kisun'ko resulted from a tragic incident involving a close collaborator of Kisun'ko, an experienced specialist, Iosif I. Vol'man. As with the complaint about antennas, it had also taken place before Stalin's death in 1953.

Kisun'ko sent the overworked Vol'man on a vacation leave over the objections of Raspletin. The latter, according to Kisun'ko, engineered a recall of Vol'man through a KGB officer in an administrative position in KB-1. When Vol'man promptly returned, the KGB men roughed him up with accusations that "he had left for vacation without a proper approval by supervisors, that he had deserted [the work] at a difficult moment. They threatened that he could end up in the 'special contingent' [of employees, that is to be arrested and then continue his work as a prisoner] and in that case they [the KGB] would help him to concentrate on the work on Berkut." The next day Vol'man died from a heart attack.[44]

---

[40]Kisun'ko, 1996, p. 11.
[41]Courtois et al., 1999.
[42]Kisun'ko, 1996, p. 21.
[43]Belous, 2009, p. 58.
[44]Kisun'ko, 1996, pp. 215–218.

Kisun'ko described his growing tensions with Raspletin,

> ... There remained [from the story with Vol'man] smoldering coals of the interpersonal [tense] relations [between us], ready to flare up at an appropriate moment. It was impossible to predict when and how this flare-up would occur and who would be destined to burn in it. I knew however who would be fanning the smoldering coals and adding fuel to the fire. But why would he [Raspletin] need it?
>
> At that time, when I had been asking myself that question, I was not mature [experienced] enough yet to understand that, in any business, one could have a situation of the type of Mozart and Salieri.[45] ...
>
> I hoped that the [spinning up] flywheel of our relations with Raspletin would not reach a destructive phase. Shouldn't Alexander Andreevich [Raspletin] change his attitude after the tragedy with Vol'man? ...
>
> But perhaps as the deputy chief designer [for Berkut] he sees in me a competitor, crossing his [career] path, and he tries to degrade me—a promising young doctor of science—in the eyes of the bosses. But I cannot tolerate being somebody's deputy. Better to be small and independent rather than a big deputy, especially a deputy [as Raspletin] clearly focused on implementation [only] where all principal questions are being settled by chief designers [S. Beria and Kuksenko] directly with the theoreticians and German specialists.[46]

What Kisun'ko meant by referring to himself as a "young doctor of science" was that he had received his second DSc degree in 1951 at a relatively young age. He was then only 33 years old, and this advanced degree unmistakably marked him as a rising star in science. Similarly to some European countries, the Soviet Union had a system of two post-Master-of-Science scientific degrees, the first (kandidat nauk) was equivalent to a PhD and the second (doktor nauk or Doctor of Science, DSc) roughly corresponding to the habilitation qualification in, for example, Germany and France.

In 1946, Kisun'ko published important articles on the emerging theory of microwave guides in leading physics journals.[47] Then, in 1949, he wrote a monograph on this subject, "Electrodynamics of Hollow Systems,"[48] making him a recognized authority in the field. One can gauge the quality and impact of the text from the fact that it has been periodically referred to in Soviet and now Russian scientific literature to this day.

---

[45]The rivalry between two prominent 18th-century composers, Wolfgang Mozart and Antonio Salieri, became a historical cliché. Popular culture accepted a simplified and not exactly accurate story of the jealous and evil Salieri hindering the genius Mozart.
[46]Kisun'ko, 1996, pp. 218, 219.
[47]Kisun'ko, 1946a, 1946b.
[48]Kisun'ko, 1949.

Raspletin defended his kandidat nauk equivalent of the PhD degree thesis at the science council of Berg's TsNII-108 in 1947. He then received his DSc degree in 1956 when he was 48 years old. In contrast to Kisun'ko, the latter degree was bestowed on him for the "totality of his work" without his having to present and defend the thesis describing his work results at a scientific council. In this peculiar and irregular way, the Soviet Union rewarded with prestigious advanced science and engineering degrees some of its important individuals such as chief designers and developers of advanced weapon systems. Preparing a thesis and publishing original results showing personal scientific contributions would have taken their valuable time away from the tasks considered as more important by the government.

Although some such degree recipients were truly qualified and deserving specialists, many others distinguished themselves as highly positioned and successful managers rather than as scientists and engineers. Justified or not, popular science opinion often separated "true" scientists from those who got their degrees by managing important programs. Similarly, many draw a distinction between academicians in the prestigious USSR Academy of Sciences who had been elected to the academy for scientific accomplishments and then rewarded with directorships of academic institutes and those who became members of the academy after serving as institute directors for many years.

Kisun'ko told his side of the story in a book of memoirs published in 1996, where he offered an unflattering view of political scheming and fights among leaders of Soviet missile defense. The assessment of the book's factual accuracy varied. Practically all main participants in Soviet missile defense programs crossed paths with Grigorii Kisun'ko, many as direct competitors, rivals, and antagonists at one time or another. Kisun'ko became embittered by his removal from being the chief designer of the first operational missile defense system, A-35, in 1975. His publication certainly ruffled the feathers of many key individuals, chief designers, and top government officials, and their responses, consequently, were not necessarily factual and sometimes turned personal.

One of Raspletin's close associates, Pivovarov, called the book a slander.[49] One should note here, however, that Kisun'ko described Pivovarov and his actions on some occasions very negatively.[50] Another accomplished missile defense specialist, who had worked with both Kisun'ko and Raspletin, characterized the book as written by somebody "removed for a long time from our business of missile defense and having a grudge against everybody and everything."[51] In another publication, he also described Kisun'ko and Raspletin as "outstanding personalities," with their relations being "not

---
[49]Semenov, 2008, p. 331.
[50]Kisun'ko, 1996, pp. 418, 419.
[51]Golubev, 2003, p. 578.

simple."[52] One active participant in the events noted, however, that Kisun'ko was "often correct" in his critical description of major personalities and their actions.[53] Kisun'ko clearly took on the missile defense establishment in his book. There would be very few impartial, independent, and at the same time knowledgeable individuals outside those who had clashed with Kisun'ko and had scores to settle with him.

Alexander Raspletin passed away in 1967, and he never publicly told his story of the events. In its glory days, the tightly controlled Soviet Union would have never allowed any information about the inner workings of sacred defense matters to be made public.

Karl Al'perovich had been close to Raspletin both professionally and personally for many years. In his memoirs, he chose not to dwell at all on the relations between Raspletin and Kisun'ko, in spite of their being so consequential to the development of early Soviet air and missile defenses. Nobody else among the direct and knowledgeable participants in those events lived long enough or had the energy and desire to offer factual details, even when a period of relative openness in Russia made it possible to do so in the 1990s and early 2000s. In fact, openly challenging the establishment views even in those times would have required unusual personal courage, independence, and integrity. Moreover, such action would have likely carried an unbearable and devastating personal price of breaking relations and friendships with many lifetime associates.

Al'perovich was certainly in a unique position to provide an important eyewitness story should he have decided to do so. He avoided the topic altogether in his books. My guess is that he was probably not willing to touch the subject, given his close personal attachment and loyalty to Raspletin and firsthand knowledge of many unpleasant details. My own attempts to reach Al'perovich by mail in 2011 in order to arrange a telephone conversation with him about the events and relationship between Raspletin and Kisun'ko did not succeed. (Al'perovich was 89 years old in 2011.) Today, the window of openness in Russia that would have allowed the appearance of an unvarnished account of the development of early air and missile defenses, a sensitive topic and a subject of national pride, has been shut.

The recently published celebratory book on the occasion of the 100th anniversary of the birth of Alexander Raspletin clearly suggests that the story would be unlikely to ever be told factually and without bias. The book described that Raspletin had been awarded the highest state decoration, the title of the Hero of Socialist Labor, for the S-25 development in 1956. It did not say a single word, however, that the second and only other specialist in

---

[52]Semenov, 2008, pp. 429–430.
[53]Drozdov, 1998, Chapter 7.

KB-1, Raspletin's colleague and rival Kisun'ko, had also received this highest honor for his contributions to the same program.[54]

Moreover, the book mentioned the infamous meeting dealing with the antenna problems in Lavrentii Beria's office in 1953, noting—casually and in a dismissively light manner—that "to the happy luck of the antenna developers [Kisun'ko and Zakson] they got away with only being frightened." It did not say anything about the role played by Raspletin and Kalmykov that had led to this episode. The description clearly downplayed the grave personal danger and obvious severity of possibly literally life-threatening consequences for Kisun'ko and Zakson, with consequences that certainly extended beyond mere demotion or loss of jobs.[55]

In any event, as an old Russian saying goes, two bears cannot share the same lair. The paths of the two strong and competitive leaders and individuals in KB-1, Raspletin and Kisun'ko, would soon part, for better or worse, with initiation of the missile defense program.

---

[54]Semenov, 2008; p. 158.
[55]Semenov, 2008; p. 130.

Chapter 5

BEGINNING OF MISSILE DEFENSE

IN RESPONSE TO A NEW THREAT

In 1944, national-socialist Germany deployed the first mass-produced ballistic missile, the A-4.[1] The public knew it as the V-2 Vengeance weapon (Vergeltung in German) (Fig. 5.1). The German army, the Wehrmacht, launched 3000 of these rockets against London, Antwerp, and other Allied targets (Fig. 5.2). A supersonic missile, the V-2 provided no audible warning of its approach. Although radar detected the rocket, nothing could stop it. People felt helpless in front of this silent, essentially impersonal and indiscriminate threat.

The advent of long-range ballistic missiles, especially when coupled with atomic warheads, promised to revolutionize warfare. Visionary military officers, scientists, and engineers began to contemplate possible defenses against the emerging weapon. British officials already were considering what could be done after the first V-2s hit London in 1944. Leading U.S. thinkers were also looking into the possibilities of missile defense in 1945 (Appendix A).

In the Soviet Union, as early as in 1945 Georgii M. Mozharovsky[2] initiated the first study of possible defenses against missiles—"A missile against a missile with radar support"—at the N.E. Zhukovsky Air Force Engineering Academy in Moscow.[3] He continued his work in 1948 at the Scientific-Research Institute No. 4 (NII-4) of the Ministry of Defense in Bolshevo near Moscow. This military institute belonged to the growing cluster of missile and space technology research and development organizations and design bureaus around Podlipki. (Bolshevo is 1.5 miles, or 2.5 km, away from Podlipki.)

Simultaneous with Mozharovsky, Anton Y. Breitbart[4] began work in 1945 on a radar for detecting long-range ballistic missiles in the scientific-research

---
[1] Kennedy, 1983; Gruntman, 2004, pp. 137–165, and references therein.
[2] Georgii Mironovich Mozharovsky, 1901–1975.
[3] Golubev et al., 1994, p. 10; Kamensky, 2003, pp. 600, 601; Pervov, 2003, pp. 13–16.
[4] Anton Yakovlevich Breitbart, 1901–1986.

Fig. 5.1 First mass-produced German ballistic missile A-4 (V-2) being prepared for test launch, ca. 1944. The rocket paint pattern facilitates observation of its attitude in flight. A technological marvel for its time, the missile spectacularly demonstrated the potential of rocket technology to revolutionize future warfare. Photo courtesy of NASA.

institute NII-20 of the Ministry of Armaments. This was the same NII-20 that would volunteer in 2 years to assist Sergo Beria in developing his antiship missile Kometa.

In the 1930s, Breitbart rose to prominence as a leading specialist in television in the Soviet Union and authored the first textbook in this field in 1935.

Fig. 5.2 German ballistic missile A-4 (V-2) hits Antwerp, Belgium, on November 27, 1944. Major cities made attractive targets for this rather inaccurate weapon. The A-4 carried a high-explosive warhead with a mass of 2200 lb (1000 kg) and had a range of up to 180 miles (300 km). Photograph courtesy of National Archives and Records Administration.

He also played an important role in building Soviet radar during World War II. In 1942, Breitbart became chief designer of the fire control radar SON-2ot, or Stantsiya Orudiinoi Navodki No. 2 (Gun Laying Station No. 2), at Plant No. 465. This was a reverse-engineered analog of the British gun-laying radar GL MkII, received through the lend-lease program. The letters *ot* in the SON-2ot nomenclature stood for otechestvennaya (national or homemade), in contrast to the British prototype.

Breitbart's study of missile detection by radar, code-named Pluton (Pluto in English), led to a concept of two pulse radars, one (in the meter-wavelength range) for surveillance and finding targets and the other (in the centimeter-wavelength range) for precise tracking. In 1946, the scientific-technical council of NII-20 concluded that radar detection of ballistic missiles "is an exceptionally complex problem, the most complex among all known radar problems here [in the USSR] or abroad." The institute froze further work on the Pluton program as impractical at that time, but supported research into radar for "high-altitude high-velocity targets."[5]

---

[5]Davydov, 2009, pp. 59–60.

Other organizations also joined the exploratory effort in missile defense. In 1948, Lev R. Gonor,[6] director of NII-88, initiated a major study of antiballistic missile systems in cooperation with the aforementioned NII-20. Gonor served as the first director of NII-88, which was organized in Podlipki following the 1946 decree Matters of the Rocket Weapons. During World War II, the Soviet government decorated him with three Orders of Lenin and the gold star of the Hero of Socialist Labor for organizing the mass production of artillery pieces and mortars. In the late 1940s, Gonor's NII-88 concentrated on guided missiles for air defense in addition to being the home of national ballistic missile development. There, Sergei Korolev worked on Soviet analogs of the German A-4 and more capable follow-on models.[7]

The antiballistic missile program at NII-88 and the institute director himself did not last long, however. Soon, the state anti-Semitic campaign led to purges of Jews in the rocket establishment.[8] The government fired Gonor in July 1950. He was arrested in January 1953 but survived the ordeal. In any event, Stalin's order in 1950 to build the high-priority air-defense system Berkut put uncertain missile defense projects on the back burner. Nevertheless another organization, the Scientific-Research Institute No. 885 (NII-885), also started its theoretical and system studies of missile defenses in 1951. (The Moscow-based NII-885 would later gain prominence by building inertial guidance systems for ballistic missiles and space launchers.)

Perhaps the most important event that ultimately led to a large-scale national missile defense program occurred in August 1953. Six months after the death of Stalin, Chief of the General Staff of the Soviet Army Marshal Vasilii D. Sokolovsky (Fig. 5.3) sent a letter to the Central Committee of the Communist Party of the Soviet Union. Six other marshals in very prominent positions also signed the letter: Georgii K. Zhukov (first deputy minister of defense), Alexander M. Vasilevsky (deputy minister of defense), Mitrofan I. Nedelin (commander-in-chief of artillery), Konstantin A. Vershynin (commander-in-chief of air defense), Nikolai D. Yakovlev (first deputy of the commander-in-chief of air defense), and Ivan S. Konev (chairman of the military council and commander of the Carpathian Military District).[9]

The letter described the growing threat of ballistic missiles: "It is expected that the probable adversary will have in the near future long-range ballistic missiles as the main means of delivery of nuclear charges to strategic objects of our country. Air Defense systems, currently deployed and under development, cannot defend against ballistic missiles. . . ."[10]

---

[6]Lev Ruvimovich Gonor, 1906–1969.
[7]Semenov, 1996; Gruntman, 2004, pp. 280–282.
[8]Kostyrchenko, 2005, p. 386 ; Gruntman, 2007, pp. 45–47; 2010, p. 49.
[9]Georgii Konstantinovich Zhukov, 1896–1974; Alexander Mikhailovich Vasilevsky, 1895–1977; Mitrofan Ivanovich Nedelin, 1902–1960; Konstantin Andreevich Vershynin, 1900–1973; Nikolai Dmitrievich Yakovlev, 1898–1972; Ivan Stepanovich Konev, 1897–1973.
[10]Belous et al., 2009, p. 86.

Fig. 5.3 General Vasilii D. Sokolovsky (right) with the Allied Supreme Commander General Dwight D. Eisenhower (center) at the Tempelhof airfield in Berlin in late May or early June 1945. Sokolovsky commanded Soviet forces in the occupied East Germany from 1946 to 1949 and then served as deputy defense minister. He was chief of the general staff from 1952 to 1960. In 1962, Sokolovsky was the editor of an authoritative treatise on the Soviet military strategy (Sokolovsky, 1963) that emphasized missile defense. Photograph courtesy of ITAR-TASS.

The highly influential marshals, all famed veterans of the recent World War II, urged the government to begin development of national missile defense.

A TIME OF CHANGES IN THE MILITARY-INDUSTRIAL COMPLEX

The year 1953 witnessed changes in the power structure and uncertainties in the decision-making authority in the Soviet Union. The communist dictator Stalin died in March after ruling the land with an iron fist for almost 30 years. Lavrentii Beria, who had been directing the development of advanced weaponry since World War II, was arrested in June in the ensuing power struggle in the Kremlin and then executed in December. New leaders of the country embarked on a reorganization of the vast military-industrial complex.

The Soviet government first split the Third Chief Directorate (TGU) into two directorates, Glavspetsmash and Glavspetsmontazh, and then folded them together with the nuclear programs into a newly created Ministry of Middle Machine Building (Ministerstvo Srednego Mashinostroeniya, commonly called in its shortened form Minsredmash). This ministry would oversee the nuclear complex until the dismantling of the Soviet Union in the 1990s.

Rocketry, guided missiles, and ballistic missiles separated from the Ministry of Middle Machine Building into administratively independent entities on April 14, 1955. The new Ministry of General Machine Building (Ministerstvo Obshchego Mashinostroeniya, usually called by its acronym MOM) assumed control of ballistic missile and space programs. It would manage them until the early 1990s.

A sequence of state committees on radioelectronics took over and directed research, development, and industrial facilities engaged in air and missile defenses and related areas, to be eventually reorganized into a new Ministry of Radio Industry in 1965. Valerii D. Kalmykov headed these organizations for 20 years, from 1954 to 1974.

The new highly influential administrative body, the Commission on Military-Industrial Matters of the Presidium of the USSR Council of Ministers, after a lengthy administrative evolution, established itself in its final form in December 1957. This body, commonly known as the Military-Industrial Commission and by its acronym VPK in Russian, would direct and coordinate the research and development of new weapons and their manufacturing. Dmitrii F. Ustinov first chaired the commission from 1957 to 1963. Then, Leonid V. Smirnov[11] headed it for the next 22 years, from 1963 to 1985. The commission oversaw the Soviet military-industrial complex until the disintegration of the Soviet Union in the early 1990s. The new assertive Russia reestablished the commission in 2006.

After Stalin's death, Nikita S. Khrushchev (Fig. 5.4) emerged as the winner in the internal power struggle in the Kremlin. He led the Soviet Union until 1964, when he was ousted and replaced by Leonid I. Brezhnev. Khrushchev reformed the Soviet military and focused the armed forces and defense industry on nuclear weapons, ballistic missiles, and missile defense. Under his leadership the Soviet Union achieved the first intercontinental ballistic missile (ICBM); operationally deployed intermediate-range ballistic missiles (IRBMs) on a large scale, as well as first ICBMs; put into orbit the first artificial satellites of the earth; established space reconnaissance; launched the first cosmonauts into space; fielded operational antiaircraft guided missile systems; demonstrated the first intercepts of long-range ballistic missile warheads; and started antisatellite weapons programs. Nikita Khrushchev remained an enthusiastic proponent of new military technologies and was directly

---

[11]Leonid Vasil'evich Smirnov, 1916–2001.

Fig. 5.4 Soviet leader Nikita S. Khrushchev enthusiastically advocated and spearheaded advancement of military capabilities into nuclear weapons, ballistic missiles, space systems, antisatellite weapons, and missile defense. Photograph (18 September 1959) courtesy of Franklin D. Roosevelt Library, Hyde Park, New York.

involved—until his last days in power—in decision making that advanced nuclear, missile, and space weapons.

Missile Defense Challenges

In 1953, missile defense looked impossible to many. The first antiaircraft guided missile system, S-25, was still in development. Actually, the military units and engineers had scored the very first successful intercepts of aerial targets in tests at Kapustin Yar only a few months prior to the marshals' letter urging development of antiballistic missile capabilities. The acceptance of the S-25 for operational deployment was two long and uncertain years away in the future.

The first applications of radar during World War II showed that antiaircraft projectiles and artillery shells could occasionally be seen for short periods of time on radar screens. Detecting and reliably tracking ballistic missile warheads, however, presented enormous technical difficulties.

Effective radar cross-sections (RCS) of warheads were expected to be about 0.1 m^2 or 100 times smaller than those of typical aircraft. Ballistic missiles flew at velocities up to 10 and even 20 times faster than airplanes.

Consequently, in order to provide warning sufficiently in advance, an antiballistic missile defense radar would have to detect approaching warheads at distances of many hundreds of kilometers, and even thousands of kilometers for future ICBMs. Very high velocities of missiles translated into a 1-km miss distance for a timing error of only 0.1 s for detonating interceptor charges.

In addition, engineers built nuclear warheads that were particularly rugged and specially designed them to withstand the rough mechanical environment of ballistic missile launch and the high thermal and dynamic loads of atmospheric reentry. Reliable destruction of such hardened targets without detonating their nuclear charges presented another major challenge.[12]

The intercept of a warhead would last only a few minutes. Therefore, antimissile missiles had to be launched on short notice in a highly automated, computer-controlled process with precision, speed, and guidance accuracy far beyond the state of the art. At that time, electronics relied almost exclusively on vacuum tubes; digital computing was still in its infancy.

MEETING AT TGU

The Central Committee of the Communist Party of the Soviet Union (CPSU) instructed the TGU's Scientific-Technical Council, headed by Alexander Shchukin, to evaluate the missile defense proposal of the military leaders at a special meeting. This consequential event took place in September 1953 in a new building of the council half a mile from the Kremlin. High-level government officials, many chief designers of weapon systems, and top defense scientists were in attendance.

The council assigned the task of recording the comments of meeting participants to a TGU specialist, Nikolai Ostapenko. The notes were to be used later for preparation of a report to the Central Committee. Consequently, council chairman Shchukin instructed Ostapenko to record the exact words of the speakers because of the importance of this sensitive gathering of highly influential participants. The preserved notes helped to reconstruct the event.

Director of the Radiotechnical Laboratory (RALAN) of the Academy of Sciences, "the [grand] master of radio science and technology," Alexander L. Mints (Fig. 5.5) spoke first and gave a harsh assessment of the proposal of the marshals. Ostapenko recorded Mints's words that the proposal "does not have any science foundation" and that "it is such a stupidity as shooting an [artillery] projectile at another projectile." He concluded that "the proposal cannot be technically implemented . . . and this should be the answer to an illiterate proposal of 'military dreamers.'"

Then, KB-1's Alexander Raspletin, without sparing words, also blasted the famed military leaders by saying that "the marshals [had] proposed

---
[12]e.g., Golubev et al., 1995.

Fig. 5.5 Alexander L. Mints, director of the Radiotechnical Laboratory, RALAN (Radiotechnical Institute RTI from 1957), of the USSR Academy of Sciences. Photograph courtesy of the A.L. Mints Radiotechnical Institute.

unfathomable nonsense, stupid fantasy."[13] Chairman of TGU's Scientific-Technical Council Shchukin then had to remind the gathered specialists and officials that the powerful positions and distinction of the authors of the letter made it "impossible to duck the raised [by them] issue."

Some other meeting participants offered less hostile assessments and even showed interest in the proposal. For example, Leonid Leonov, the designer of the A-100 early warning radar of the S-25 system, expressed interest in beginning studies of detecting targets with small reflecting surfaces such as warheads. "Whether we want it or not," concluded Leonov, "[warheads] will become real targets for our radars."

However, it was KB-1's Grigorii Kisun'ko (Fig. 5.6) who made a statement, as the meeting progressed, that had the "effect of an exploded bomb." Kisun'ko came well prepared and, in contrast to others, brought to the discussion specific scientific and engineering details of the challenge. Ostapenko recorded Kisun'ko's words, "The proposal [of the marshals] has been brought at the right time. Its essence is of practical importance. . . . Americans are developing missile weapons of various types. Missile warheads will become targets in the not far distant future. . . ."[14]

---

[13]Belous et al., 2009, p. 92.
[14]Belous et al., 2009, p. 93.

Fig. 5.6 Grigorii V. Kisun'ko, chief designer of the experimental system that achieved the first intercept of a long-range ballistic missile warhead. Photo (1966) courtesy of ITAR-TASS.

Here Mints interrupted the speaker, "There are no intercontinental ballistic missiles yet alive!" At this moment the leader of the Soviet ICBM program, the powerful Sergei Korolev, also present at the meeting, joined the fray defending his turf, "There will soon be [operational ICBMs] and not only as test articles. Yes! Yes!, respected corresponding member [of the Academy of Sciences Mints]."

Kisun'ko continued:

> Preliminary calculations show that our radars will be able to detect and track ballistic missile warheads with effective RCS [radar cross-sections] two orders of magnitude smaller than those of modern perspective aircraft if one increases the power of radar transmitters by a factor of 20, builds large antenna systems 15–20 m [50–65 ft] in diameter, and develops radar receivers with a sensitivity of about 10^{-13} W.
>
> All mentioned parameters of radars are quite achievable by the comprehensive focused scientific-research [program]. I state this with the full responsibility.[15]

---

[15]Belous et al., 2009, p. 93.

The meeting ended with a recommendation to form a committee to initiate feasibility studies of missile defense in both KB-1 and RALAN (Fig. 5.7).

DEVELOPMENT LEADERS

Fedor V. Lukin[16] assumed the top technical manager position as chief engineer of KB-1 in November 1953. He directed the design bureau during the period of administrative turbulence and reorganizations. Lukin saw the promise of the new initiative in missile defense and supported feasibility studies in this powerful organization.

Two rising stars and rivals in KB-1, Raspletin and Kisun'ko, could take on this major new program of missile defense. Only these two in the design bureau would soon receive the highest USSR decorations and title of the Hero of Socialist Labor for development of the operational Moscow air defense system S-25, reflecting their status and accomplishments. Both were ambitious and young. In 1954, Raspletin was 46 years old and Kisun'ko was only 36.

Raspletin remained skeptical about missile defense. He decided at that time to concentrate instead on perfecting air defenses. Already in November 1953 the USSR Council of Ministers had authorized development of his new mobile S-75 (SA-2) antiaircraft system. Raspletin advanced air defenses with new, increasingly capable designs until his death in 1967. Boiling with energy, Kisun'ko jumped on the new opportunity offered by the emerging area of missile defense. This endeavor required the talents of both a scientist and a manager, a combination that he possessed.

Kisun'ko directed a large leading technical department in KB-1 with more than 1000 employees.[17] In 1954, he assigned selected groups of engineers to conduct preliminary studies of specific technical issues important for missile defense such as powerful transmitters, large antennas, waveguides for high-power signals, and precise tracking. Some specialists of the former TGU also contributed their expertise.[18]

In addition to Kisun'ko, the former chief designer of Berkut, Pavel Kuksenko, got interested in the new area as well. In December 1953, KB-1 established a special theoretical laboratory, headed by Nakhim A. Livshits, for evaluation of missile defense. Eight months later Livshits issued a feasibility study report.

In Alexander Mints's RALAN, a group under Mikhail M. Veisbein[19] began work on missile defense in January 1954. Many other initially skeptical and vocal opponents of the concept would also soon join the rapidly growing effort, lavishly funded and supported by the government.

---

[16]Fedor Viktorovich Lukin, 1908–1971.
[17]Kisun'ko, 1996, p. 311.
[18]Belous et al., 2009, p. 95.
[19]Mikhail Mikhailovich Veisbein, b. 1906.

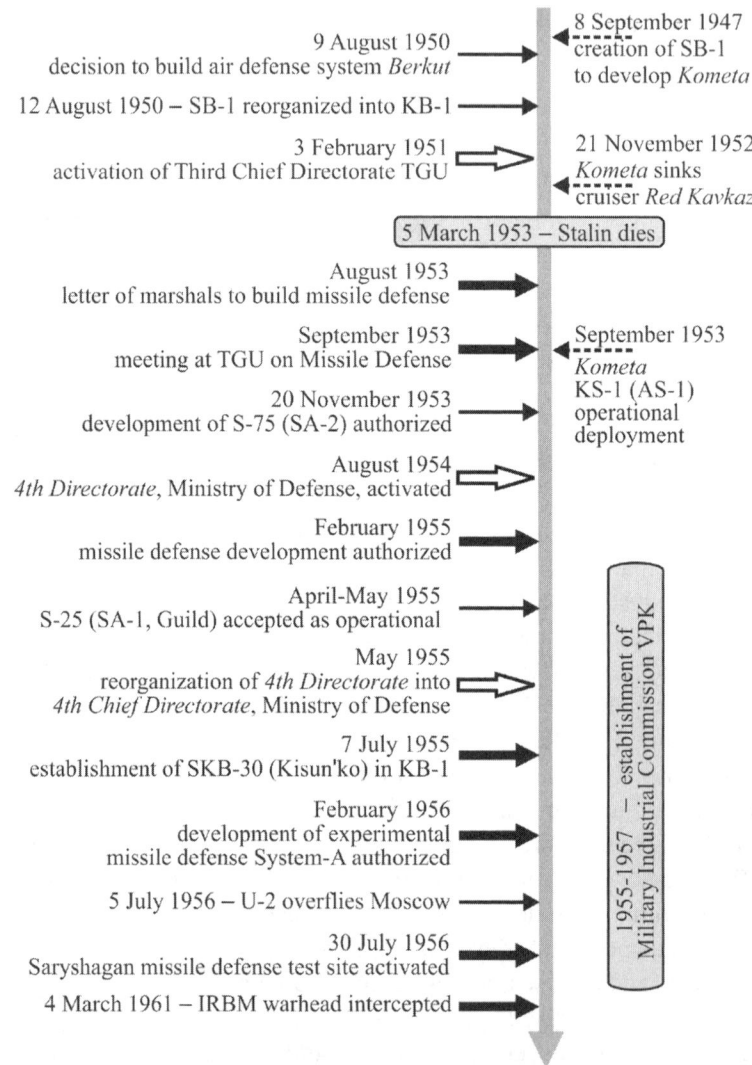

Fig. 5.7 Timeline of the emerging Soviet guided missile air defense and missile defense. Arrows: dashed line: development of Kometa; narrow solid black: air defense systems S-25 (Berkut; SA-1) and S-75 (SA-2); thick solid black: missile defense events; thick white: evolution of overseeing government organizations. Also shown is the establishment of the Military-Industrial Commission (VPK).

So, by the end of 1954, three groups had proposed concepts of missile defense, those of Livshits and Kuksenko at KB-1 and Mints at RALAN. The Scientific-Technical Council of TGU's successor organization, Glavspetsmash, declined Kuksenko's proposal, approved Livshits's concept, and was ambivalent about Mints's proposal.[20]

---

[20]Pervov, 2003, p. 26.

The Central Committee of the CPSU issued a decree on 2 February 1955 (followed by a decree of the Council of Ministers the next day) authorizing the development of missile defense.[21] Subsequently, Minister of Defense Industry Dmitrii Ustinov reorganized KB-1, at first forming two semi-autonomous special design bureaus (SKBs) within the organization, SKB-31 and SKB-41, with Raspletin and Andrei A. Kolosov[22] as chief designers, respectively. (Government orders transferred Kolosov, along with Kisun'ko and Livshits, from the Military Academy of Communications in Leningrad to support development of Berkut. Kolosov first worked at a military institute in a Moscow suburb, Mytishchi, 3 miles from Podlipki, and then joined KB-1.) These two SKBs consolidated ongoing KB-1 projects in "legacy" areas of air defense and air-launched missiles.

Understanding the radar properties of incoming warheads was indispensable for firming up the concept of missile defense. Sergei Korolev, who led the development of long-range ballistic missiles, adamantly opposed opening his test launches to observation by outsiders. After all, the entire concept of missile defense sought ways of destroying "his" ballistic missiles. Only the personal interference and insistence of Minister Ustinov forced Korolev to let Kisun'ko and Mints attend tests at Kapustin Yar in June 1955. (A RALAN group under Veisbein conducted the first radar observations of ballistic missiles from 1956 to 1957. Engineers under Kisun'ko began extensive tests in 1957 at the new proving ground at Saryshagan.)

After the trip to Kapustin Yar, Kisun'ko was ready. He reported to Minister Ustinov his plan for the development of missile defense. Consequently, on 7 July 1955, Ustinov ordered—over the objections of Raspletin—the creation of another new special design bureau within KB-1. Kisun'ko became head of this new SKB-30 (Spetsial'noe Konstruktorskoe Byuro No. 30) and chief designer for missile defense.[23]

The reorganization of KB-1 had thus been completed: Kisun'ko's SKB-30 concentrated on missile defense; Raspletin's SKB-31 designed air defense systems; and Kolosov's SKB-41 led work on air-launched guided missiles (Fig. 5.8). In 1960, Kolosov left KB-1 and Anatolii I. Savin[24] became chief designer of SKB-41. Savin then turned the focus of his bureau to development of space defenses and antisatellite weapons. In 1973, he separated his organization, forming an independent Central Scientific-Research Institute Kometa (TsNII Kometa).[25]

Raspletin and his part of KB-1 continued the highly successful development of advanced air-defense systems, many exported to foreign countries and serving as an important instrument of Soviet policies abroad. Nevertheless, he

---

[21] Yakovlev, 1999, p. 174; Pervov, 2003, p. 27; Belous et al., 2009, p. 142.
[22] Andrei Alexandrovich Kolosov, b. 1907.
[23] Kisun'ko, 1996, p. 310; Pervov, 2003, p. 28.
[24] Anatolii Ivanovich Savin, b. 1920.
[25] Vlasko-Vlasov, 2002.

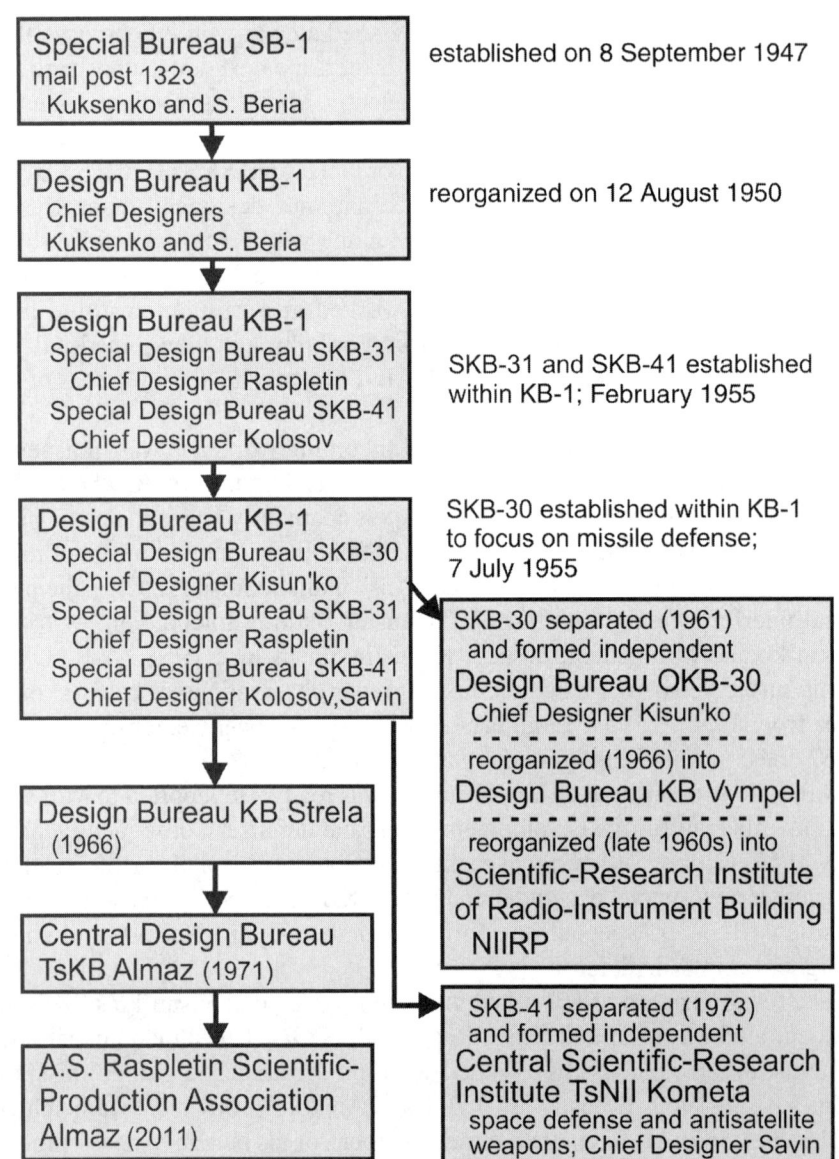

Fig. 5.8 Evolution of the Design Bureau KB-1 from 1947 to 2011 and its branched-off independent organizations. Today, many leading former air and missile defense establishments have been folded into the new business group Almaz-Antei.

did not entirely shun missile defense and could not avoid later entering the rapidly growing, important field.

Raspletin remained skeptical about the ability of defensive systems to overcome penetration aids, particularly possible decoys accompanying attacking intercontinental ballistic missiles at high altitudes. Consequently, in the early

1960s he authorized studies of a new missile defense system, S-225, against shorter range IRBMs. Vladimir I. Markov headed this effort. In contrast to ICBMs, IRBMs flew at lower altitudes where atmospheric drag made inflatable decoys and dipole reflectors less effective by separating them from the warheads.

At that time, Kisun'ko was already actively working on the operational missile defense system of Moscow, A-35. So Markov tried, without the knowledge of his boss Raspletin, to suggest this system to Kisun'ko as a terminal defense component for the A-35. He then dropped this idea out of fear that Raspletin, in Markov's words, would "tear his head off" for interactions with his "enemy" Kisun'ko.[26] In addition, Raspletin also joined forces in the early 1960s with Vladimir N. Chelomei[27] in proposing the development of antisatellite weapons and a national strategic missile defense system, Taran.[28] (Taran means a battering ram, as well as the act of one aircraft ramming another.)

Sometime during initial studies in the unsettled new field of missile defense, Grigorii Kisun'ko had a consequential meeting with the powerful Alexander Mints. Mints suggested Kisun'ko should join his organization, but the latter declined the alliance. Adversarial relations with Mints followed afterwards. As Kisun'ko described it:

> In the meantime the study of a concept, suggested by me, of the initial problem of missile defense had begun in KB-1, which planned establishing a missile defense test site as the base for experiments in order to lay the science foundation for creating an operational military system of missile defense. The concept was reinforced by industrial technical solutions and theoretical calculations, confirming the concept feasibility and suggesting principles of its implementation.
>
> At that time, I had a memorable conversation with [Corresponding Member of the USSR Academy of Sciences Alexander L.] Mints. He started it in this manner, "It will be too crowded for your work [on missile defense] in KB-1. But if you split from KB-1, taking its half [with you], and join forces with our [scientific-research institute] NII, then we will be able to do big things. Also, I am already old and you are young.[29] At first, you will [practically] run all [my] NII [in my name] and then take my position formally both in the NII and in the Academy of Sciences."
>
> I politely thanked Alexander Lvovich [Mints] for the flattering proposal but said that I could not accept it because we had essentially diverged on the [technical] principles of missile defense. He replied [using a popular saying], "How to build a bridge, along the river or across it, could be

---

[26]Markov, 2003, p. 400.
[27]Vladimir Nikolaevich Chelomei, 1914–1984.
[28]Semenov, 2008, pp. 225–231.
[29]Mints and Kisun'ko were 58 and 36 years old, respectively, at that time.

sorted out later. The main task now is to stake the area and to make it ours." Then I responded that it was not in my character to partition such a unique organization as KB-1.

Listening to my response, Alexander Lvovich darkened, looked at me without kindness, and said, "You will regret one day that you did not understand me. . . ."

We coldly parted and never again revisited the subject of our conversation. And I did not understand at first that this cold parting was the beginning of a cold war that Alexander Lvovich Mints waged [against me] since that time until the end of his life. And [he did it] not alone but with his mighty allies.[30]

By the end of 1955, Kisun'ko had prepared a specific proposal for the experimental system of missile defense. Interception of a warhead required its precise tracking. Kisun'ko's concept relied on the method of three distances, sometimes loosely and erroneously called "triangulation."[31] Three widely separate radars had to accurately measure their distances to a target at the same moment of time, which allowed one to determine its absolute position. (In essence, there were three unknowns, the coordinates x, y, and z of the target. Measuring distances to three widely separated radars provided three simple equations that could be solved easily for the unknown coordinates.) Because the warhead velocity significantly exceeded that of a practical intercepting missile, the interceptor was to meet the approaching warhead head-on. A high-explosive charge on the interceptor would then detonate and create a field of fragments, destroying the target on impact.

A competing proposal from Alexander Mints's RALAN included three vertically pointed radars, or radar "fences." He planned to determine the warhead trajectory by measuring moments of time and coordinates of fence crossings. The predicted system accuracy of 4–5 miles (6–8 km) in distance and 1.2 miles (2 km) in azimuth did not make the intercept realistic.

A joint meeting of the Scientific-Technical Councils of KB-1 and RALAN took place on February 1, 1956, to evaluate the competing proposals. Representatives of the Ministry of Defense and overseeing government bodies attended the gathering. With many powerful interests involved, the councils came to a compromise decision to recommend the practical development and evaluation of Kisun'ko's experimental system and further study of Mints's system.

The organizational structure of the early Soviet missile defense effort began to emerge. The new special design bureau SKB-30, headed by Kisun'ko, spearheaded development of the approved experimental system. For many

---

[30]Kisun'ko, 1996, p. 299.
[31]e.g., Golubev et al., 1992.

years he would be fighting contentious battles with rivals and opponents. Mints's competing RALAN continued its own development of related technologies, looking for opportunities to initiate major programs. Both organizations, SKB-30 and RALAN, would succeed. A number of other powerful establishments also joined the effort that expanded into many new areas such as early warning of ballistic missile attack, above-the-horizon and over-the-horizon radars, space situational awareness, antisatellite weapons, and missile defense against shorter range ballistic missiles.

SYSTEM A AUTHORIZED

In February 1956, the Presidium of the CPSU Central Committee and the USSR Council of Ministers formally authorized development of the experimental System A, proposed by Kisun'ko, and its demonstration at a special test range. The obvious complexities of missile defense and the challenges of achieving the desired characteristics of its key elements made the need to begin with an experimental system inevitable. The initial effort was to eventually pave the way for a missile defense system of the capital, Moscow.

The first operational German ballistic missile, A-4, did not separate its warhead. The United States and Soviet Union introduced warhead separation in early post–World War II missile programs.[32] Kisun'ko's experimental system thus had to demonstrate defense against one ballistic target, a pair consisting of a warhead and the separate body of the rocket accompanying it in flight.

System A included a long-range search and acquisition radar to detect the incoming missile as early as possible, determine its coordinates, and pass this information to three precise tracking and guidance radars (Fig. 5.9). These latter radars formed an equilateral triangle and accurately measured their distances to the target, which determined the warhead position by the method of three distances. The radars tracked the approaching warhead with high precision and guided the intercepting missile toward it. System A had to identify the warhead in the target pair and destroy it by exploding the fragmentation interceptor.

For the originally envisioned operational missile defense system, Kisun'ko planned to place three precise tracking radars on the existing S-25 outer ring road built around Moscow. This design determined the 95-mile (150-km) distances between the radars. The development plan included a similar experimental prototype to be built first at a new test range in Kazakhstan to prove the concept.

Even before the experimental System A achieved its first intercepts in 1961, the government had issued a decree on 8 April 1958, authorizing work on an

---

[32]Gruntman, 2004, pp. 213–215.

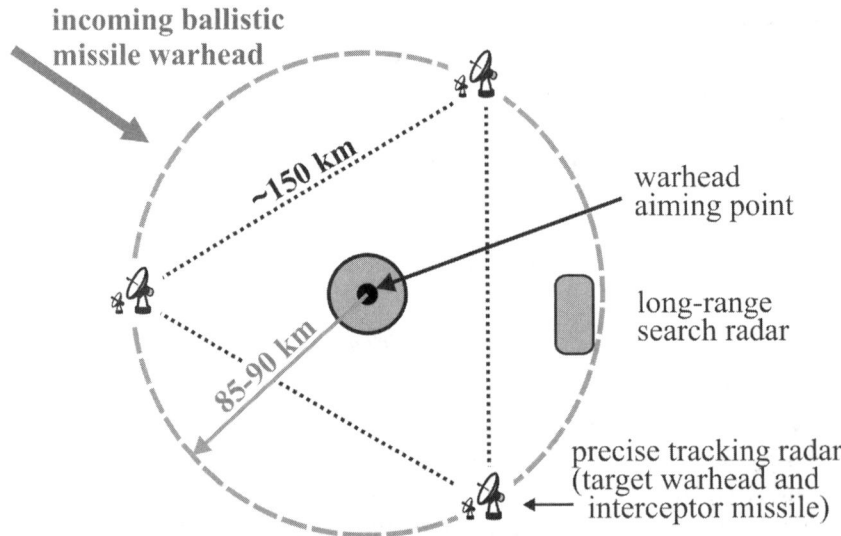

Fig. 5.9 Original concept of System A for the missile defense of Moscow. A long-range search radar detects the incoming ballistic missile warhead and the accompanying rocket body and passes the information to three precise tracking and guidance radars. The latter radars form an equilateral triangle and measure their distances to the target. The system then calculates the absolute position of the warhead, tracks it, and guides the interceptor missile. For the operational system, Kisun'ko initially planned to place the three precise tracking radars at the Moscow ring road built to support bases of the outer circle of the S-25 air defense system. This design determined the distance of 95 miles (150 km) between the radars. The development program called for building a similar experimental system at a test site in Kazakhstan to prove the concept.

operational missile defense system for Moscow, designated the A-35, with Kisun'ko as chief designer. The A-35 requirements went beyond the original concept of System A and called for intercepting up to eight warheads. A highly accurate method of three distances would have then required eight groups of three radars each. Simultaneous operation of 24 similar but independent radars in shared space presented significant technical difficulties.

Consequently, the logic of science and engineering led to switching to a traditional, but less accurate, approach for determining the distance and elevation and azimuth angles of a target by a single radar. This resulted in lower precision of warhead tracking and hence larger miss distances in interception. The latter, in turn, required more destructive interceptors than the fragmentation charges planned for System A. In addition, as it would soon become clear, reliable destruction of hardened warheads by interceptors with conventional blast high-explosive charges presented a major challenge. Consequently, the interceptors of the operational missile defense system A-35 would be nuclear armed.

Kisun'ko realized early the inevitability of nuclear intercepts for the operational system, even as he had been working on the more challenging nonnuclear concept of System A. A leading missile defense specialist, Oleg Golubev, gave the full credit to Kisun'ko for understanding the issue and choosing nuclear-armed interceptors,

> To compensate for the worsening accuracy of coordinate determination [due to a different detection approach], it was necessary to increase [destructive] power of a special (nuclear) warhead of the antimissile missile. Therefore, one could not consider use of a high-explosive fragmentation interceptor warhead in the [operational missile defense] system A-35. G.V. Kisun'ko clearly understood it. He had even considered [earlier] a possibility of using a special (nuclear) warhead on interceptor missiles of System A. Such a special warhead was built and underwent preparations. Therefore, statements periodically appearing in the media that the use of nuclear warheads in the national missile defense contradicted the views of G.V. Kisun'ko are absurd. It was he who was and remains "the father" of nuclear warheads for interceptor missiles, and this was his accomplishment. While defending a city, a missile defense system had to provide a "clean" destruction of the incoming warhead, that is without detonating the [target] warhead's nuclear charge at the altitude of intercept and without nuclear fallout from the destroyed warhead. Only interceptor missiles with such special (nuclear) warheads can guarantee destruction of approaching warheads.[33]

The complications of building and then fielding the A-35 operational system, with the changing requirements, delays, and interruptions in funding of the early 1960s; the system upgrade (known as the A-35M) after its deployment; and rivalry and turf battles in missile defense would come years later. In 1956, Kisun'ko and his collaborators faced an immediate unprecedented technical challenge of the nonnuclear intercept of a single ballistic missile target warhead by an experimental system at a special, yet to be built, new test range in a desert.

The Fourth Chief Directorate had been overseeing air defense systems in the Ministry of Defense since 1954. In 1956, the government formed a new body, the Fifth Directorate, within the Fourth Chief Directorate to specially serve as the procuring and coordinating arm for emerging missile defense, on behalf of the Military-Industrial Commission (VPK) and the Ministry of Defense.

The Fifth Directorate later expanded its responsibilities to early warning of ballistic missile attack (by both space-based sensors and ground radars), space surveillance, and antisatellite weapons. Colonel, later Lieutenant General,

---
[33] Golubev, 2003, p. 577.

Mikhail G. Mymrin[34] became the first commander of the Fifth Directorate and served in this position from 1956 to 1965. The directorate also oversaw construction of the new missile defense proving ground.

System A called for major advancements in many areas of science, engineering, and manufacturing. A number of leading research and development organizations took part in the endeavor. In addition, a large new test range was built in a short period of time in the middle of a harsh and desolate desert.

The decree of the Central Committee of the Communist Party and the Council of Ministers on 18 August 1956, set in motion the development of System A and established demanding schedules for the program. As Kisun'ko described it,

> by that time the survey of the location of the test site had been completed and planning units of the Ministry of Defense worked full speed on design of installations of the future test site; first trains with military construction units began to arrive at the nearest railroad station [Saryshagan in Kazakhstan] in July [1956]; the General Staff issued a directive to activate a new command for the test site, military unit No. 03080. The decrees of the Central Committee and the Council of Ministers assigned codes "Test Site A" to the test range and "System A" to the experimental complex of missile defense.[35]

The Soviet government apparently succeeded in keeping details of this new major program secret. One year later, in March 1957, a Central Intelligence Agency National Intelligence Estimate (NIE) guessed about the possible development of missile defense in the Soviet Union:

> We estimate that anti-ICBM defense would receive the higher priority, and that the USSR could probably develop a missile system of some capability against the ICBM for first operational use during the period of 1963–65. We are unable to estimate with confidence the characteristics of such a system. It might carry a 1,000-pound payload to a horizontal distance about 40 n.m. and an altitude 200,000 feet. An extension of these range and altitude capabilities would require advances in radar design which we believe are not within Soviet capabilities during this period... . Development of antimissile defense systems will undoubtedly be continued beyond the period of this estimate.[36]

Development Team

The high velocity of long-range ballistic missile warheads compressed the intercept into a very short period of time. Consequently, the missile defense

---

[34]Mikhail Grigorievich Mymrin, 1918–1984.
[35]Kisun'ko, 1996, p. 315.
[36]Central Intelligence Agency, 1957, p. 16.

system had to operate to a significant degree automatically, precisely synchronize its various elements spread over hundreds of kilometers, and rely on real-time computing supported by reliable communications.

To demonstrate interception, the experimental System A included the following main components:

- Long-range surveillance and acquisition radar for detection of approaching ballistic missiles
- Three radars for precise tracking of warheads by the method of three distances and for final guidance of interceptor missiles
- Additional dedicated radar to initially guide the interceptor missile from its launch site toward the target and then hand it over to precise tracking radars for the final phase of the intercept
- Guided interceptor missiles with a high-explosive fragmentation warhead
- High-capacity, low-error communication lines
- A digital control computer with complex algorithms and codes

Grigorii Kisun'ko and his OKB-30 took the responsibility for designing and integrating the entire System A and for developing precise tracking and guidance radar systems. Each such Precise Guidance Radar (Radar Tochnogo Navedeniya, or RTN) consisted of two collocated radars: one for warhead tracking and another for tracking the intercepting missile during the final phase of the intercept.

Two organizations engaged in the development of competing variants of the long-range surveillance and acquisition radar, Aksel' Berg's NII-108 and Alexander Mints's Radiotechnical Institute (RTI). (RTI emerged from the reorganization of RALAN in 1957.) One missile defense veteran viewed the decision to build two competing systems as politically driven and unjustified wasteful duplication that continued for many years and also later led to redundant systems for tracking of orbiting space objects.[37] Some others attributed duplication to necessary hedging against technological uncertainty.[38]

Early on Kisun'ko chose the radar design of Vladimir P. Sosul'nikov[39] in NII-108, who based it on his earlier radar Dunai-1. (Dunai is the Russian name for the Danube river.) The new continuous wave sector radar with electronic beam scanning, Dunai-2, became part of System A. In his competing development, Alexander Mints changed the approach from the initial fence concept to a sector pulse radar, Central Finding Station—Preliminary (Tsentral'naya Stantsiya Obnaruzheniya—Predvaritel'naya, or TsSO-P).

Sosul'nikov would later build the next variant of his long-range radar, Dunai-3, for the first operationally deployed A-35 missile defense system for Moscow. Although the RTI's TsSO-P radar had never been used in System A,

---

[37]Drozdov, 1998.
[38]Ivantsov, 2003, p. 473; Belous et al., 2009, 112.
[39]Vladimir Panteleimonovich Sosul'nikov, 1921–2008.

its technology evolved with time into the successful long-range radars Dnestr, Dnestr-M, and Dnepr, which played prominent roles in the Soviet antisatellite weapons program and the above-the-horizon radar systems for early warning of ballistic missile attack. They also provided technical capabilities for tracking orbiting space objects.

The Mints's RTI remained a main participant, perennial competitor, and powerful rival of Kisun'ko in missile defense. Later, in addition to the next generation of long-range surveillance radars Daugava and Daryal, the institute also built multipurpose fire-control radar systems for missile defense, culminating with Don-2N in the current A-135 operational missile defense system for Moscow.

The bitter rivalry between Kisun'ko and Mints that had begun in the mid-1950s continued unabated for many years. The first commander of the Missile and Space Defense Forces, General Yuri V. Votintsev, recalled the meeting, prior to his appointment in 1967, with the Secretary of the Central Committee of the CPSU for defense issues Dmitrii F. Ustinov. Ustinov told Votintsev that he

> had an earnest request [to Votintsev]. Because of the nature of your appointment, you will be working with general designers [of missile defense systems] comrades Mints and Kisun'ko. Each of them is now working on development of his own weapon system, and your objective is to try to combine their efforts ... [T]his will shorten the time for system development and [decrease] expenses of the state.[40]

A new independent special design bureau, Fakel led by Petr D. Grushin (Fig. 5.10) in Khimki, built another major component of System A, the V-1000 interceptor missiles. Several key organizations in air and missile defenses—KB-1, Fakel, Lavochkin's OKB-301, Mints's RTI, and the Vympel Association (see Chapter 9) that consolidated national missile defense development under one administrative roof in the 1970s—were located within 7 miles (10 km) of each other in the same general area in northwestern Moscow (Fig. 5.11).

Samuil P. Rabinovich[41] from NII-20 of the Ministry of Defense Industry designed the Antimissile Missile Initial Guidance Radar (Radiolokatsionnaya Stantsiya Vizirovaniya Protivorakety, or RSVPR) for the initial part of the interceptor trajectory immediately after its launch. This was the same NII-20 that had first appeared in this text when taking part in development of young Beria's Kometa antiship missile, and was later "cannibalized" and finally evicted from its premises. The institute would once again lose its most accomplished specialists who built the RSVPR for System A, including Rabinovich.

---

[40]Votintsev, 1993, part 2, p. 30.
[41]Samuil Pavlovich Rabinovich, 1909–1989.

Fig. 5.10 Tombstone (left) of Petr D. Grushin at Moscow's elite Novodevich'e Cemetery, showing his contributions to air and missile defense systems (a missile launcher at the bottom-right of the stone is shown magnified on the right). One can see a reflection (left) of the author, taking this photograph, in the polished marble of the tomb. Photo (2006) courtesy of Mike Gruntman.

Large distances between radars, the interceptor launch site, and the command and control computing center of the missile defense system made highly reliable high-capacity communications especially important. In March 1956, the government established a special Scientific-Research Institute No. 129 (NII-129) to design and build this critical part of System A. Chief engineer of the institute Frol P. Lipsman[42] headed the effort. Since 1966, NII-129 has been known as the Moscow Scientific-Research Radiotechnical Institute (Moskovskii Nauchno-Issledovatel'skii Radiotekhnicheskii Institut). The socialist state would unceremoniously force the highly accomplished and active Lipsman into early retirement in 1979 in one of its perennial purges of Jews.[43]

---

[42]Frol Petrovich Lipsman, 1915–2008.
[43]Nemirovskaya and Shnitser, 2011.

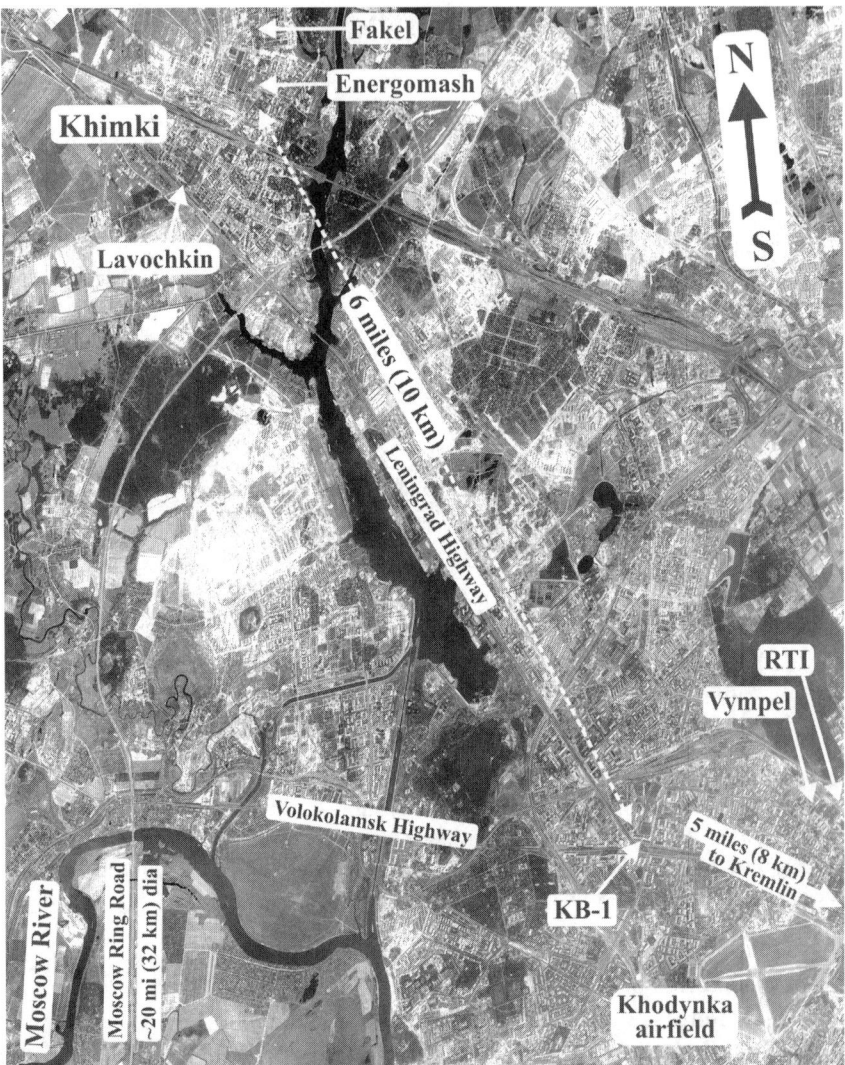

Fig. 5.11 Main contributors to early air and missile defenses in northwestern Moscow. The ring road formed an administrative boundary of Moscow. Kisun'ko's OKB-30 separated administratively from KB-1 in 1961. Mints's RTI moved to its location in 1957–1958. Formed in 1970, Scientific-Industrial Association Vympel consolidated the national missile defense effort. Lavochkin's KB and Grushin's Fakel were in the town of Khimki, and the nearby Energomash, headed by Valentin Glushko, built large liquid-rocket engines. Original photograph (KH-7; Mission 4030; 16 July 1966) available from the U.S. Geological Survey; photograph identification, interpretation, and processing by Mike Gruntman.

Sergei A. Lebedev and Vsevolod S. Burtsev[44] headed the development of powerful computers and complex codes at the Institute of Precision Mechanics and Computer Engineering (Institut Tochnoi Mekhaniki i Vychislitel'noi Tekhniki, or ITMVT) of the USSR Academy of Sciences. System A's computer performed real-time processing of a large amount of data, developed intercept solutions, and worked out guidance commands for the interceptor missile. A pioneer of Soviet computers, ITMVT director Lebedev had been playing a leading role in the field since the 1940s. Burtsev succeeded him as director of ITMVT in 1973 and later built Soviet supercomputers for operational missile defense systems and other defense applications.

Grigorii Kisun'ko put together the core team for System A development in 1956. Dozens of other scientific, research, and applied institutions and numerous design bureaus and manufacturing plants contributed to the program by advancing technology and building various subsystems and components. The military would then test the experimental missile defense system at a remote secrete location in Kazakhstan where the new proving ground had to be built from scratch.

---

[44]Sergei Alexeevich Lebedev, 1902–1974; Vsevolod Sergeevich Burtsev, 1927–2005.

Chapter 6

SARYSHAGAN TEST SITE

DESERT IN KAZAKHSTAN

The government decision to build a missile defense system for Moscow included first demonstrating the experimental System A at a special new test site. This proving ground grew with time into one of the largest in the Soviet Union. The military used it to evaluate experimental prototypes Aldan and Amur-P as part of the development of operational missile defense systems A-35 and A-135, respectively. The installation also played an important role in testing numerous other advanced antiaircraft (S-75, S-200, S-300, Dal'), antimissile (S-225), and antisatellite systems as well as laser weapons.

The ordinance of the USSR Council of Ministers from 3 February 1956 ordered the Ministry of Defense to establish a new proving ground for the emerging missile defense program. General Sergei F. Nilovsky, commander of the air defense test range GNIIP-8 at Kapustin Yar, led a survey group. They worked from 28 February until 11 April 1956 and selected a large desolate area in the Betpak Dala desert west of the Balkhash (Balqash) lake in Kazakhstan in Soviet Central Asia (Figs. 6.1 and 6.2).

A small settlement with a railroad station, Saryshagan, near the lake's shore became the jumping off point for building the new installation. It also gave the name to Saryshaganskii Poligon (Saryshagan Test Range, or simply Saryshagan), sometimes called Balkhashskii Poligon after the lake.

The Ministry of Defense ordered an experienced 45-year-old engineering officer, Colonel Alexander A. Gubenko,[1] to assume command of the Directorate of Engineering Works No. 32 (Upravlenie Inzhenernykh Rabot No. 32, or UIR-32) to oversee construction of the test range. (Gubenko would later be promoted to the rank of one-star general, or major general.) In the Soviet Army, UIR directorates played the role of engineering headquarters, executing construction

---

[1] Alexander Alexeevich Gubenko, 1911–2004.

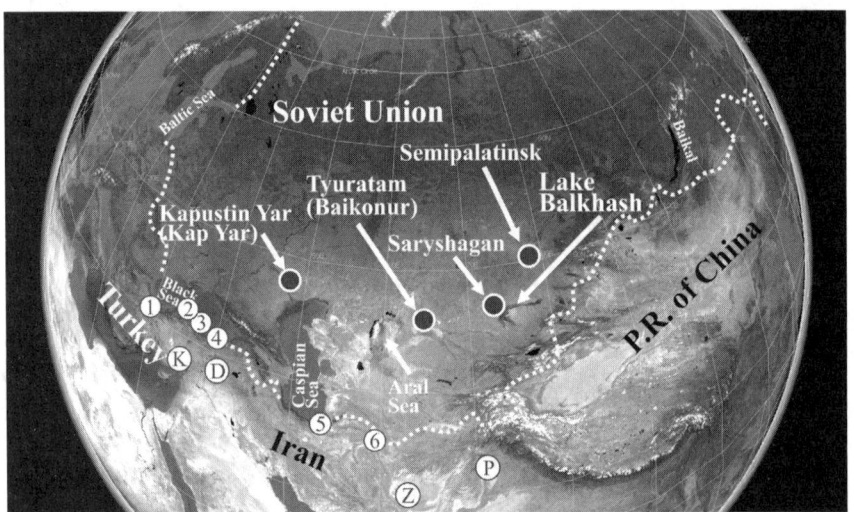

Fig. 6.1 Rangeheads (black filled circles) of three major missile development sites in the Soviet Union: Kapustin Yar (Kap Yar), Tyuratam (Baikonur), and Saryshagan. Also shown is the main nuclear test site near Semipalatinsk. Tyuratam, Saryshagan, and Semipalatinsk are located in the present-day independent Kazakhstan; Kapustin Yar is in Russia near the border with Kazakhstan. All three missile sites are at similar geographic latitudes.

The numbered white circles show a string of U.S. electronic intelligence (ELINT) listening stations in Turkey and Iran that intercepted telemetry during missile tests in the 1950s and 1960s. The military sites Karamursel (1), Sinop (2), Samsun (3), and Trabzon (4) in Turkey and the CIA station at a private palace of the Shah of Iran in Behsher (5) operated beginning in the 1950s. The CIA added a particularly important listening post (6) in a remote area 40 miles (65 km) east of Meshed in Iran in 1965 to improve the reach to the Tyuratam and Saryshagan missile tests. In addition, an outpost in Peshawar (P) in Pakistan also engaged in electronic intercepts.

In 1955, the U.S. Air Force activated an AN/FPS-17 radar at Diyarbakir (D) in eastern Turkey to observe Soviet missile launches at Kapustin Yar. A similar radar was also installed at Samsun (3). In 1964, the military added an advanced tracking radar, AN/FPS-79, at Diyarbakir. Later, satellites in geostationary orbit enabled intercepts of electronic signals from proving grounds in the Soviet Union and People's Republic of China.

U-2 reconnaissance planes of the CIA's Detachment B took off and landed at airfields at Incirlik (K) in southern Turkey, Zahedan (Z) in Iran, and Peshawar (P) in Pakistan during overflights of Soviet Central Asia in 1957–1960.

Original composite satellite photograph (Landsat, mid-1990s to early 2000s) courtesy of NASA WorldWind; image processing, callouts, schematic USSR border (white dotted line), and markings by Mike Gruntman.

projects by using brigades, regiments, and battalions of construction troops and other specialized engineering units assigned to them.

In April 1956 the military recalled Gubenko to Moscow from his assignment in Siberia and briefed him on the new important job. Gubenko then visited Kisun'ko at his design bureau. The chief designer of System A gave

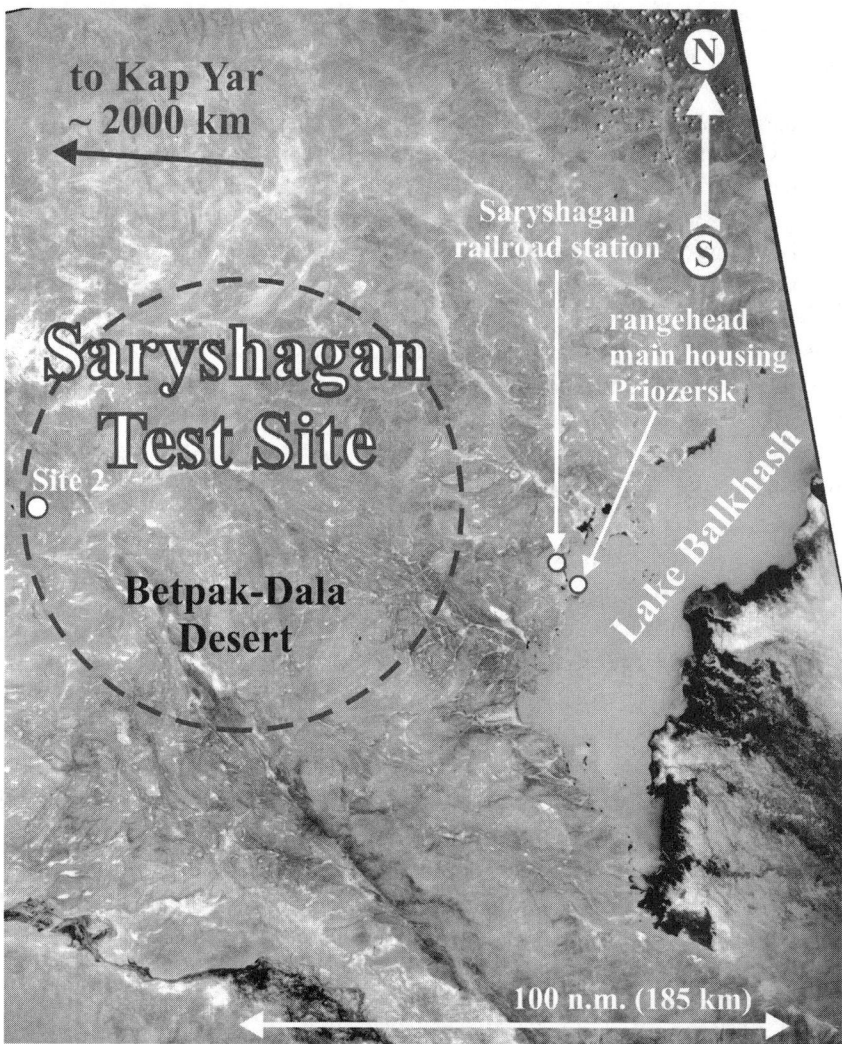

Fig. 6.2 Satellite photograph of the large desert area of the Saryshagan missile defense test range west of Lake Balkhash activated in 1956. The dashed-line circle corresponds in size to the outer ring of the S-25 air defense sites around Moscow, the planned location of three precise guidance RTN radar units of System A. The westernmost Site 2 for one radar and the range headquarters and main housing area at the lakeshore (the future town of Priozersk) not far from the Saryshagan railroad station were among the highest priority construction objectives in 1956. Original satellite reconnaissance photograph by KH-5 Argon mapping camera (Mission 9058A; 29 August 1963) available from the U.S. Geological Survey; photograph identification, interpretation, and processing by Mike Gruntman.

Gubenko the impression of a "powerful, tall ... [and] simply handsome man." He recalled that his first conversation had shown

> that [Kisun'ko] was a highly intellectual, well-mannered man and [that] his speech was very polite and [grammatically] correct ... I was struck by the confidence projected by all appearances of Kisun'ko. He did not seem to show any doubt that all objectives would be achieved. I felt that he knew thoroughly the field of his work, with great erudition prominently showing in his every word. ...[2]

From the very beginning, Kisun'ko and Gubenko established a relationship of mutual respect, and successfully worked together as partners for many years. They became personal friends.

Saryshagan represented the third major missile test facility established in Soviet Central Asia and the adjacent area of Russia (Fig. 6.1). All three centers—Kapustin Yar, formed in 1946; Tyuratam (later known as Baikonur), activated in 1955; and now Saryshagan—had exceptionally large territories. The new missile defense range in the Betpak Dala desert stretched 155 miles (250 km) from north to south and 370 miles (600 km) from east to west and occupied an area larger than 30,000 miles^2 (80,000 km^2).

The area was a dry rocky plateau with small hills and wide depressions at 1300–1600 ft (400–500 m) altitude above sea level. A Saryshagan veteran, Colonel Alexander F. Kulakov,[3] described the harsh continental climate as "dry with cold winters and hot summers. Air temperature varies from maximum +50°C [+122°F] down to minimal –45°C [–49°F]. Soil freezes [during winters] to the 2-meter [6.5 feet] depth."[4]

Annual precipitation did not exceed 4.7 in. (120 mm). Gusts of winds reached 50–65 ft/s (15–20 m/s), with sudden powerful sandstorms. Evgenii V. Zhadeiko[5] served at Saryshagan as a young lieutenant in the late 1950s. He remembered how he had forgotten to close a small window pane in his room when he left for work on a beautiful quiet sunny morning without any wind. Later in the day a sudden sandstorm arrived and, within a couple hours, deposited a 2-cm layer of sand everywhere in his apartment.[6] On the positive side, there were more than 250 sunny days each year, and colorful tulips covered the desert in April and May.

Economic activity in the desert was limited to shepherds using the area for seasonal movement of sheep. Time seemed to be frozen there (Fig. 6.3). One railroad line ran along the western shore of the lake. There was practically no population, no roads, and no industry. This desolate region thus offered the

---

[2] Gubenko, 2003, p. 668, 669.
[3] Alexander Fedorovich Kulakov, b. 1924.
[4] Kulakov, 2006, p. 9.
[5] Evgenii Vladimirovich Zhadeiko, b. 1937; retired as colonel in 1990.
[6] Zhadeiko, n.d. p. 11.

Fig. 6.3 Frozen in time. Top: indigenous nomads and their yurt, also called a kibitka, in the lower Syr Darya river area in 1871–1872. Bottom: local people in the same area near the Tyuratam ballistic missile range in the mid-1950s.

"Their [nomad] habitation, both in winter and summer, is a kibitka, a circular tent made of felt spread over a light wooden frame. The frame is easily taken apart and put together, and is so light as to form a load for a single camel only. The broad pieces of felt are easily stretched over it.... The kibitka forms a most comfortable abode, being cool in summer and warm in winter" (Schuyler, 1876, p. 35).

Photographs from (top) Kun, 1872, p. 34 and (bottom) collection of a Baikonur veteran, Colonel Yu. V. Bonchkovsky, courtesy of Alexander Yu. Bonchkovsky.

safety and security desired for secret weapons programs. Such considerations tended to place missile proving grounds in uninhabited areas where the harsh living conditions kept away even hardened native inhabitants. The Soviet Union was not unique with such locations for weapons and missile development centers. The French rocket sites at Hammaguir in Algeria and the Australian site at Woomera were not unlike those at Tyuratam and Saryshagan in this respect.[7]

Lake Balkhash stretched for 370 miles (600 km) from east to west, with the Sarymsek Peninsula dividing it into two distinct parts (Fig. 6.4). The narrow and deep eastern part of the lake contained salty water, whereas fresh water from the large Ile river filled the wider western part near Saryshagan. The river fell into the lake opposite from the Saryshagan side, about 40 miles (60 km) to the south. Fortunately for the military, the lake water near the proving ground was potable after filtering.

In a typical year, the lake froze from late November to early April. Abundant sturgeon, bream, pike, and sazan populated the lake's waters. Fish offered important recreation and some diversity of diet in this otherwise God-forsaken area. Soon pike-perch were introduced into the lake, which displaced some other fish and became popular among anglers throughout the year, including in winters. Reed-covered shores also offered excellent duck hunting, especially on the eastern side. The quality of water began to deteriorate in the 1970s after the opening of a copper processing plant in the growing industrial

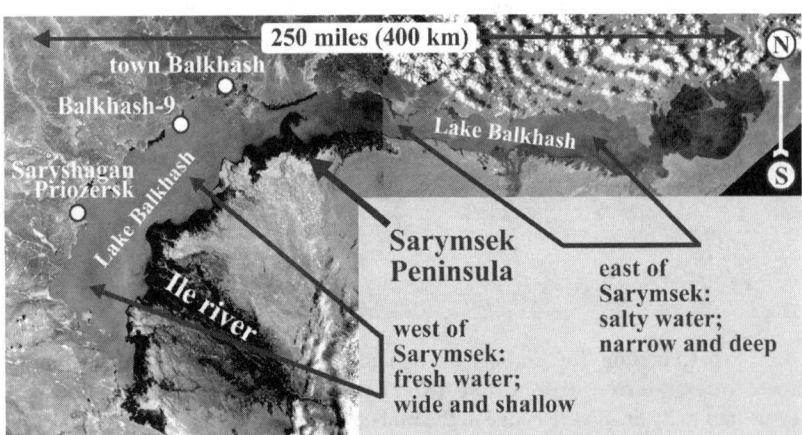

Fig. 6.4 Composite satellite photograph of Lake Balkhash in Kazakhstan at the eastern edge of the Saryshagan test site in 1964. White circles show the railroad station Saryshagan and the nearby main residential area Priozersk, Balkhash-9 (the future site of Dnestr radar of the antisatellite weapon system and for early warning of ballistic missile attack and satellite tracking), and the town of Balkhash. A large coal-burning plant in the latter provided electric power for the Saryshagan range. The Sarymsek peninsula divided the fresh and salty waters of the western and eastern parts of the lake, respectively. Original satellite reconnaissance photographs by KH-5 Argon mapping camera (Mission 9066A; 21 August 1964) available from the U.S. Geological Survey; photograph identification, interpretation, and processing by Mike Gruntman.

---

[7]Gruntman, 2004, pp. 313–315, 418–420, 423–425.

town of Balkhash (Fig. 6.4) on the northern shore of the lake 80 miles (130 km) from the range headquarters.

CONSTRUCTION IN THE DESERT

The Ministry of Defense assigned the engineering directorate UIR-32 to build the Saryshagan test range. The directorate, also known as military unit v/ch 19313, enjoyed comfortable quarters in a large regional center, the town of Odessa on the Black Sea shore in Ukraine. The prospect of transfer to Kazakhstan terrified many of its officers and civilian employees, who did everything possible to avoid being sent to the new desolate location with a harsh climate. By this time, Soviet leader Khrushchev had begun relaxing the severe communist restrictions and control of people lives, giving at least civilians some freedom of choice in their employment. Consequently, only 12 officers and 1 civilian nurse out of the 357 directorate personnel in Odessa joined Gubenko when they stepped down from a train at the tiny Saryshagan station (Fig. 6.2) on 5 July 1956.[8]

A few days later the first construction battalion with more than 500 men and various machines disembarked from trains, beginning a continuous stream of servicemen and equipment sent to this top priority national effort (Fig. 6.5). With time, the Ministry of Defense also transferred a number of engineering officers to staff UIR-32.

Within three months trains had brought 5000 automobiles to the area, which lacked even primitive roads. Gubenko established a special school to retrain soldiers for driving in the treacherous desert and opened major repair and maintenance facilities. The military also redeployed some experienced engineering units who had been working under similar harsh conditions at Tyuratam constructing the ballistic missile range and future space launch base. The work there had started a year earlier, in 1955, and was now in full swing.

Two widely separated locations, the main residential area Site 4v and a technical position at Site 2, became the highest construction priorities in 1956. In addition, engineering units quickly established a support base near the Saryshagan railroad station. The words *ob"ekt* and *ploshchadka*, meaning *site* or *technical position*, were common names for various installations and areas of the range. The military assigned identifying numbers to the sites, by which they were commonly referred. These locations varied from small facilities such as a radio relay or an instrumentation outpost with a few engineering huts and a small barrack for a couple dozen soldiers to a major airfield or main residential quarters housing many thousands of servicemen and civilians.

Soldiers began building the main residential area at Site 4v (the letter *v* stood for vremennaya, or temporary) at the "waist" of a peninsula protruding

---

[8]Gubenko, 2003; Belous et al., 2009, pp. 149–151.

Fig. 6.5 Officers and enlisted men of construction troops and other specialized engineering units wore a badge like this with a prominently displayed bulldozer on their collar tabs. The badge measured 0.55 × 0.9 in. (14 × 22 mm). This branch of the USSR armed forces built numerous military installations, bases and outposts, roads and bridges, airfields, ports, missile silos, headquarters, barracks, and residential buildings and also completed many civilian infrastructure projects in remote areas.

into the lake 7 miles (11 km) southeast from the railroad station (Fig. 6.6). The site served as the initial staging location for arriving military construction units. It also accommodated the temporary range headquarters and provided housing for military families and engineers from industry. The military soon opened a movie theater and a garrison clubhouse there. People quickly nicknamed the settlement Poluostrov, or peninsula. In two years the work expanded to Site 4p (the letter *p* stood for postoyannaya, or permanent) 2 km to the east. Eventually these two areas merged, forming a new town, Priozersk (Site 4).

By mid-1957, 75 construction battalions had arrived at Saryshagan and were building 332 permanent and many temporary structures at 18 locations. The next year, in 1958, they worked on 643 permanent buildings and various structures at 31 sites scattered across a huge territory.[9]

Gubenko wrote that by the middle of 1957 more than 110,000 military personnel and technicians and fitters from industry had worked around the clock on the construction of various technical positions, buildings, and infrastructure and on installing equipment at the Saryshagan range.[10] Many other Russian publications then repeated this number, although nobody independently corroborated it.

The actual number of engaged construction troops was certainly very large, but likely lower than what was stated by Gubenko. Consider, for comparison,

---

[9]Belous et al., 2009, p. 152.
[10]Gubenko, 2003, p. 671.

Fig. 6.6 Area in the vicinity of the Saryshagan railroad station near the shore of Lake Balkhash in 1966. The station served as the disembarkation point for the arriving military construction units and equipment as well as the storage depot and support base from the earliest days of the proving ground in July 1956. Among the first two top priority construction objectives of the new range was the main residential and headquarters area at the peninsula 7 miles (11 km) southeast of the station. Its construction began first at Site 4v and soon expanded to Site 4p. The two areas then merged into the town of Priozersk (Site 4). Original satellite reconnaissance photograph by KH-7 camera (Mission 4027; 21 April 1966) available from the U.S. Geological Survey; photograph identification, interpretation, and processing by Mike Gruntman.

Fig. 6.7 Saryshagan railroad station with spurs, ramps, storage depots, and residential areas in 1966. The first military construction units found only a small settlement when they had disembarked here 10 years earlier. The support base near the station served construction troops and engineering units of the range for many years. Original satellite reconnaissance photograph by KH-7 camera (Mission 4027; 21 April 1966) available from the U.S. Geological Survey; photograph identification, interpretation, and processing by Mike Gruntman.

work on the Tyuratam ballistic missile range conducted by the 130th Directorate of Engineering Works (UIR-130).[11] The directorate's chief engineer, Alexander Yu. Gruntman,[12] oversaw engineering for Tyuratam's construction from 1955 to 1963. He estimated later that the number of officers and soldiers of construction troops had not exceeded 20,000 during those early years of building this top priority large ballistic missile range and space launch center.[13] The number of servicemen would double at Tyuratam in the 1980s during a major expansion of engineering facilities to prepare for the launch of the reusable space vehicle Buran by the new and largest Soviet rocket Energia.

The support area near the Saryshagan station (Fig. 6.7) straddled the railroad and became the main disembarkation point for military units and supplies. It also had some temporary barracks; a sawmill and lumberyard; garages for trucks, automobiles, bulldozers, scrapers, road rollers, excavators, and various other heavy construction equipment; and storage areas for pipes, cement, steel, and building materials. Construction units quickly built ramps and established temporary depot areas for unloaded materials, machines, and equipment. The military expanded the station and branched off railroad spurs directly to some technical positions of the range. They also opened a locomotive repair shop and maintenance facilities as well as a fueling station. A diesel engine generator provided some electric power.

---

[11]Gruntman, 2004, pp. 312–314.
[12]Alexander Yulievich Gruntman, 1912–1975.
[13]Sergei Gruntman, private communications, 2003, 2014; Gruntman, 2004, p. 317.

The other highest priority construction objective of the early days of 1956, Site 2, lay about 150 miles (240 km) away to the west (Fig. 6.2). Construction troops crossed the roadless desert and reached Site 2 several weeks after the arrival of the first units to Saryshagan.

Kisun'ko described, somewhat poetically (he occasionally wrote poetry), the harsh conditions and difficulties and dangers of navigating in the roadless treacherous desert.

> The Betpak Dala desert astonishes by its striking monotony. Roads are there everywhere and nowhere. Small hills, mounds, depressions, and cavities monotonically transition from one to another and look so similar to each other, like waves at sea. Surprisingly, driving in the [roadless] desert somehow turns, imperceptibly [for the driver], into moving along the same circular path.
>
> Monotonous scenery becomes even greater in winter, when carried by the wind snow flows through the steppe, like milk, filling all cavities and holes, [making] traps for vehicles, and instantaneously covers tracks behind the car. When wind is strong, a shroud of a snow blizzard covers everything, making headlights powerless to penetrate it; the sound of wind becomes louder than engines; and automobiles may get lost even following one another in a convoy....
>
> [Commander of construction troops] Gubenko spent ten railroad cars of timber to install wooden landmark poles, with branches ... [of brush] and [bunches of] grass attached to them, every half a kilometer on all the 300-km east-to-west way [from Site 4 and Saryshagan railroad station] to Site 2. These landmarks served as good aids for drivers, both as road marks and as wood for making fires at stops. They did not burn the poles indiscriminately, but first every other one then every other one again until they began to wander in between the remaining poles.[14]

UIR-32 would complete a network of permanent roads with hardened surfaces at the Saryshagan proving ground by 1964.

Other infrastructure demands for the range were also pressing. The military required reliable high data rate communications based on microwave radio relay towers for forthcoming missile intercepts. Some remote sites consumed significant electrical power. Just one precise guidance RTN radar used 650 kW of power when engaged in active tracking of missiles. A power plant in the town of Balkhash became an important source of electricity, connected to the test site by high-voltage lines (Fig. 6.8). Many technical positions operated independent diesel power generators. For example, six diesel engines (designed for sea-going ships) provided power at Site 2.

---

[14] Kisun'ko, 1996, pp. 352, 353.

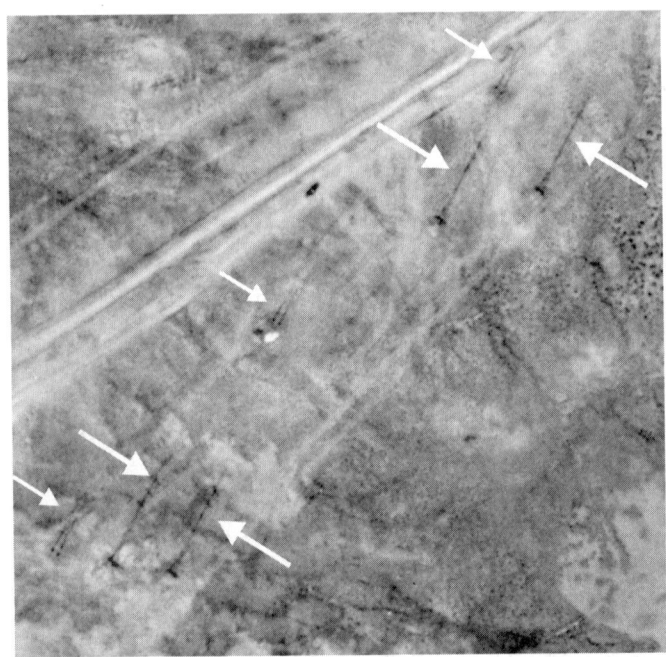

Fig. 6.8 Aerial reconnaissance photograph of high-voltage (110 kV) electric power lines leading to and from the main distribution station near Site 4v in 1960. The larger and smaller arrows point to larger and smaller towers, respectively. The main line carried electric power from a power plant at the town of Balkhash to the distribution station. Original photograph (U-2 Mission 4155; 19 April 1960) courtesy National Archives and Records Administration; photograph identification, interpretation, and processing by Mike Gruntman.

In spring 1958, the military combined two hospitals, one for the construction troops and the other (opened in September 1957) for the test range, into the main garrison hospital with 300 beds.[15] The medical facility also served the growing civilian population, including children. Major remote sites had their own medical posts.

Clean water supply to the growing number of technical positions, barracks, and living quarters was especially urgent because of the looming danger of disease. Gubenko assigned the task of building the first pumping station at the lake and connecting water pipes to the nearby areas to an engineering unit of Lt. Col. Levon S. Arzanov.[16] Arzanov and his 40 officers and 600 men disembarked on a cold December night in 1956 at the Saryshagan station, where they had arrived from the Tyuratam missile range. In April 1957, water began

---

[15]Kulakov, 2006, pp. 152, 153.
[16]Levon Sumbatovich Arzanov, 1912–1992.

flowing to many buildings, a major celebration for everybody in the remote garrison.[17]

The scarcity of clean water and difficult living conditions took a toll, however. Already in 1958, the first major epidemic of dysentery broke out. Fifteen thousand soldiers from the construction troops got sick, but hospital facilities could accommodate only 1000 men.[18] It took a concerted effort of commanders and medical officers to avoid such disruptive events in the future.

GNIIP-10

On 30 July 1956, the General Staff of the Soviet Army activated (by order ORG/6/40258) the State Scientific-Research Test Range No. 10 (Gosudarstvennyi Nauchno-Issledovatel'skii Ispytatel'nyi Poligon No. 10, or GNIIP-10), also known as military unit v/ch 03080. This day is now celebrated as the date of the creation of the Saryshagan test site. Major General Stepan D. Dorokhov[19] assumed command of GNIIP-10 and promptly arrived at the location. His main three deputies were chief engineer Col. Mikhail I. Trofimchuk, chief of staff Col. Alexei I. Isaev, and first political officer and head of the political department Col. Vladimir L. Subbotin.[20] After Dorokhov's sudden death in 1966, Trofimchuk served as commander of GNIIP-10 from 1966 to 1969. Colonel Arzanov transferred to GNIIP-10 and became Dorokhov's deputy representing the range, the "customer," in its interaction with the construction troops of UIR-32.

Two weeks after activation of the new proving ground, on 15 August 1956, the USSR Council of Ministers approved highly demanding schedules for the development of the experimental System A. In addition to the rapidly growing number of construction troops at Saryshagan, the first specialized air defense units and freshly minted young officers graduating from military colleges and academies began to arrive at GNIIP-10. They formed numerous testing and operational organizations at the range.

Already by November 1956, mobile radio relay stations had established communications among a few key widely spread locations. Then, an Air Force mixed squadron reached Saryshagan in January and February 1957 with a number of planes to support the growing activities. Many remote technical sites of the range built airstrips that could receive liaison aircraft as well as light transport planes delivering personnel, equipment, and supplies.

Subsequently, the military reorganized aviation units into the new 60th Mixed Aviation Test Division. It included a large variety of transport, fighter,

---

[17]Arzanov, 2003, pp. 696, 697.
[18]Kulakov, 2006, p. 23.
[19]Stepan Dmitrievich Dorokhov, 1913–1966.
[20]Mikhail Ignatievich Trofimchuk, 1922–1985; Alexei Ivanovich Isaev; Vladimir Lukyanovich Subbotin.

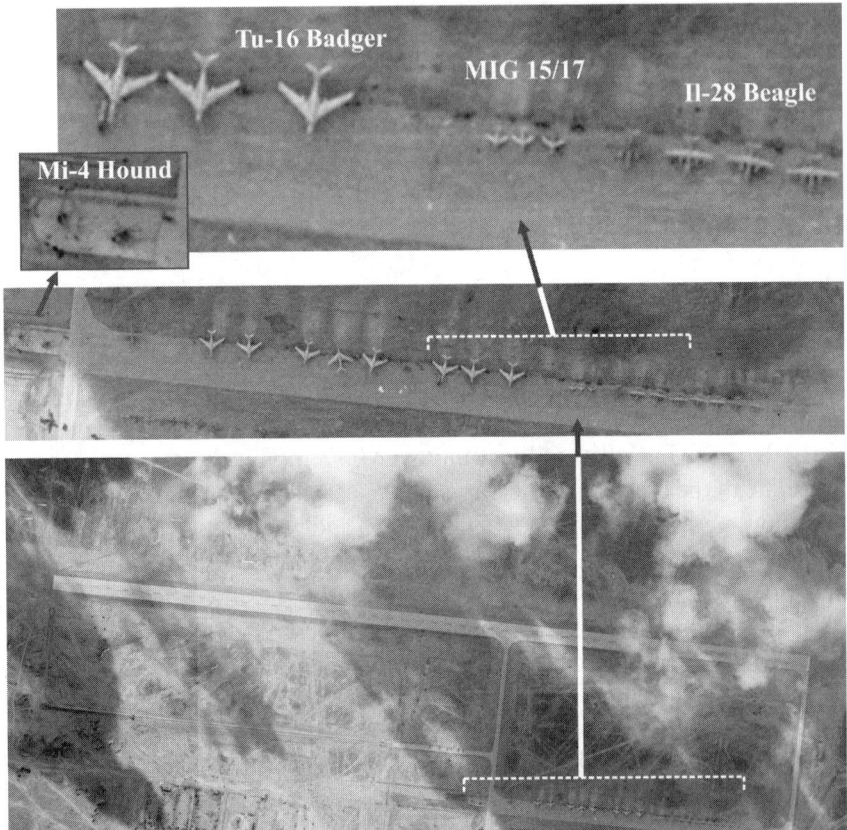

Fig. 6.9 Aerial reconnaissance photograph of the main airfield (Site 7), partially covered by clouds, 10 miles (16 km) to the west of Priozersk at the Saryshagan test range in April 1960. The runway (bottom) stretched for 9850 ft (3000 m) with an alternating black and white strip in the center. Landing aids provided all-weather capabilities. There were 70 aircraft on the field at the moment when a U-2 took this photograph, including 9 Tu-16 (Badger) and 14 IL-28 (Beagle) bombers, 14 MiG-15 (Fagot) and MiG-17 (Fresco) fighters, 5 Li-2 (Cab; modified license-built Douglas DC-3) passenger and transport planes, various other transport and liaison airplanes, and 5 Mi-4 (Hound) helicopters (Central Intelligence Agency, Photographic Intelligence Center, 1961, pp. 28, 30). Regular flights by turboprop Il-14 and jet Tu-104 (a civilian version of the Tu-16) airliners would soon connect Saryshagan with Moscow's Vnukovo airport. Original photograph (U-2 Mission 4155; 19 April 1960) courtesy National Archives and Records Administration; photograph identification, interpretation, and processing by Mike Gruntman.

bomber, liaison, and other airplanes and helicopters. The total number of deployed aircraft sometimes reached 300. In addition, the division included supporting radio, engineering, and maintenance units. It also flew radio-controlled target airplanes and deployed parachute targets.

A 1960 aerial photograph (Fig. 6.9) showed the main airfield [Site 7 of the range located 10 miles (16 km) to the west from Priozersk] with 70 aircraft.

The field had a 185-ft (55-m) wide runway that stretched for 9850 ft (3000 meters). In addition, there were two taxiways and one parking ramp, with "another taxiway and two parking ramps ... under construction." The adjacent area included various technical support sites, fuel tanks, communications facilities, hangars, and the living quarters consisting of "35 barracks-type buildings capable of housing 1,200 persons."[21] Regular flights connected the main airfield with a Moscow airport starting in the early 1960s.

Officers and enlisted men of GNIIP-10 followed construction units and spread across the vast area to newly established technical positions. By the end of 1957, the missile range personnel (excluding many thousands of construction troops) exceeded 500 servicemen and civilians.[22] Some bases housed large numbers of personnel and had numerous engineering and support buildings, structures, and equipment. There were also small outposts where the everyday life of servicemen differed from the routine at larger installations overseen by many senior officers.

Lieutenant Zhadeiko served in the late 1950s at a large base, Site 1, working at one of the precise guidance radars, RTN-1. He described a small radio relay post nearby commanded by his friend from military college days. There, another young lieutenant, Vladimir Bocharov, and a small detachment of fewer than two dozen soldiers maintained a lone 100-m-high radio relay tower. Zhadeiko wrote that

> [i]n the fall, in the month of November, ... [Bocharov together] with best shots [from his unit] hunted saigas with carbines, and they had meat [in his unit] during the winter. Only knowledgeable people knew how to cook the saiga meet. First, it was soaked in vinegar and then cooked with special herbs and spices.[23]

The saiga is a type of antelope that inhabits large stretches of Kazakhstan including the Betpak Dala desert and the Tyuratam ballistic missile test range area, as well as some territories near the northern shores of the Caspian Sea in Russia. Hunting saigas was one of the very few recreational activities enjoyed by officers at remote military posts in the desert in the 1950s and 1960s. Today, saigas are an endangered species. In the Betpak Dala and Tyuratam areas, the population of these antelopes dropped from 400,000 in 1980 down to 15,000 in 2000.[24]

The life of servicemen at Saryshagan was not easy, especially during the early years. One veteran officer who had worked at the range from 1959 to 1983 recalled that

---

[21]Central Intelligence Agency, Photographic Intelligence Center, 1961, pp. 28, 30.
[22]Belous et al., 2003, p. 152.
[23]Zhadeiko, n.d., p. 5; also quoted in Kulakov, 2006, p. 55.
[24]Lagrot, 2009.

[i]n the end of the 1950s and early 1960s, the living conditions of officers [at Saryshagan] were far from ideal. They spent years waiting for apartments, and sometimes several families lived in one unit. At large sites, officers lived 10–12 men in one room on two-level bunks in poorly heated dormitories. They waited years in order to buy [scarce basic goods such as] refrigerators, laundry machines, and carpets. A photographic camera was considered a luxury item. Not every officer could afford a decent [civilian] Sunday suit.[25]

As the range expanded its activities and grew, the military reorganized it in 1960, creating four directorates.[26] The first and second directorates conducted tests and analyses of missile defense and air defense systems, respectively. The third directorate focused on processing of test results, and the fourth directorate operated instrumentation sites. A number of units provided security, communications, logistics, supplies, housing maintenance, and medical care. During just one year (1961), transport units, for example, moved 120,000 tons of cargo and more than 1.5 million passengers.[27]

Test Site "Put on the Map"

The expansion of weapon development in Soviet Central Asia could not stay secret for too long. U.S. electronic intelligence (ELINT) listening stations (Fig. 6.1) at Karamursel, Sinop, Samsun, and Trabson in Turkey and at Behsher in Iran had been intercepting telemetry transmitted by missiles during tests at Kapustin Yar since the mid-1950s.[28] In addition, the U.S. Air Force activated an AN/FPS-17 radar at Diyarbakir in eastern Turkey and began tracking missile launches in 1955. In 1965, the ELINT capabilities further improved after the CIA added another listening outpost in a remote area 40 miles east of Meshed in Iran, which proved particularly important for observing tests at Tyuratam and Saryshagan. The Air Force also activated a new tracking radar, AN/FPS-79, at Diyarbakir in 1964.[29]

The first series of deep-penetration reconnaissance flights over the European parts of the USSR by U.S. U-2 aircraft in July 1956 had put to rest the uncertainties of a possible bomber gap between the two superpowers. President Eisenhower then suspended further incursions. In the mid-1950s, electronic intercepts revealed the expansion of Kapustin Yar activities to new areas to the east, but could not determine their precise locations. Major concerns were now increasing about the scale and pace of advancing Soviet programs in long-range ballistic missiles and nuclear weapons that threatened to upset the

---

[25]Gusarov, 2012, part 5.
[26]Kulakov, 2006, p. 88.
[27]Kulakov, 2006, p. 152.
[28]Richelson, 2002, pp. 33, 88; Wheelon, 2004.
[29]Zabetakis and Peterson, 1964.

strategic balance. Consequently on 6 May 1957, Eisenhower authorized a new series of overflights for the summer of that year. This operation, SOFT TOUCH, focused on missile and nuclear production, test, and development facilities in Central Asia and Siberia.[30]

At first, Mission 4035 searched for a new ballistic missile development center on 5 August 1957. The U-2 aircraft reached Tashkent and then proceeded northwest along the main Turkestan railroad line to Orenburg. The pilot noticed and made oblique photographs of a spur track that led north to a sprawling launch complex. He passed directly over and photographed a new garrison town with the range headquarters and main residential area near a small railroad station. Chief information officer of the CIA's National Photographic Interpretation Center (NPIC), Dino A. Brugioni,[31] named the newly discovered missile development center Tyuratam after the railroad station.[32] Four years later, Soviet officials publicized a different name, Baikonur, a settlement 200 miles (320 km) away from Tyuratam, as the space launch site in a clumsy and unnecessary attempt to hide the real and well-known by that time location.[33]

Then on

> its way to [a nuclear test site near] Semipalatinsk [in northeastern Kazakhstan], the 21 August [1957 U-2] mission flew a search pattern over the western end of Lake Balkhash looking for another Soviet missile-related installation and made the first photographs of what was later determined to be the new missile test center at Saryshagan. This facility was used to test radars against incoming missiles fired from Kapustin Yar, 1,400 miles [2,250 km] to the west.[34]

The next day, on 22 August, another mission "photographed a well-planned, modern community of 20,000 people not previously known of on the north shore of Lake Balkhash. This turned out to be the headquarters of the Sary Shagan antimissile test range, a real find."[35]

These 1957 U-2 missions "put on the map" the new weapon proving ground west of the Balkhash lake. The test facilities were in the initial phases of construction, and many installations did not exist at that time. Nevertheless Brugioni later noted that the photo interpreters had been "surprised to find a very large antimissile test station ... replete with all kinds of trails, cable lines, and instrumentation sites."[36]

---

[30]Pedlow and Welzenbach, 1998, p. 135.
[31]Dino A. Brugioni, b. 1921.
[32]Brugioni, 1984; Gruntman, 2004, pp. 318–322; Brugioni, 2010, p. 229.
[33]Gruntman, 2004, p. 318.
[34]Pedlow and Welzenbach, 1998, p. 139.
[35]Lowenhaupt, 1968, p. 9.
[36]Brugioni, 2010, p. 234.

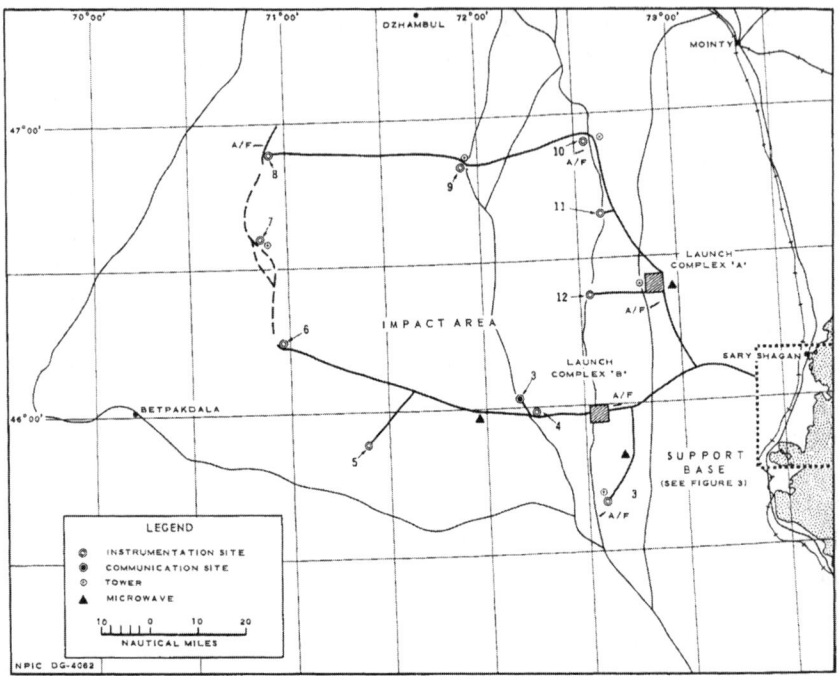

Fig. 6.10 The CIA's map (1961) of the Saryshagan test range with radar, instrumentation, communications, and missile launch sites; airfields; and the support base (Fig. 6.11) near the lakeshore. The range rapidly grew, providing development and test grounds for various air defense, missile defense, and antisatellite weapons programs. Figure from Central Intelligence Agency, Photographic Intelligence Center, 1961, p. 11.

No more U-2 missions flew in this area until the last series of overflights of Central Asia almost three years later. Then another U.S. U-2 aircraft took detailed photographs of expanding activities at the Saryshagan range that allowed reconstruction of the site layout and its many technical capabilities (Figs. 6.10 and 6.11). This was the last successful U-2 deep-penetration overflight of the USSR, Mission 4155, on 9 April 1960.

A staff officer of the Air Defense Forces at that time recalled many years later that Soviet leader Nikita Khrushchev had been "indignant" about the failure to intercept the intruder and that "many generals and other officers were penalized."[37] Khrushchev wrote that "in April [1960] we'd had an opportunity to shoot down a U-2, but our antiaircraft [guided missile] batteries were caught napping and didn't open fire soon enough."[38] The new antiaircraft S-75 missiles referred to by Khrushchev would bring down a U-2 only three weeks later during the next, and last, deep-penetration overflight of the USSR on 1 May 1960. The military also deployed S-75 batteries at Saryshagan (Fig. 6.12).

---

[37]Orlov, 1998/1999, p. 10.
[38]Khrushchev, 1974, p. 443.

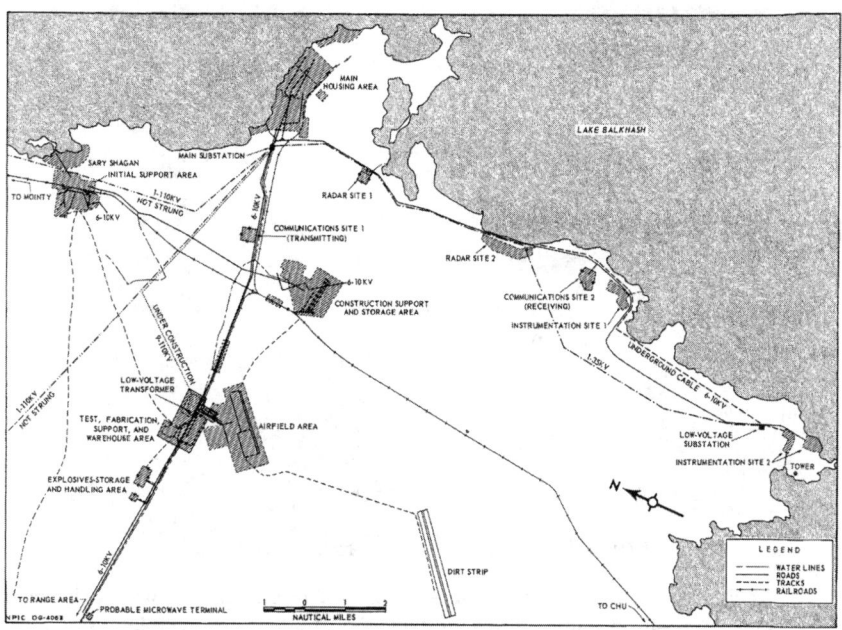

Fig. 6.11 The CIA's map (1961) of the support base at the Saryshagan test range. The Saryshagan railroad station (on the left) initially served as the main disembarkation area. The main housing area on the lakeshore at the peninsula became the town of Priozersk in 1963. The map also shows the main airfield (Fig. 6.9), storage areas, communication sites, high-voltage power lines, and two long-range surveillance radars, Dunai-2 of System A (marked as *radar site 2* on this map) and TsSO-P (*radar site 1*). The proving ground continued to grow as GNIIP-10 expanded its support for various air defense, missile defense, and antisatellite weapons programs. Figure from Central Intelligence Agency, Photographic Intelligence Center, 1961, p. 15.

The U-2 overflight in April 1960 provided detailed aerial photographs of the expanding Saryshagan range for the first time. U.S. photo interpreters clearly saw the profound importance of the observed installations, in spite of the area being partially covered by clouds. The operational debrief of the mission noted that

> [a]n area of some 8,000 square miles [20,700 km^2] between lake Balkhash and [the town of] Dzhezkazgan is seen for the first time and discloses many missile-associated installations. Included are interferometer-type instrumentation sites, numerous communications and tracking facilities, missile launch and support areas, a large dirt strip (16,300 × 260 ft [4970 × 80 m]), and other identified installations....
>
> Numerous guided-missile-associated activities have been located.... The complexity of this entire area complicated by partial cloud cover, will necessitate considerable detailed analysis, using all-source intelligence, to put together a composite intelligence picture....

Fig. 6.12 Satellite photograph of a surface-to-air S-75 (SA-2) battery with a characteristic "hexadic" pattern of launch positions deployed 14 miles (23 km) inland to the west of Priozersk. Original satellite reconnaissance photograph by KH-7 camera (Mission 4027; 21 April 1966) available from the U.S. Geological Survey; photograph identification, interpretation, and processing by Mike Gruntman.

> Detailed study of this area could result in the most provocative and possibly the most productive results from any TALENT [U-2 photography] coverage of Soviet missile activities to date. ...
>
> Not until the final readout has been made will the roles and interrelationships of the installations in this area be understood.[39]

The U-2 photography alone had not conclusively revealed the nature and technical details of the development programs at Saryshagan, which required correlation of information from various sources, particularly electronic intelligence. Immediately after the loss of a U-2 aircraft over the Soviet Union on 1 May 1960, Director of Central Intelligence Allen Dulles[40] issued a memorandum summarizing the accomplishments of the U-2 program. (The official U-2 history quotes parts of this document verbatim, allowing its attribution to Dulles.[41]) Dulles noted, apparently referring to Mission 4155, that "[p]reliminary analysis of photography collected on one of our most recent flights indicates that the Soviets may be engaged in research concerning anti-ballistic missile radars and tracking. It is too early, however, to determine whether or

---

[39]Central Intelligence Agency, 16 April 1960, pp. 1, 2.
[40]Allen Dulles, 1893–1969.
[41]Central Intelligence Agency, Office of Special Activities, 1960, p. 14; Pedlow and Welzenbach, 1998, p. 316.

not these developments include an actual anti-ballistic missile ... development program."[42]

By this time, Kisun'ko's team was only several months away from its first attempts in fall 1960 to intercept target warheads of long-range ballistic missiles.

Less than four months after termination of U-2 overflights, in August 1960, a Corona satellite successfully brought back to Earth the first reconnaissance photographs from space.[43] Since then, U.S. intelligence began to follow the developments at Saryshagan with continuously improving quality of images.

Actually, Corona was not the first reconnaissance satellite that peeked into the Soviet Union and Saryshagan. A few months earlier, on 22 June 1960, the Thor-Able rocket had launched the first electronic intelligence satellite, GRAB. Its name, GRAB, was a cover that innocuously stood for Galactic Radiation Background satellite, intended to perform astrophysical measurements. It shared the launch (a dual launch) with the U.S. Navy's much larger prototype navigational satellite Transit-2A.

In addition to its science instruments, the GRAB spacecraft carried a secret electronic intelligence payload. It turned on the sensors to "listen" to air defense radars over the Soviet Union for the first time on 5 July 1960, six weeks before the first successful Corona space mission delivered its photographs. GRAB findings included electronic signals from the rapidly growing antiballistic missile defense installations at Saryshagan. The satellite remained operational until August 1962.[44]

The emissions of Saryshagan radars Dunai-2 of System A and TsSO-P being developed by RTI would soon also be detected and characterized in some detail by using an unconventional method... measuring their signals scattered from the surface of the moon.[45] At the same time, satellite-based intelligence systems rapidly advanced. In the 1960s, radar emission sensors flew on satellites in low-Earth orbit. Later, payloads on geostationary satellites began intercepts of telemetry transmitted during missile trials at Soviet and Chinese proving grounds.[46]

GROWING INSTALLATION

The advent of satellite reconnaissance led to periodic coverage of the Saryshagan test range by Corona photography during various seasons and times of the day. It revealed minute details of expanding facilities, infrastructure, and operations. Photo interpreters found strung power lines and

---

[42]Central Intelligence Agency, Office of Special Activities, 1960, p. 14.
[43]Ruffner, 1995; Wheelon, 1997; Gruntman, 2004, pp. 403, 404.
[44]Moynihan, 2000; Gruntman, 2004, pp. 409–411; McDonald and Moreno, 2005.
[45]Eliot, 1967; Brown, 1969; Wheelon, 2004.
[46]Ball, 1988; Dyer, 1998, p. 269.

underground cables and estimated their voltages. They determined the types of road surfaces, the sizes and functions of buildings at technical sites, the capacity of barracks for soldiers and living quarters for officers and civilian personnel; identified waste disposal areas; and mapped buried pipes. Barbed wire fences with patrol zones pointed to specially guarded areas such as storage areas for explosives or particularly sensitive technical positions. After snowfalls in the winters, soldiers first cleared roads and paths to the headquarters of military units as well as to outdoor latrines near barracks, which helped to determine the layout of military bases, the deployment of units, and the number of servicemen.

U.S. intelligence analysts developed their own numbering system to identify technical sites and various areas scattered across the vast territory of the range (Figs. 6.10 and 6.13). Obviously, these numbers differed from those of GNIIP-10. The two largest long-range radar systems also received "personal" nicknames, Hen Roost for Dunai-2 and Hen House for TsSO-P. The box "Site Designations at Saryshagan Range" relates the CIA and Soviet nomenclature.

Although Kisun'ko's System A dominated the early work at Saryshagan, the proving ground expanded its activities into numerous other weapon programs in air and missile defenses. The range also actively engaged in advancing new applications that branched off from missile defense, such as tracking and cataloging orbiting artificial satellites and other space objects, early warning by radar of ballistic missile attack, antisatellite weapons, and later laser weapons.

Alexander Mints's RTI institute began building its long-range surveillance radar TsSO-P and an accompanying smaller fire-control radar for warhead intercepts. Already in 1956–1957, the institute's engineers had tested a simplified 12-m-diameter antenna prototype of the latter at a temporary position near the town of Aral'sk, observing ballistic missiles launched from Kapustin Yar. Mints's deputy, Mikhail Veisbein, oversaw this development.[47]

The RTI soon terminated work on its fire-control radar but continued development of the long-range TsSO-P. This program led to construction of one of the largest structures at Saryshagan (Fig. 6.14). The TsSO-P pulse radar operated at the 150-MHz frequency (2 m wavelength); its enormous aperture was 820 ft (250 m) long and 49 ft (15 m) in height. The radar emitted a narrow (in the horizontal direction) fan beam that extended 20 deg vertically. The beam scanned the sector in azimuth by frequency modulation of the signal and determined target elevation by comparing signals received by its two antennas.[48]

Operations of the TsSO-P radar required real-time computational capabilities. Consequently, in late 1957 Alexander Mints brought in a prominent

---

[47]Ivantsov, 2003, pp. 474, 475; Pervov, 2003, p. 26.
[48]Ivantsov, 2003, pp. 474, 475; Pervov, 2003, pp. 75–77; Belous et al., 2009, pp. 332, 333.

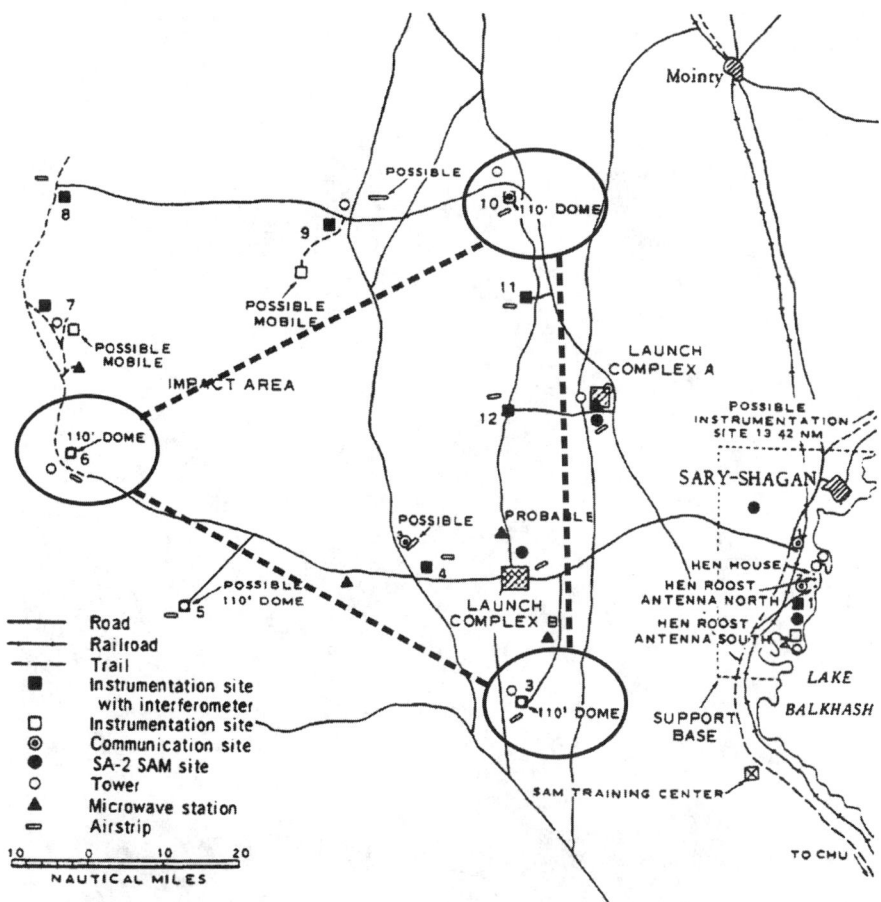

Fig. 6.13 Schematic map of the Saryshagan test range from a CIA 1963 intelligence report. The map shows radar, instrumentation, communications, and missile launch sites as well as airfields and main roads. The range grew rapidly, serving as testing grounds for various air defense, missile defense, and antisatellite weapon programs. CIA reports also provided detailed layouts of various technical positions and installations of the range. Three overlaid ellipses, connected by straight dashed lines, show the locations of three precise guidance RTN radars of System A, forming an equilateral triangle. Launch Complex B housed the System A interceptors. Initial plans had called for a backup launch site to the north, Launch Complex A, that was subsequently refocused on testing advanced antiaircraft missiles. The support base area (outlined by dotted straight lines on the right) showed north and south Hen Roost antennas of the long-range surveillance radar Dunai-2 of System A and the Hen House long-range radar (TsSO-P) being developed by RTI. Original figure from Ruffner, 1995, p. 219; System A geometry overlay by Mike Gruntman (Gruntman, 2004, p. 408).

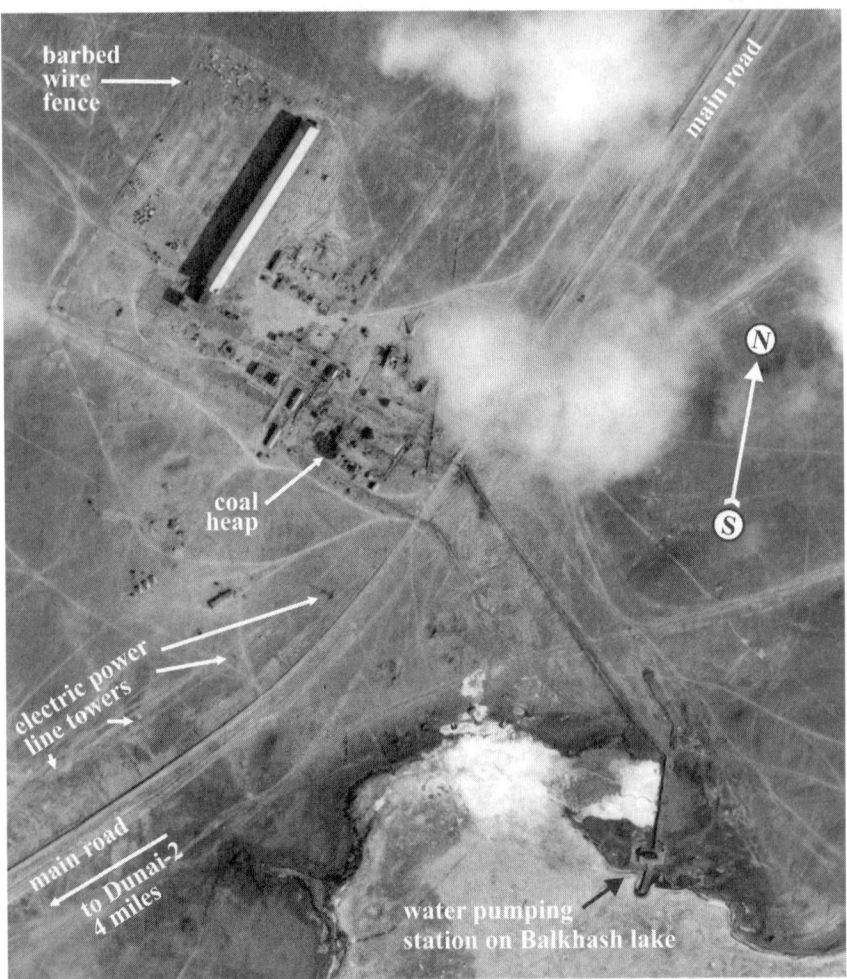

Fig. 6.14 Aerial reconnaissance photograph of the largest 890-ft (270-m) long structure at the Saryshagan proving ground housing the long-range early-warning radar TsSO-P in 1960. The front of this windowless building at Site 8 pointed in the general direction of Kapustin Yar. U.S. intelligence analysts nicknamed this huge building and its radar the Hen House. A small connected structure with windows, at a right angle to the main building, housed control electronics. The barbed wire fence surrounded this technical position next to a road and electric power lines leading southwest to the Dunai-2 antennas at Sites 14 and 15. A pumping station on the lakeshore nearby supplied the facility with water. Original photograph (U-2 Mission 4155; 19 April 1960) courtesy National Archives and Records Administration; photograph identification, interpretation, and processing by Mike Gruntman.

Site Designations at Saryshagan Range

GNIIP-10	CIA	Facility/Area
1	3	Precise guidance radar RTN-1
2	6	Precise guidance radar RTN-2
3	10	Precise guidance radar RTN-3; also later known as Site 53
4v		First barracks of construction troops and first residential area
4p		Expansion of residential area
4		Town of Priozersk, combining Sites 4v and 4p
6	Launch Complex B	V-1000 missile launch positions; also later known as Site 52 (target and interceptor radars of Aldan testbed of A-35)
7	Airfield and nearby area	Technical position for storage, preparation, and maintenance of V-1000 missiles; main airfield and aviation division
8	Radar site 1	TsSO-P radar (Hen House)
9		OS-2 node with Dnestr radars, near Gulshad settlement; also known as Site 54
14	Radar site 2	Receiving antenna of Dunai-2 radar (Hen Roost, North)
15	Radar site 2	Transmitting antenna of Dunai-2 radar (Hen Roost, South)
16, 17, 20, 21, 22	Instrumentation sites	
35	Launch Complex A	Test site for antiaircraft missiles S-75, Dal', S-200, etc.; test site for missile defense system S-225
38		Radar RE-4; later known as Site 51 (radars of Argun' testbed); laser weapons
40		Missile range headquarters; computer center

specialist, computer pioneer Isaak S. Bruk.[49] Bruk had built the first Soviet computers M-1, M-2, and M-3, and in 1958 became the founding director of the Institute of Control Computers (Institut Elektronnykh Upravlyayushchikh Mashin, or INEUM) of the USSR Academy of Sciences. He then developed a specialized computer, M-4, for TsSO-P. Although RTI specialists did not

---

[49] Isaak Semenovich Bruk, 1902–1974.

consider the M-4 particularly successful, it laid the foundation for the next generation of INEUM computers for operational early warning systems of ballistic missile attack.

In September 1961, TsSO-P demonstrated tracking of a ballistic missile launched from Kapustin Yar and soon afterwards tracking of an orbiting satellite. The radar performed numerous experiments detecting artificial satellites, which led to the selection of its successor designs for the antisatellite weapon system.[50]

The TsSO-P radar operated at frequencies somewhat lower than those of System A's Dunai-2. Consequently, the former was more susceptible than the latter to disturbances in the troposphere and ionosphere. In addition, it was easier to "blind" it through high-altitude nuclear explosions either detonated on purpose by the attacking side or produced by the defender's nuclear-tipped interceptors. On the other hand, the lower frequency of the TsSO-P made it harder, even impractical, to diminish the radar efficiency by deploying chaff because dipole reflectors would have to be 1 m long and thus unacceptably heavy.

The arguments between Kisun'ko and Mints and their respective supporters about the best long-range acquisition radar included considerations extending beyond technical merits and thus never reached a conclusion. The interests of powerful competing organizations and government officials made the issue intractable, and concurrent development of both systems continued.

Site 2

A main construction priority in 1956, Site 2, represented a technical position about 150 miles (240 km) to the west of Saryshagan (Figs. 6.15, 6.16, and 6.17). The designers placed it on an imaginary circle (Fig. 6.2) corresponding in size to the outer ring of the S-25 system around Moscow. In addition to housing the future RTN-2, Site 2 had another urgent role to play.

The entire concept of missile intercept relied on accurate tracking of approaching target warheads. To demonstrate such capabilities, Kisun'ko initiated a crash program to build a dedicated Experimental Radar (Radiolokator Eksperimental'nyi, or RE).[51] With the introduction of its modernized variant RE-2, the original RE would be referred to as RE-1. The Kisun'ko's SKB-30 design bureau led development of the RE and RTN radar systems.

As military construction units began to arrive at Saryshagan in summer 1956, the first convoy of 500 automobiles with 2500 servicemen and supplies departed to Site 2 across the desert. Without roads, it took more than four days

---

[50]Ivantsov, 2003, p. 475.
[51]Tolkachev, 2003; Pervov, 2003, pp. 60–63; Belous et al., 2009, p. 153; Aitkhozhin and Gantsevich, 2010a.

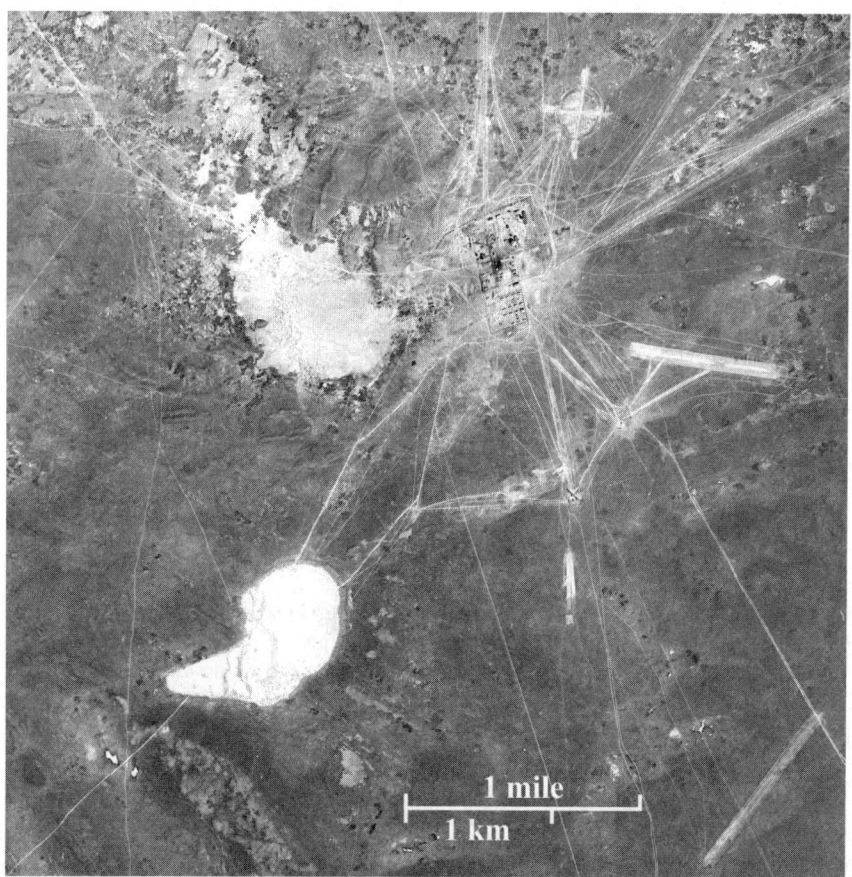

Fig. 6.15 Satellite photograph of Site 2 in 1966. This remote area 150 miles (240 km) west of Priozersk was among the highest construction priorities when the government activated the Saryshagan test range in 1956. Site 2 included a technical position with radar, electric power generators, communications facilities, and a residential area. The site housed one of the three precise guidance radars of System A, RTN-2, under the 115-ft (35-m) radome (Fig. 6.16). In the late 1950s, the RE-1 and RE-2 experimental radar at Site 2 confirmed the feasibility of detection and tracking of target warheads and accompanying missile bodies. CIA analysts spent significant effort trying to determine the purpose of a cross-type interferometer installation outside the fenced area, also photographed at some other sites of the test range. As a large remote base, Site 2 had an airstrip for light transport and liaison airplanes (Fig. 6.17). Original satellite reconnaissance photograph by KH-7 camera (Mission 4032; 20 September 1966) available from the U.S. Geological Survey; photograph identification, interpretation, and processing by Mike Gruntman.

to reach the destination. The soldiers brought fuel, food, tents, mobile kitchens, electric power generators, uniforms, medical supplies, spare parts, and construction materials. The remote area did not have any water but was abundant in scorpions and desert phalanx spiders as well as poisonous snakes.

Fig. 6.16 Fragment of a satellite photograph (Fig. 6.15) showing a fenced-off area at Site 2 in 1966. The large 115-ft (35-m) dome of the precise guidance RTN-2 radar of System A casts a clearly seen shadow. Original satellite reconnaissance photograph by KH-7 camera (Mission 4032; 20 September 1966) available from the U.S. Geological Survey; photograph identification, interpretation, and processing by Mike Gruntman.

Gubenko recalled that "water had to be brought from 40 km [25 miles]"[52] to Site 2 and that there had been no human settlements around. The soldiers drilled a number of wells; however, the water that was found did not meet quality requirements for mixing with cement to make concrete. After desperate testing in primitive conditions, military engineers determined that although the water from some wells did not satisfy strict government standards, it yielded concrete of acceptable strength.

In addition to the RE-1 radar, the remote installation also included smaller radar systems, optical kinetheodolites, and radio communications equipment. The narrow pencil beam of the experimental radar was less than 1 deg wide and could only determine the range (distance) to the target, not the direction. Therefore, the RE-1 relied on external pointing to follow moving targets. An optical kinetheodolite, KT-50 (a small telescope capable of recording images on film), which was in turn guided by another small radar, provided required accurate pointing. Establishing the uniform standard time system presented another challenge. Clock synchronization turned out to be especially important for complex experiments with components separated by hundreds and sometimes thousands of kilometers.

---

[52] Gubenko, 2003, p. 674.

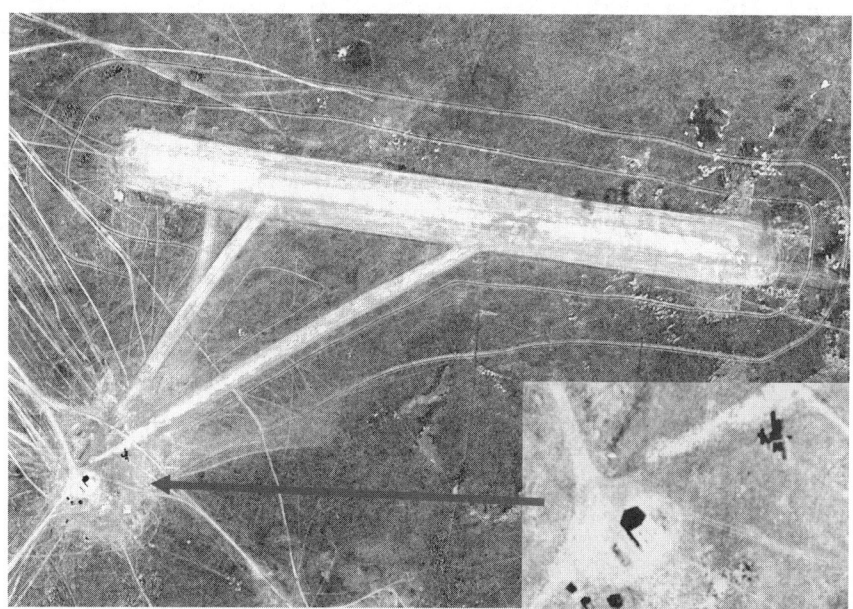

Fig. 6.17 Fragment of a satellite photograph (Fig. 6.15) showing an airfield at Site 2 with a parked airplane (insert on the right) in 1966. Original satellite reconnaissance photograph by KH-7 camera (Mission 4032; 20 September 1966) available from the U.S. Geological Survey; photograph identification, interpretation, and processing by Mike Gruntman.

The RE-1 radar had an agile 15-m-diameter antenna, RS-10, in the Cassegrain configuration with a horn emitter. (The two-mirror Cassegrain design is widely used in optical telescopes and radio antennas. It includes two reflectors, the primary large concave parabolic mirror and the secondary small convex hyperbolic mirror.) The mechanisms with two degrees of freedom could point this large antenna in any direction. Because of the program urgency, the experimental radar borrowed the 2-MW power transmitter operating at the 10-cm wavelength (frequency 3 GHz) from the B-200 radar of the S-25 air defense system. A spherical radome protected the radar from the elements.

The operation of high-power radar required adherence to strict safety measures because of intense microwave radiation. Exposure limits were not well known and understood in the 1950s. In addition, a general sense of urgency led to neglect in enforcing safety regulations, especially in the early days of the new test range.

Officer Alexander Kulakov arrived at Saryshagan in May 1957. GNIIP-10 assigned him to a new safety office that did not have any staff yet. (Two years later he transferred into a computer unit and became, with time, a leading specialist in the development and application of complex computer codes supporting missile defense tests and operational systems.) Kulakov recalled that in 1957

> superpowerful electromagnetic and especially microwave radiation was thought to be dangerous but everyone had to protect himself as one could. I went [to get guidance] on this problem to a medical academy in Leningrad. They promised to study the subject but I have never received any results from them. Only God would know how many people suffered from this radiation. But rumors [and anecdotal evidence] are very sad.[53]

Another officer, Zhadeiko, had begun his service at the RTN radar at Site 1. He later recalled that

> [t]he installation of the large [15-m diameter] antenna was completed in November 1958. During mechanical tests it moved, with screeching, from pointing to the horizon at the north to [the horizon at the] south, and from east to west, trying to cover all possible directions in the hemisphere. When the [microwave transmitter] power was turned on, a special signal sounded and the red lamp lighted. It was quickly explained to us that the invisible [microwave] rays were deadly to our [ability to have] children in the future. Therefore, when we walked to the radar, we moved in zigzags, dashing to its location during time intervals when it pointed into the zenith.
>
> ... [O]ne engineer from industry [for example] covered [private parts of his body] with a white-enameled basin...
>
> [Safety] specialists visited us to measure [levels of] the microwave radiation and said that they were ten times higher than [the safe] limits. After one year, when the [RTN] radar began continuous operations, several men ended up in the hospital. The guys working at the receiver unit and with the transmitter got the largest [microwave] radiation doses.[54]

Providing realistic target warheads for missile defense intercepts also presented a challenge. In 1956, the largest operational ballistic missile was the R-2 (article 8Zh38 in the Soviet military nomenclature). The military had accepted it in 1952 and then deployed it in rocket units. Sergei Korolev's design bureau in Podlipki produced this significantly improved version of the first Soviet ballistic missile R-1 (8A11), the latter, in turn, based on the German A-4.[55]

The West referred to the R-1 and R-2 as SS-1 (Scunner) and SS-2 (Sibling), respectively. The R-2 later earned the distinction of jump-starting the national ballistic missile program of the People's Republic of China in the late 1950s.

---

[53]Kulakov, 2006, p. 29.
[54]Zhadeiko, n.d., pp. 3, 4; also quoted in Kulakov, 2006, p. 54.
[55]Gruntman, 2004, pp. 283, 284.

The Soviet Union had transferred the R-2 technology and helped the Chinese communists to organize the manufacturing of this missile, designated Model 1059, before cutting aid off in 1960 when the relationship between the two totalitarian Marxist giants deteriorated.[56]

The 360-mile (580-km) range of the single-stage R-2 was insufficient for reaching Saryshagan when launched from Kapustin Yar 1200 miles (2000 km) away. Correspondingly, the construction troops built a temporary launch position closer to Saryshagan near the rapidly expanding ballistic missile proving ground at Tyuratam. The RE-1 radar observed for the first time an R-2 fired from that location on 7 June 1957; more launches followed in August. The experiments demonstrated detections of both the separated target warhead and the accompanying body of the ballistic missile; the measured radar cross-sections were about 0.3 m^2 and a few dozen square meters, respectively.

The operational frequency being designed for the RTN radar of System A differed from that of the RE-1 because the latter relied on a transmitter borrowed from the B-200 radar. Therefore, as the first RE-1 tests were progressing, SKB-30 built a new variant of the experimental radar, RE-2, operating at the RTN nominal frequency 2 GHz (wavelength 15 cm) and increased the radar pulse power to 10 MW by combining four horn emitters.

The RE-2 began numerous tests in summer 1958, detecting and tracking the warheads of R-2 ballistic missiles as well as newer R-5 missiles (8A62, SS-3, Shyster) and later R-12 (8K63, SS-4, Sandal) as they became available. Also developed by Korolev, the single-stage R-5 used a cryogenic oxidizer (liquid oxygen) similar to its predecessors R-1 and R-2 and achieved a range of 750 miles (1200 km). Its variant R-5M became the first Soviet ballistic missile designed to carry a nuclear warhead. The military demonstrated this capability with the successful launch of a live nuclear warhead and its detonation over the target in February 1956. The Soviet Union had operationally deployed more than 30 such missiles by 1960.[57]

The R-5 also could not reach the target point at the Saryshagan range when fired from established positions at Kapustin Yar. Therefore, the military built a temporary launch site for R-5 missiles 30 miles (50 km) north of the small town of Chelkar located about 50 miles (80 km) north of the Aral Sea (Fig. 6.18). Already in 1960, a U.S. intelligence report identified this Chelkar field facility as a launch site operated by the command at Kapustin Yar and described that it was "apparently used for the training of operational crews on missiles launched in conjunction with the antiballistic missile (ABM) program being conducted at Saryshagan."[58]

---

[56]Gruntman, 2004, pp. 440, 441.
[57]Gruntman, 2004, p. 284.
[58]Guided Missiles and Astronautics Intelligence Committee, 1960a, C-1; Guided Missiles and Astronautics Intelligence Committee, 1960b, p. 4.

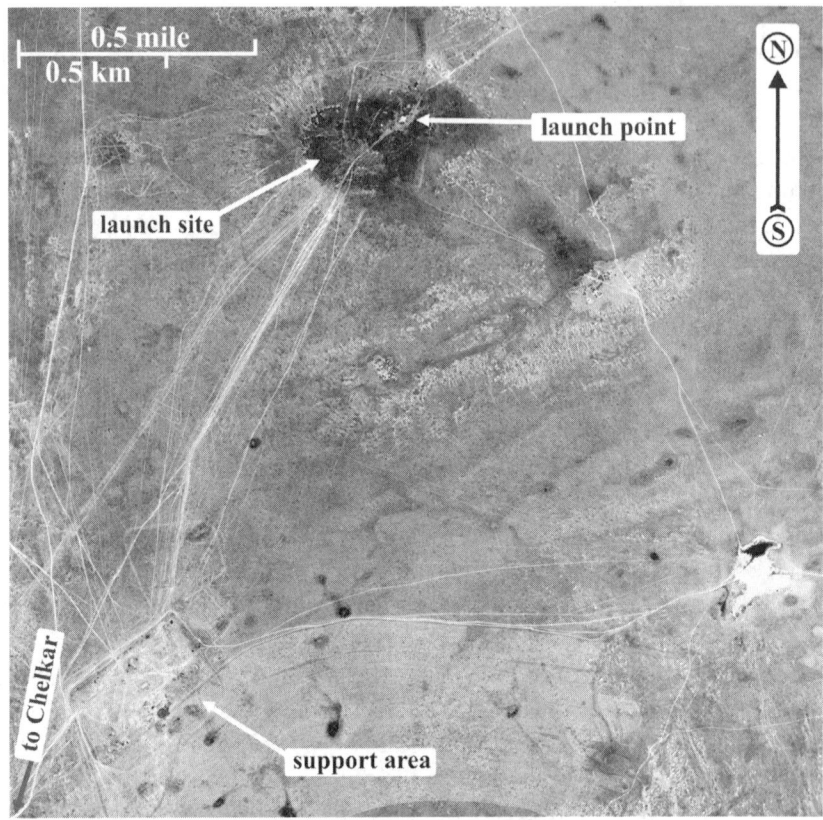

Fig. 6.18 Field launch site of R-5 ballistic missiles 30 miles (50 km) north of the small town of Chelkar, located 50 miles (80 km) north of the Aral Sea in Kazakhstan in 1966. Units of the strategic rocket forces fired ballistic missiles from this position toward the Saryshagan range starting from 1958. The CIA called this facility "the launch point ... SP-5" and determined that the Kapustin Yar missile range operated it (Guided Missiles and Astronautics Intelligence Committee, 1960a, C-1). In 1962, the analysts described that this position "consists of a fenced launch site with seven associated vehicle revetments, a probable communications area, and a possible baseline instrumentation and guidance facility.... The support area contains nine buildings and a possible power generation station. There are no external power lines serving this installation" (Guided Missiles and Astronautics Intelligence Committee, 1962, III-C-1). Original satellite reconnaissance photograph by KH-7 camera (Mission 4029; 4 June 1966) available from the U.S. Geological Survey; photograph identification, interpretation, and processing by Mike Gruntman.

Korolev's competitor Mikhail K. Yangel[59] developed the IRBM R-12 in the Yuzhnoe Design Bureau in Dnepropetrovsk (Dnipropetrovsk), Ukraine. This new-generation, single-stage, intermediate-range ballistic missile (IRBM) had a fueled mass of 92,600 lb (42,000 kg) and could carry a 3500-lb

---

[59]Mikhail Kuz'mich Yangel, 1911–1971.

(1600-kg) nuclear warhead to a range of 1250 miles (2000 km). In contrast to Korolev's missiles, which used cryogenic liquid oxygen, the liquid-propellant engine of the R-12 burned storable noncryogenic propellants—kerosene as fuel and nitric acid with the addition of nitrogen tetroxide as oxidizer. The military favored storable propellants because it could now maintain IRBMs in launch readiness for extended periods of time. The new ballistic missile was also the first in the Soviet Union to rely on an entirely autonomous guidance system, another important feature for a weapon.

The R-12 successfully flew for the first time in June 1957. It became the first truly mass-produced IRBM in the Soviet arsenal, with 172 and 458 missiles operationally deployed by 1960 and 1962, respectively. Soviet leader Khrushchev famously described these rockets as coming from factory lines "as sausages" in his speech in the United Nations in 1959.[60] The fielded nuclear-armed SS-4 missiles could effectively reach all North Atlantic Treaty Organization (NATO) countries in Europe.

Tracking warheads by experimental radar provided a wealth of information indispensable for the design of missile defenses; for example, it revealed relationships between the characteristics of scattered radar pulses and the rotation of the observed objects. Such data directly supported development of missile defense techniques dealing with possible penetration aids such as decoys. In addition, the observations confirmed that the wakes behind approaching warheads did not affect the accuracy of their tracking by radar at selected operational frequencies.

Also in 1958, for the first time the RE-2 conducted multiple observations of an artificial satellite, Sputnik-3. An optical instrument, a kinetheodolite, again pointed the radar. The detection of space vehicles demonstrated the feasibility of precise measurements of coordinates of high-velocity space objects in orbit[61] and laid the foundations for future satellite tracking and antisatellite weapons programs. Kisun'ko's RE-2 radar had achieved satellite tracking three years ahead of the TsSO-P of his rival Mints.

The design of effective missile defenses relied on knowledge of radar cross-sections and emitted and received radar power. Initially, engineers performed radar calibrations by measuring signals scattered by metal spheres flown on meteorological balloons. Later, development of the first operational missile defense system, A-35, expanded these techniques by placing into orbit dedicated artificial satellites with specific sizes and shapes as radar targets.

Saryshagan installations were too close to Tyuratam to conduct important tracking of warheads of the first Soviet ICBM R-7 (8K71, SS-6, Sapwood) launched from there. Consequently, engineers placed the new upgraded experimental radar RE-3 at the Kamchatka peninsula in the target area of

---

[60]Gruntman, 2004, pp. 286–289.
[61]Aitkhozhin and Gantsevich, 2010a.

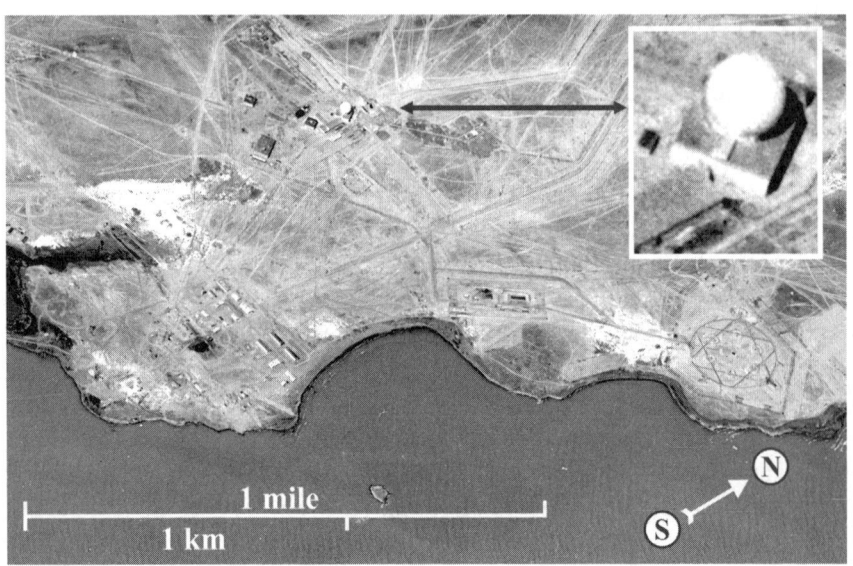

Fig. 6.19 Satellite photograph of Site 38, also known later as Site 51, with the dome of the RE-4 experimental radar at the Saryshagan proving ground in 1966. The insert shows the RE-4, which was built to test new technologies for the operational A-35 missile defense system. The range also deployed an S-75 (SA-2) antiaircraft battery at the lakeshore nearby (bottom right in the photograph). Within a couple years, the military expanded the facilities of the site (Chapter 9) by adding in the immediate vicinity of RE-4 two new prototype radars (target radar RKTs-35TA *Istra* and interceptor missile radar RKI-35TA) of the testbed Argun' for the modernized A-35M. From the 1970s, the site also engaged in evaluating the Terra prototype laser weapons. Original satellite reconnaissance photograph by KH-7 camera (Mission 4027; 21 April 1966) available from the U.S. Geological Survey; photograph identification, interpretation, and processing by Mike Gruntman.

intercontinental ballistic missile (ICBM) test launches. They also built and deployed another new experimental radar, RE-4, in 1962 at Site 38 (later known as Site 51) of the Saryshagan range (Fig. 6.19). The RE-4 tested new tracking technologies for the evolving operational A-35 missile defense system. It also demonstrated reduction of noise in signal receivers by cooling them down to liquid nitrogen temperatures. This technical position later expanded its activities to evaluating the testbed Argun' for the modernized A-35M and the development of laser weapons.

PRIOZERSK

From the first days of its arrival at the Saryshagan railroad station in July 1956, the military focused its attention on building the main residential area. It began work on Site 4v on the shore of Lake Balkhash at the peninsula 7 miles (11 km) south-southeast of the railroad station (Fig. 6.6). In addition to the UIR-32 directorate and numerous military construction units, this area housed growing numbers of officers of GNIIP-10. It also accommodated visiting civilian specialists from Kisun'ko's design bureau and his contractors.

Two years later, the military expanded construction of living quarters to Site 4p 2 miles (3 km) to the east. The 1960 aerial photograph showed significant housing at Site 4v and several blocks of new buildings at Site 4p (Fig. 6.20).

Fig. 6.20 Growing main housing area of the Saryshagan range, in what would be Priozersk, in April 1960. The U-2 aircraft obtained this oblique photograph on the last successful deep-penetration flight over the Soviet Union. Thawing and cracking ice fields still covered the lake. A few small clouds cast their shadows on the ground. One such cloud hung directly above Site 40 (at the bottom of the photograph) and concealed the main headquarters of GNIIP-10 and the computer center. The area showed 326 housing and 20 administrative buildings as well as more than 230 trucks and cars in motor pools (Central Intelligence Agency, Photographic Intelligence Center, 1961, p. 14). The town would rapidly expand to Site 4p on the right and fill the empty space between Sites 4v and 4p. The planned street pattern showed the main Transportnaya Street on the left connecting to the future Lenin Street on the right. The first buildings already appeared in the area between Transportnaya Street and the lakeshore, housing the main hospital of the range. Original photograph (U-2 Mission 4155; 9 April 1960) courtesy National Archives and Records Administration; photograph identification, interpretation, and processing by Mike Gruntman.

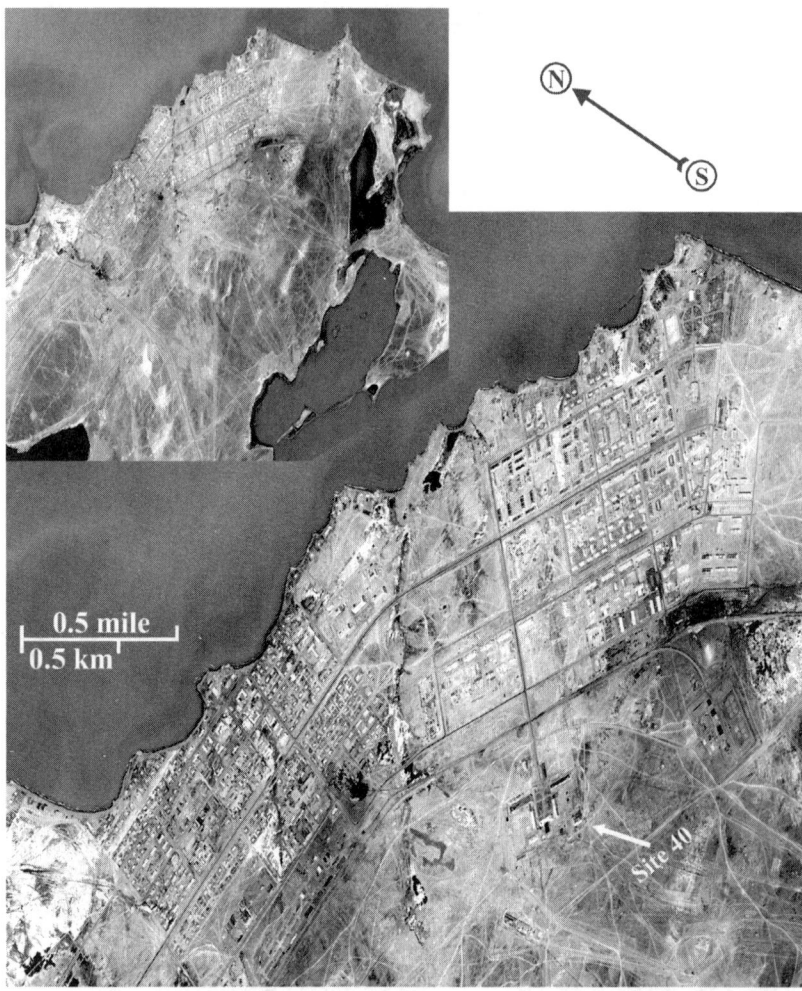

Fig. 6.21 Satellite photographs of the peninsula (top left) and Priozersk (bottom) in 1966 showing expanded residential areas, with new streets, housing blocks, and administrative buildings. The area of the Saryshagan range headquarters and the computer center (Site 40) is clearly seen in the photograph, in contrast to the 1960 U-2 photograph (Fig. 6.20) where a cloud had obscured the view. The road pointing from Site 40 directly to the lakeshore ends at a new television tower (Fig. 6.22) erected in 1963. More buildings also appeared in the hospital area. Original satellite reconnaissance photograph by KH-7 camera (Mission 4027; 21 April 1966) available from the U.S. Geological Survey; photograph identification, interpretation, and processing by Mike Gruntman.

A medical facilities area between these two sites already revealed the first hospital structures. Six years later a satellite photograph showed many new buildings on the streets of Site 4p and the growing hospital (Fig. 6.21). By that time Sites 4v and 4p had combined into one Site 4, with the settlement first known as Koktas and then the town Priozersk.

By 1962, this main residential area had 25,000 inhabitants.[62] It boasted several stores, three preschools/kindergartens, and three schools. (A typical school enrolled pupils of all grades, from the primary first year through 10th or 11th year of graduating classes. The number of pupils in a school could have ranged from several hundred to more than a thousand.)

The permanent main garrison clubhouse, commonly called Dom Ofitserov, or House of Officers, opened in 1963. It replaced the club in a temporary building at Site 4v that had functioned since 1959. The chief political officer and the political department of the garrison oversaw the club that, in addition to being a tool of communist ideological control and propaganda, served as a cultural center in a remote military settlement. It included a movie theater, doubling as a venue for official celebrations of state holidays and other important events in the life of the garrison; housed various sports, cultural, and hobby groups; and conducted political and cultural activities reaching families of officers and civilians. Sports played an especially important role as recreation at Saryshagan. The main stadium opened its doors in 1964 and gradually expanded its facilities.

Site 40 lay a couple kilometers inland from Site 4 and the lakeshore (Fig. 6.21). It included the GNIIP-10 headquarters and the main computing facilities. A number of remote technical positions and outposts of the range also grew into large settlements with barracks for soldiers, residential housing for military families, and accommodations for visiting engineers and technicians. Hundreds to thousands of servicemen and civilians worked and lived, for example, at Sites 1, 2, 6, 7, 9, and 35. A garrison club at Site 35 was similar to that of the main residential area.

A young engineer of SKB-30, Leonid G. Khvatov,[63] spent extended periods of time at the range working on RTN radar systems. He described that Site 3, for example,

> represented a small settlement surrounded by a barbed wire fence. It had the RTN radar (one-story building with the target and article [interceptor missile] antennas [in radomes] on its sides), a hotel for representatives [specialists] from industry, a hotel for officers, barracks for soldiers, headquarters of the military unit, houses for the commanders [and senior officers] of the unit, and other buildings supporting the installation. There was an airstrip for planes four kilometers away.[64]

Some enterprising military commanders of units at remote locations opened small farms at their sites. There, soldiers tended to a few dozen cows and pigs,

---

[62]Kulakov, 2006, p. 158.
[63]Leonid Georgievich Khvatov, b. 1931.
[64]Khvatov, 2003, p. 586.

supplementing the rations of the servicemen and providing milk to children.[65]

In spite of a large number of residents, Site 4 did not have the usual municipal services. People had to travel to the town of Balkhash 60 miles (100 km) away, or even much farther to a regional center, Karaganda, for such mundane tasks as registration of marriages and births. Finally, in spring 1963, the government of the Soviet Kazakh Republic granted the status of a town to the settlement and approved its new name, Priozersk. (Priozersk literally means near the lake.) Now, the town received a local government and municipal law enforcement, and could establish essential services.

Built on a rocky peninsula, Priozersk lacked vegetation. Everybody took part in nurturing precious green areas. People excavated holes in the ground, filled them with soil brought from distant locations, and planted trees. With care and a lot of water, Priozersk boasted large green areas. The town also built a television center with a transmitting tower (Fig. 6.22) in 1963 and began broadcasting its own programs.

After some bureaucratic maneuvering and lobbying, the Saryshagan range had succeeded in getting on a government list of towns with a "special status," which improved its supply of food and other goods. Such special treatment was among the essential privileges granted in the Soviet Union to the so-called

Fig. 6.22 Satellite photograph of the television tower erected in 1963 in Priozersk near the lakeshore. A straight road connected the tower with Site 40. Original satellite reconnaissance photograph by KH-7 camera (Mission 4027; 21 April 1966) available from the U.S. Geological Survey; photograph identification, interpretation, and processing by Mike Gruntman.

---

[65] e.g., Egorov, 2009, p. 29.

"closed towns" engaged in important defense research, development, and manufacturing.

The main residential area, Priozersk, continued to grow. A veteran of Saryshagan, Yuri G. Venediktov,[66] began his service in UIR-32 in 1957 as a young lieutenant. He worked at the range for 17 years and became chief engineer of the directorate. Venediktov summarized that by 1975 Priozersk had 354 residential houses with 10,200 apartments, 12 hotels and dormitories with 2500 beds, 22 stores with 180 employees, 4 schools with 4300 pupils, 11 kindergartens for 2500 children, 4 medical clinics, and one civilian hospital and one military hospital with 50 and 650 beds, respectively.[67] There were also the garrison clubhouse, television center, and numerous administrative buildings. In addition, military construction units built, starting from scratch, 30 miles (50 km) of railroads and 900 miles (1500 km) of automobile roads connecting various sites of the range at the lakeshore (Fig. 6.23) and inland.

The initial rush of construction of the range installations and the town resulted in problems, however. Creating the pleasant and praised green areas and planting trees undermined building foundations. Growing trees in a desert required a lot of water. Surveyors had not done comprehensive geological studies of the area prior to construction of the range. The military had built Priozersk on a rock with, as was discovered later, multiple fractures filled with gypsum. Watering green areas caused a gradual rising of the water bed, which affected the structural strength of the underlying fractured rock. Consequently, many buildings required strengthening of their foundations, using reinforced concrete plates and deep piles.[68]

Completion of a hydroelectric dam and filling of the reservoir on the Ile river reduced the flow of fresh water into Lake Balkhash in the 1970s. As a result, the mixing of salty and fresh waters accelerated. Pumped lake water became increasingly saline, causing the growth of deposits in pipes in Priozersk. It became necessary to replace many water lines in the town, increase their diameter, and even in some cases use stainless steel pipes.

NUCLEAR EXPLOSIONS IN SARYSHAGAN'S SKIES

As the work on the nonnuclear System A progressed, the military and Kisun'ko's designers had firmly arrived at the decision to employ nuclear-armed interceptors in the operational missile defense system. Consequently, establishing the destruction effectiveness of nuclear explosions on target warheads emerged as a top priority. In addition, it became clear that the attacking side could detonate nuclear charges at various altitudes to temporarily interfere with and blind missile defense radars, thus allowing other warheads

---

[66] Yuri Georgievich Venediktov, b. 1934.
[67] Venediktov, 2003, p. 684.
[68] Venediktov, 2003, pp. 684, 685.

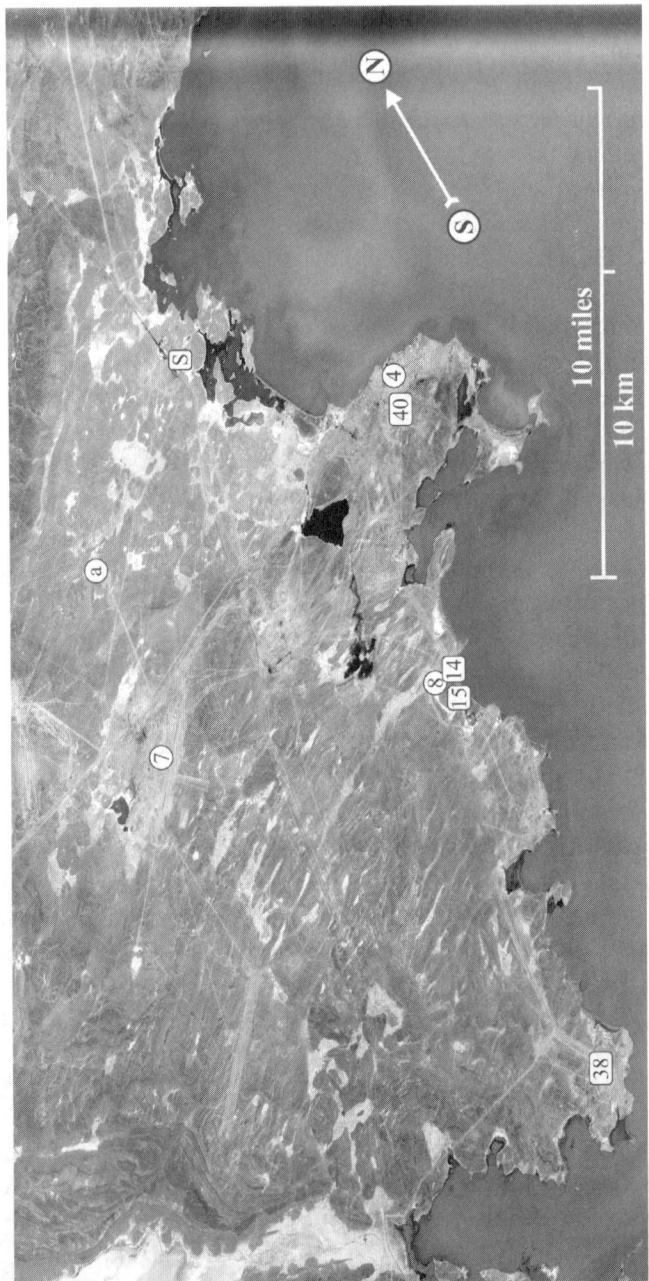

Fig. 6.23 Photograph of the test range area near the Balkhash lakeshore in 1966. Sites: S: Saryshagan railroad station; 4: Site 4 (Priozersk); 7: main airfield; 8: TsSO-P; 14, 15: Dunai-2; 38: Site 38 (also 51); 40: missile range headquarters and computer center; a: S-75 battery (Fig. 6.12). Original satellite reconnaissance photograph by KH-7 camera (Mission 4027; 21 April 1966) available from the U.S. Geological Survey; photograph identification, interpretation, and processing by Mike Gruntman.

TABLE 6.1 MISSILE DEFENSE NUCLEAR EXPLOSION TESTS IN 1961–1962

Date	Yield	Altitude
October 27, 1961	1.2 kton	93 miles (150 km)
October 27, 1961	1.2 kton	186 miles (300 km)
October 10, 1962	300 kton	180 miles (290 km)
October 28, 1962	300 kton	93 miles (150 km)
November 1, 1962	300 kton	37 miles (59 km)

to break through defenses. Therefore, it was necessary to understand the effects of nuclear explosions on radar.

Consequently, the government prepared a series of dedicated nuclear tests, known as Operation K. The objectives included determining the effects of nuclear explosions on ballistic missile warheads; the atmosphere; various radioelectronic units, especially the radars of missile defense System A; and interceptor missiles and their guidance system.

The Soviet Union conducted a total of five nuclear tests for Operation K in 1961 and 1962. Publications by missile defense participants[69] listed identical altitudes (80, 150, and 300 km) of explosions and no exact dates and yields, apparently repeating the first publication on the matter. A recent well-researched study[70] unrelated to missile defense provided details of the nuclear explosions. It showed a significantly different altitude (37 miles or 59 km) of the lowest detonation (Table 6.1).

Here, the yield of 1 kton (1 kiloton or 1000 t) means the energy equivalent to that released in an explosion of 1 kton of common chemical high-explosive TNT. One kton of TNT corresponds to energy of 10^{12} calories or 4.186×10^{12} joules. (For comparison, the first two atomic bombs operationally detonated over Japan in 1945 were in the 10–20 kton range.)

In Operation K, rocket units simultaneously launched two ballistic missiles along similar trajectories toward the Saryshagan range. The detonation of a nuclear charge carried by one missile allowed researchers to study the effects of the explosion on the other warhead mockup nearby that was equipped with numerous sensors. System A had to track the latter warhead and guide the interceptor missile toward it despite the nuclear explosion. In addition, ground facilities and instruments on sounding rockets, also launched during the test, simultaneously performed numerous measurements of the conditions in the upper atmosphere and ionosphere. As the CIA's deputy director for science and technology from 1963–1966 Albert "Bud" D. Wheelon[71] described it,

---

[69]Kisun'ko, 1996, pp. 429, 430; Kamensky, 2003, p. 605; Pervov, 2003, p. 100; Belous et al., 2009, pp. 184–190.
[70]Greshilov et al., 2008, pp. 266, 267.
[71]Albert D. Wheelon, 1929–2013.

"two sounding rockets carried instruments aloft during each test. Thus four vehicles arrived at roughly the same point in space at the same time: target, nuclear carrier, and two sounders."[72]

Saryshagan officer Kulakov recalled that "specialists and members of [their] families in Priozersk did not feel special anxiety about nuclear explosions because of complete illiteracy about possible consequences and lack of dosimeters and information about levels of radiation."[73] As another officer noted, "safety [procedures were] slightly above the zero level. The same was the level of information [about nuclear explosions and their effects] available to the participants of the tests."[74]

Kulakov wrote that

> [e]verybody wanted to observe them [nuclear explosions] visually. The effect of observation depended on the altitude of the nuclear detonation. ... I succeeded to observe one of them by leaving my working place in the building. I did not have time to put soot on a glass plate for observation. Therefore, I saw [by the naked eye] a small flash far away at a distance of 300 km. The majority of those looking through soot-covered glass plates could not see even that.
>
> A completely different effect was produced by explosions at [lower] altitudes of 80 and 150 km [as observed] not only at technical positions closer to the explosions but also in the vicinity of Priozersk where family members of servicemen had been [temporarily] sent to from those sites. At the technical positions, windows were covered by black paper. All those who remained there were issued special protective means.[75]

An eyewitness to the explosions described that "the sky was showing [colored] stripes, ominously beautiful and extraordinary" after the nuclear explosion.[76] An officer serving at the RTN radar on Site 2, the closest to the detonation, recalled that

> to the northwest, we saw a huge [ominous] poisonous-greenish rugged cloud against the background blue sky. Everybody had unsettling feelings. At home, I turned on the radio and could hear only static noise on all frequencies. Perhaps we were under a powerful electron cupola. ...
>
> The sky was clear the next day. ... However, during the next several months one could see blind saigas in the steppe.[77]

---

[72]Wheelon, 2004, p. 253.
[73]Kulakov, 2006, p. 76.
[74]Pervov, 2003, p. 101.
[75]Kulakov, 2003, pp. 76, 77.
[76]Belous et al., 2009, p. 190.
[77]Tsukov, 2003, p. 374.

Another Saryshagan officer noted that the range issued only standard gas masks as protection. He added that he could read a book outside his house at night after the test because a luminous formation hung in the sky above.[78]

Operation K provided important insights into the effects of nuclear explosions on target warheads and missile defense radar. The Soviet government assigned a major nuclear weapons center at Chelyabinsk-70 in the southern Ural mountains to build nuclear warheads optimized for missile defense. Intercepts at high altitudes precluded reliance on blast waves, so the design of nuclear charges maximized their neutron and x-ray output and also tried to focus it in the desired direction.[79]

The nuclear tests of Operation K in 1961 and 1962 took place after the first successful intercepts of target warheads. Prior to that, in the late 1950s, the engineers from industry and officers of GNIIP-10 at the Saryshagan test range focused on building the highly complex experimental System A.

---

[78]Gusarov, 2012, part 2.
[79]Golubev et al., 1994, pp. 44, 45; Kamensky, 2003, p. 610.

Chapter 7

Experimental System A

System Concept

The experimental System A aimed to prove the feasibility of defense against a single pair target, a warhead and the accompanying ballistic missile body. Its main components were scattered across a vast territory of the Saryshagan range and included the long-range Dunai-2 radar with antennas at Sites 14 and 15 on the lakeshore; three precise guidance RTN radars at Sites 1, 2, and 3 at the corners of an equilateral triangle; V-1000 interceptor missiles and their initial guidance RSVPR radar at Site 6; and the main control center at Site 40 near Priozersk (Fig. 7.1). High data rate microwave communication channels connected system elements.

Enormous scientific and engineering challenges for System A included development of radar, interceptor missiles and their blast fragmentation warheads, and real-time control of the entire system by a digital computer. The duration of the entire event was so short that the system operated automatically, with operator contributions limited to early stages of the intercept. Computers, algorithms, and codes had to synchronize all system components, automatically collect and process diverse information received through high data rate communications channels, identify and separate the warhead from the accompanying rocket body, predict its trajectory in the gravitational field of the Earth and account for the effects of drag in the atmosphere, launch and guide the missile to the intercept point, and detonate the interceptor warhead with precise timing.

The complexity of System A called for mobilizing the resources of many of the country's leading scientific, research, development, and industrial organizations.

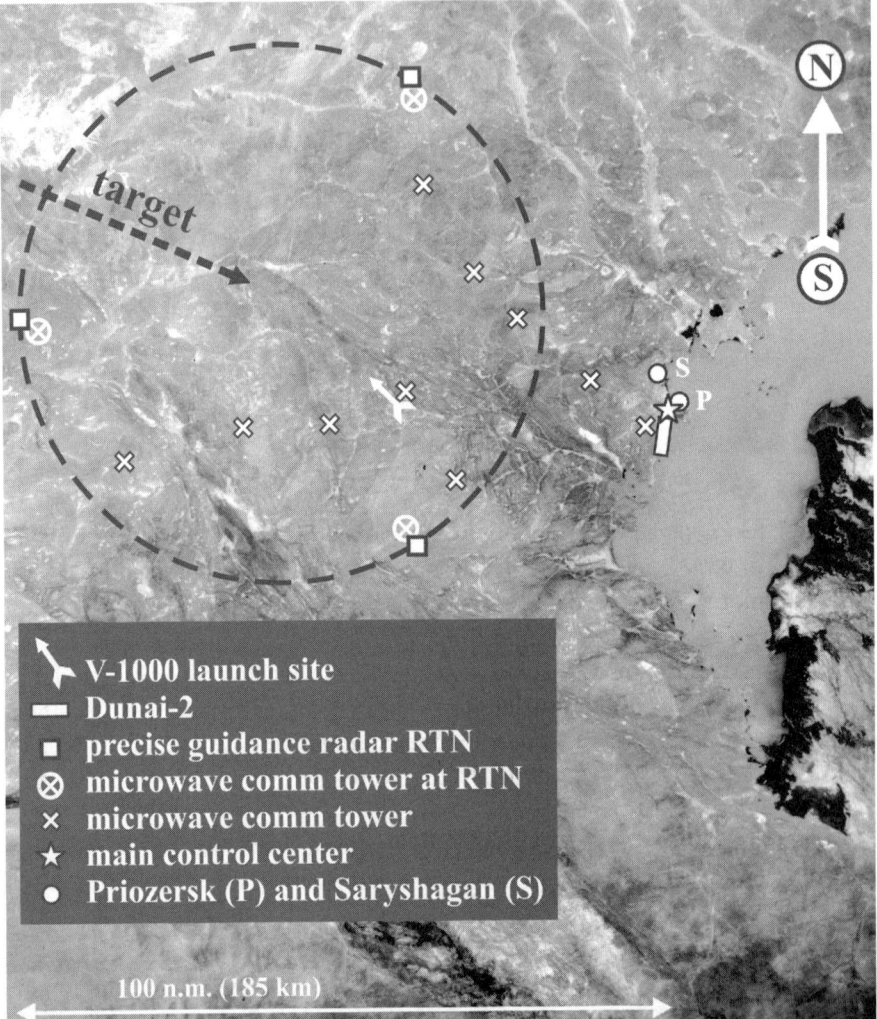

Fig. 7.1 System A at the Saryshagan test range. The dashed black arrow shows the trajectory of approaching ballistic missile targets. White squares of three precise tracking RTN radars (Sites 1, 2, and 3) form an equilateral triangle. The long-range Dunai-2 radar (white rectangle; Sites 14 and 15) is to the south from the Saryshagan railroad station (S) and town Priozersk (P) on the lakeshore. The main command center and computer facility (star; Site 40) is a couple kilometers inland from Priozersk. A white arrow (Site 6) shows the launch site of V-1000 interceptor missiles with the initial guidance RSVPR radar. Microwave communications towers (crosses and circled crosses) connected all components of System A. Original satellite photograph by KH-5 Argon mapping camera (Mission 9058A; 29 August 1963) courtesy U.S. Geological Survey; photograph identification, interpretation, and processing courtesy Mike Gruntman.

Long-Range Search Radar Dunai-2

Vladimir P. Sosul'nikov[1] led development of the long-range search radar Dunai-2 at Aksel' Berg's NII-108. Sosul'nikov and his group would separate in 1958 into a new branch of NII-108 and then form an independent institute.

Based on its prior work, NII-108 designed Dunai-2 as a continuous wave radar. A common pulse radar periodically transmits short pulses of electromagnetic radiation, which are reflected by a target and received by the radar. The measured round trip time of a pulse then determines the distance to the target. (The propagation speed of radio waves is practically constant and equals the speed of light.)

In contrast, a continuous wave (CW) radar emits electromagnetic waves at a certain frequency without interruption, and its emission power does not peak sharply, in contrast to traditional pulse radar. The CW radar continuously receives radiation reflected by a target. It determines the target velocity along the line of sight in a straightforward way by measuring a frequency shift between the emitted and received signals, described by the Doppler effect.

Such a basic CW radar cannot obtain the distance (range) to a target. However, if its frequency varies in time according to some law, then a CW radar can also determine the distance to the object. Therefore, periodic variations of the emitted signal frequency, the frequency modulation, is essential for range measurements.

In large and powerful CW radar, such as Dunai-2, a leakage of the transmitted signal into the receiver circuitry, known as the *spillover*, called for two physically separated transmitting and receiving units. Two such antenna structures were erected at the lakeshore 7.5 miles (12 km) south of Priozersk (Fig. 7.2 and Fig. 6.23 in Chapter 6) at Sites 14 (receiving antenna) and 15 (transmitting antenna). The antennas stood 1 km apart with minimal spillover. A supporting military settlement with barracks for servicemen, accommodations for engineers and technicians from industry, an athletic field, and a motor pool grew up further south from the transmitting antenna.[2]

A parabolic cylinder antenna of the radar generated a narrow fan beam oriented vertically. (Fundamental properties of an antenna such as gain, beamwidth, and polarization are identical in transmitting and receiving modes.) The transmitted fan beam steered electronically in the horizontal (azimuthal) direction. The sizes of the transmitting and receiving (Fig. 7.3) antennas of Dunai-2 were 26×490 ft (8×150 m) and 66×490 ft (20×150 m), respectively.[3] The radar could detect typical warheads at distances up to 750 miles (1200 km) with a 0.6-mile (1-km) range accuracy.[4] It determined their

---
[1] Vladimir Panteleimonovich Sosul'nikov, 1921–2008.
[2] Central Intelligence Agency, Imagery Analysis Division, 1966.
[3] Belous et al., 2009, p. 114; Khodarenok, 2010.
[4] Belous et al., 2009, pp. 112, 114.

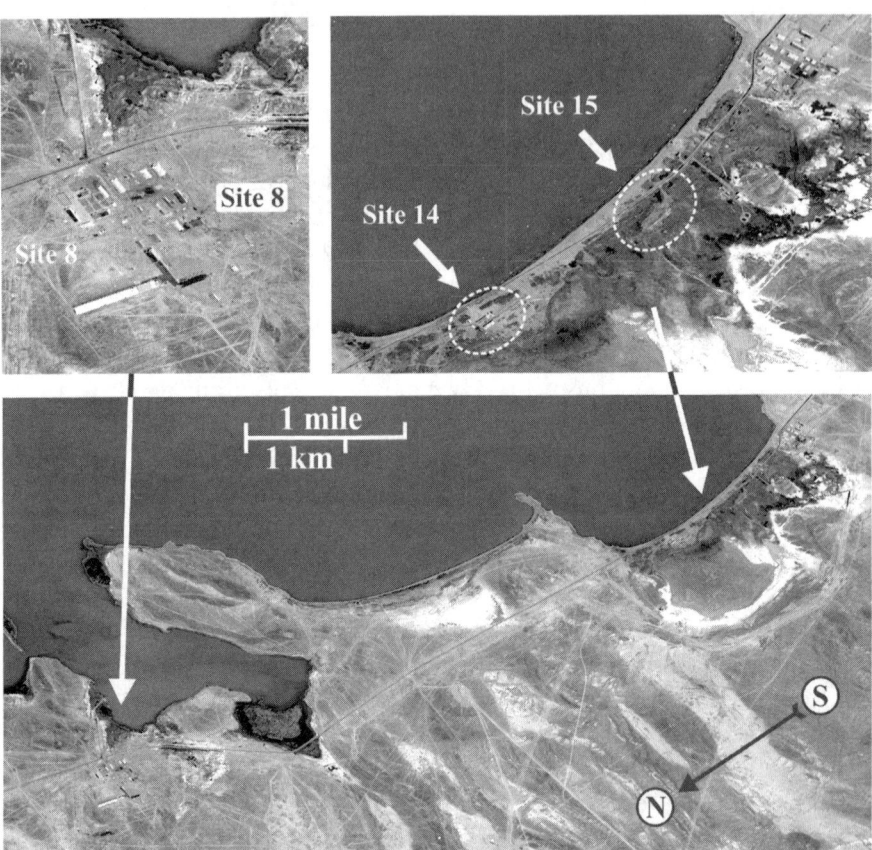

Fig. 7.2 Satellite photograph of the coast of Lake Balkhash (bottom) south from Priozersk with long-range search and acquisition radar systems in 1966. The top-right insert shows the area of the Dunai-2 radar (designed by Vladimir Sosul'nikov in NII-108) of System A with receiving (Site 14) and transmitting (Site 15) antennas separated by 0.6 mile (1 km). A military settlement with barracks for servicemen and accommodations for engineers and technicians from industry grew further south from the transmitting antenna. A road connecting these two antenna installations ran along the shore, and waves occasionally brought water over it during storms. The military raised, hardened, and paved the road in the early 1960s. The top-left insert shows the TsSO-P radar nearby (Fig. 6.14, Chapter 6) developed by Alexander Mints's RTI institute. This radar at Site 8 occupied the biggest building at the Saryshagan proving ground. U.S. intelligence analysts nicknamed the Dunai-2 antennas Hen Roost North (receiving antenna) and Hen Roost South (transmitting antenna), respectively, and assigned the name Hen House to TsSO-P. Original satellite reconnaissance photograph by KH-7 camera (Mission 4027; 21 April 1966) available from the U.S. Geological Survey; photograph identification, interpretation, and processing by Mike Gruntman.

Fig. 7.3 Receiving antenna of the long-range Dunai-2 radar (Hen Roost North) of System A at Site 14 on the shore of Lake Balkhash. Photograph (from Khodarenok, 2010) courtesy of Mikhail Khodarenok.

elevation and azimuth with an accuracy of 0.5 deg or 6.5 miles (10.5 km) in the vertical and horizontal directions at the maximum distance.

Two transmitters with 40-kW power each combined their output to produce the radar beam. A parabolic cylindrical antenna of Dunai-2 had the focal line rather than the focal point characteristic for a common parabolic dish antenna. A slotted waveguide carried the electromagnetic radiation along the antenna focal line, with radiation emitted from the slots. The operational wavelength determined the size of the slots. Such a design was similar in concept to phased array antennas but used only one or two power transmitters, as was true in this case.

From a Branch of NII-108 to NIIDAR

In 1958, the government set up a new branch of Aksel Berg's NII-108 radar institute on the premises of a design bureau of Plant No. 37 in Moscow. It assigned more than 250 specialists from NII-108, including Vladimir P. Sosul'nikov, to join the plant's engineers in the new branch.

They concentrated on development of long-range radar, first System A's Dunai-2 and then Dunai-3 and its variants for the operational missile defense system A-35 and then A-35M.

In 1960, the organization merged with Plant No. 37 and formed the new independent Scientific-Research Institute No. 37, or NII-37. Later, in 1966, the institute became the Scientific-Research Radiotechnical Institute (Nauchno-Issledovatel'skii Radiotekhnicheskii Institut, or NIRTI). Then in 1975, it received its present name, the Scientific-Research Institute of Long-Range Radio Communications (Nauchno-Issledovatel'skii Institut Dal'nei Radiosvyazi, or NIIDAR). NIIDAR designed a number of radar systems, such as Volga and Neman, over-the-horizon radar Duga for ballistic missile early warning, and radar for the space object identification system Krona.

The transmitting antenna of Dunai-2 emitted a vertically oriented 0.6-deg × 16-deg fan beam.[5] The wavelength (frequency) and the 26-ft (8-m) height of the antenna aperture fundamentally determined the 16-deg width of the beam in elevation. The large 490-ft (150-m) length of the antenna structure resulted in the narrow 0.6-deg beamwidth in the azimuthal direction. At a distance of 620 miles (1000 km), such a fan beam projected a rectangle covering a range of altitudes of 170 miles (280 km) with a horizontal width of 6.5 miles (10.5 km).

Although no available published sources provide the exact operational wavelengths (frequencies) of Dunai-2,[6] the specified beam shape constrains the wavelength to the 1.4–1.7 m range (frequencies of 175–215 MHz). (For comparison, the TsSO-P radar operated at a somewhat lower frequency of about 150 MHz.) The reflecting parabolic cylinder mirrors of both the receiving and transmitting antennas did not have to be built as solid metal surfaces, because it would have made them exceedingly heavy. A metal mesh efficiently reflected electromagnetic waves if openings in the mesh were smaller than the wavelength. Contemporary photographs showed such metal mesh structures of the antennas (Figs. 7.3 and 7.4).

The continuous wave Dunai-2 radar used linear frequency modulation to determine the distance to the warhead. The varying frequency also enabled electronic steering of the beam. When the frequency (and correspondingly the wavelength) of the electromagnetic radiation changed, the interference of emissions from the waveguide slots in the antenna focal line resulted in the shift of the transmitted beam direction, or electronic steering of the beam.

---
[5]Khodarenok, 2010.
[6]e.g., Pervov, 2003; Zavaly, 2003; Belous et al., 2009; Khodarenok, 2010.

EXPERIMENTAL SYSTEM A

Fig. 7.4 Structure of the Dunai-2 antennas from a Soviet documentary film (1961) about the first ballistic missile warhead intercepts by System A. Frames from a video uploaded to YouTube by vpknewsRU, http://www.youtube.com/watch?v=zbR_h__a5ZY&feature=youtu.be (accessed 29 May 2012); frame identification, interpretation, and processing by Mike Gruntman.

Therefore, periodic variations of the Dunai-2 frequency, the linear frequency modulation, also electronically scanned a wide sector in the azimuthal direction by the narrow fan beam. (*Linear* modulation meant that the frequency changed with time linearly.) Consequently, Dunai-2 could determine the azimuth of the target with a 0.5-deg accuracy.

The receiving component of Dunai-2 actually consisted of two antennas similar to the transmitting one and stacked one on top of the other. Therefore, its structure at Site 14 (Fig. 7.3) stood about 66 ft (20 m) tall. The beamwidth of each receiving antenna was 0.6 deg × 12 deg.[7] With these two antennas pointing in slightly different directions in elevation, one could obtain the elevation angle of a target by comparing signals received by these antennas.

In August 1958, the Dunai-2 radar for the first time detected an R-5 ballistic missile in flight at a distance exceeding 620 miles (1000 km). Then on 6 November 1958, Dunai-2 accomplished the detection and tracking of an

---

[7]Khodarenok, 2010.

incoming missile and generated target coordinates for System A.[8] Continuing extensive tests of the radar included validation of the determined coordinates by optical telescopes independently tracking targets. Use of the telescopes limited the tests to short periods of time close to local dawn and dusk. Only then could an approaching high-altitude missile be illuminated by the sun with observers on the ground being in darkness, which was favorable for optical measurements. Consequently, engineers and GNIIP-10 officers conducted many test launches in the early hours, with final preparations for the tests performed during the night.

Figure 7.5 shows the Dunai-2 receiving (northern) antenna at Site 14 photographed by a U-2 aircraft on 9 April 1960. Heavier clouds covered the transmitting (southern) antenna of the radar at the nearby Site 15. They prevented examining its design details besides finding the antenna presence and its approximate size. The CIA referred to these two antennas as *radar site 2* on its maps (Fig. 6.11, Chapter 6). A photographic intelligence report, based on the overflight, described the Dunai-2 antennas:

> This site, located at 45-56N 73-38E, contains two large linear antennas, probably radars, each with adjoining support facilities. One antenna is complete and the other, approximately 3,500 feet [1070 m] to the south, is under construction. Although the size of the second antenna cannot be determined because of cloud cover, the photographic evidence indicates that both antennas will be similar in design and size. The support facilities for the antenna under construction are also partially under clouds. The site is located along the road which runs roughly parallel to Lake Balkhash south from the major portion of the Support Base. At least three other electronic sites are located along this same road.
>
> The completed antenna appears to consist of a nearly vertical, slightly curved parabolic reflector screen about 510 feet [155 m] long, in front of which is a tubelike feed as long as the screen and 35 feet [11 m] from the ground. About 280 feet [85 m] in front of the reflector screen and parallel to it is a ground clutter screen approximately 620 feet [190 m] long and approximately [deleted] high. The top of the reflector screen is about 65 feet [20 m] long above the ground. Near the antenna is a large three-story control building 145 by 45 feet [44 by 14 m] and several other support buildings. The two antennas are parallel, and a line constructed perpendicular to them is oriented on an azimuth of [deleted] generally south of Tyura Tam rangehead.[9]

Using additional photographs from the first Corona reconnaissance satellites, another CIA report accurately concluded in February 1962 that "because

---

[8]Belous et al., 2009, p. 113.
[9]Central Intelligence Agency, Photographic Intelligence Center, 1961, pp. 54–56.

EXPERIMENTAL SYSTEM A

Fig. 7.5 Aerial reconnaissance photograph of the 490-ft (150-m) long receiving antenna (Hen Roost North) of the long-range search and acquisition radar Dunai-2 of System A at Site 14 on the shore of Lake Balkhash. Thin scattered clouds partially obscured details of the antenna design. A U-2 aircraft obtained this near oblique photograph on 9 April 1960, which also showed another antenna of similar length about 1 km to the south. Heavier clouds covered the latter, preventing its detailed analysis. Original photograph (U-2 Mission 4155; 9 April 1960) from National Archives and Records Administration; photograph identification, interpretation, and processing by Mike Gruntman.

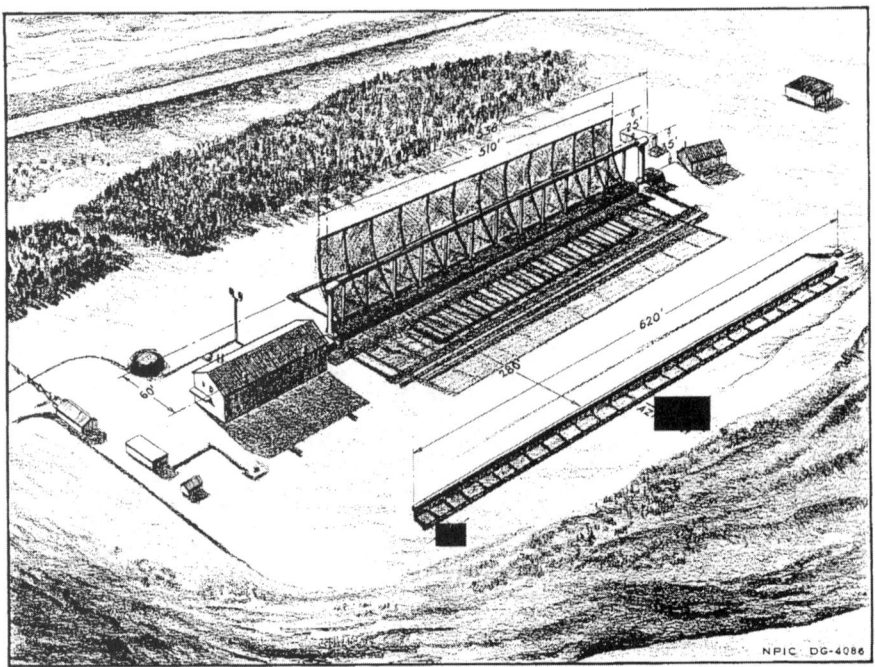

Fig. 7.6 CIA's artist rendering of a receiving antenna (Hen Roost North) of the long range Dunai-2 radar of System A described as "Concept of completed antenna at Radar Site 2. Another similar antenna is under construction 3,500 feet to the south." Such sketches protected the capabilities of photoreconnaissance systems that could have been gleaned from original photographs. Figure 26 from Central Intelligence Agency, Photographic Intelligence Center, 1961, p. 57.

of their partially similar and partially different characteristics, the two antennas appear to be separate units of one antenna system."[10] Referring to Hen Roost (Dunai-2) and Hen House (TsSO-P), CIA Deputy Director Wheelon later recalled that these observed "largest facilities [at Saryshagan] were evidently steered electronically. ... This was an emerging technology at the time, and the development of these new radars near the missile impact area was startling. Even more startling was their size."[11]

Reconnaissance photographs had revealed important design information about the Dunai-2 and TsSO-P installations (Figs. 7.6 and 7.7). However, determining radar characteristics and understanding intended applications required examining their emissions. U.S. intelligence had succeeded in establishing the radar's properties only by the mid-1960s. The electronic listening sites along the Soviet border (Fig. 6.1, Chapter 6) could not directly intercept radar emissions. The nuclear explosions of Operation K above Saryshagan

---

[10]National Photographic Interpretation Center, 1962, p. 13.
[11] Wheelon, 2004, p. 252.

Fig. 7.7 CIA's artist rendering of the largest 890-ft (270-m) long building at the Saryshagan proving ground 3 miles (5 km) north from the Dunai-2 radar. U.S. intelligence analysts assigned the nickname Hen House to this long-range TsSO-P radar, which was being developed by the RTI institute. Figure 25 from Central Intelligence Agency, Photographic Intelligence Center, 1961, p. 56.

created ionized formations at high altitudes, which allowed detection of TsSO-P emissions that were reflected from them.

The analysts missed the emissions of Dunai-2 at first and identified them only a few years later. Wheelon wrote that

> [the nuclear tests] left little doubt that the purpose of the Sary Shagan program was to develop an anti-missile defense system....
>
> ... Thirteen new microwave signals were reflected from the nuclear fireball during the test on 28 October 1962 and were recorded by the CIA site in Iran. It was clear from the geometry that these signals originated from the radars and command links at Sary Shagan, all of which would have been transmitting during such tests. On the other hand, it took considerable effort to identify these signals with the various elements of the ABM system.[12]

The CIA characterized in detail the TsSO-P radar by first examining its emissions bounced off the moon.[13] Two receivers equipped with similar 150-ft (46-m) dish antennas at Stanford, California, and the Chesapeake Bay in Maryland proved particularly useful. Detection in the continental United States of radar signals emitted on the other side of the globe at the Saryshagan range and scattered by the moon could be performed only during particular times with brief favorable positions of the moon and orientations of the Earth. For Stanford, for example, the total duration of such possible observations did not exceed 18 hours per month.

---

[12]Wheelon, 2004, p. 253.
[13]Eliot, 1967; Brown, 1969; Wheelon, 2004.

The U.S. Navy receiver at Chesapeake Bay had first succeeded in detecting Saryshagan radar emissions in early 1964. Then, in early 1965, analysts working for the CIA contractor ESL, Inc. showed that these emissions had originated from the Hen House (TsSO-P) radar and not from the nearby Hen Roost (Dunai-2).[14] When the receiver at Stanford began registering signals in 1965, it succeeded in characterizing them in significant detail. Wheelon described that

> [t]he transmitter peak power was measured to be 25 megawatts, making Hen House [TsSO-P] one of the most powerful radars in the world. It transmitted about 100 pulses per second, indicating that it was designed to operate against targets more than 800 miles [1300 km] away. It used a spread-spectrum signal so the transmitted pulses could be rapidly broadened to increase its range or range-rate resolution. The carrier frequency was changed for each pulse to provide the scanning feature for the large fixed array. The scanning could be stopped at will to provide focused tracking of a target once it was identified. All this represented a significant technical achievement in 1965.
>
> Operational deployment of the Hen House radars began in 1964 and proceeded even as the testing at Sary Shagan continued. Monthly [reconnaissance] satellite missions eventually located fifteen of these enormous radars deployed around the perimeter of the USSR.[15]

At the same time, the analysis of intercepted signals during the nuclear tests of Operation K in 1962 missed the emissions of Dunai-2 and

> [u]nderstanding the pair of Hen Roost [Dunai-2] antennas at Sary Shagan took longer [then those of Hen House], since satellite photography did not detect their deployment [elsewhere in the USSR]. On the other hand, reconnaissance photographs and military attaché reports in 1964 indicated that large phase arrays and missile launchers were being constructed around Moscow to defend the capital [against ballistic missiles] from all approaches. The first signal from the radars at Moscow was intercepted by moon bounce at the NRL [Naval Research Laboratory] facility in 1967. Its carrier frequency was changed continuously to provide beam steering. To everyone's surprise, the transmission was a continuous wave (CW) signal, not a pulsed signal like that used by Hen House and many American radars.[16]

Importantly, the new continuous wave radar near Moscow, nicknamed by analysts the Dog House (Dunai-3; Fig. 9.1, Chapter 9), also consisted of two

---

[14]Brown, 1969.
[15]Wheelon, 2004, pp. 253, 254.
[16]Wheelon, 2004, p. 254.

separate facilities as expected for CW radar in order to avoid spillover of the signal. This arrangement was similar to the still-not-understood Hen Roost (Dunai-2) radar at Saryshagan. A CIA analyst explained that Hen Roost "signals intercepted after the October 1962 nuclear test were still lying around on magnetic recording tape. It was possible that the Hen Roost signal had been there all the time and had been ignored because we [the analysts] were looking for a pulsed, rather than a CW radar."[17]

Wheelon added that "[r]etrospective analysis of the signals collected during the 1962 nuclear tests soon identified an identical CW signal. The connection of the two Hen Roost antennas at Sary Shagan then became apparent.... Similar array pairs were soon identified at Moscow, although their dimensions were different."[18]

So, a major element of System A—the long-range search and acquisition radar, Dunai-2—remained not understood by the United States until the second half of the 1960s. This story demonstrates that technical methods of intelligence collection, even ones as powerful and successful as electronic intercepts and satellite photoreconnaissance, could not fully reveal and characterize in a timely manner some technology and components behind most consequential missile defense developments of the early 1960s.

PRECISE TRACKING AND GUIDANCE RADAR RTN

Each RTN built by Kisun'ko's design bureau consisted of two separate pulse radars (Fig. 7.8). They worked at different frequencies and determined distances to the targets and interceptor missile, respectively. One radar with a larger 49-ft (15-m) diameter antenna (RS-10) tracked the target warhead and measured its range with 16-ft (5-m) accuracy (Fig. 7.9). The other radar, with a smaller 15-ft (4.6-m) diameter antenna (RS-11), tracked the interceptor missile, or izdelie (an article) as Soviet specialists often called it.[19] Then, the control computer calculated in real time their absolute coordinates by the method of three distances. The overall error in the obtained relative distance between the target and the interceptor did not exceed 30 ft (10 m).

Engineers designed both radar antennas in the two-mirror Cassegrain configuration. The shape of the 15-m antenna was manufactured with a precision of 0.1 inch (2 mm).[20] Powerful mechanisms pointed the antennas by rotating them about two axes. The requirement to intercept targets in all possible directions resulted in a fixed horizontal axis of rotation. It would have been impossible to track warheads near zenith (the vertical direction away from the center of the Earth) using the common design with a fixed vertical axis.

---

[17]Brown, 1969, p. 16.
[18]Wheelon, 2004, p. 254.
[19]Khodarenok, 2010; Aitkhozhin and Gantsevich, 2010a.
[20]Khvatov, 2003, p. 585.

Fig. 7.8 Precise guidance RTN radar at the Saryshagan range. Two radomes of the target radar (right) and interceptor missile radar (left) flank the one-story control building. A tall metal mast protects the radars from lightning. The larger dome measures 115 ft (35 m) in diameter. Photograph (from Khodarenok, 2010) courtesy of Mikhail Khodarenok.

Balancing weights reduced torques on pointing mechanisms. This made them heavy, however, with the total mass of moving structures being 92 t and 8 t for the large RS-10 and small RS-11 antennas, respectively.[21] Nevertheless, the highly agile antennas could rotate within ±90 deg about each axis with angular rates up to 13 deg/s and accelerations up to 3 deg/s^2.

The pointing mechanism of the RS-10 antenna consumed 70 kW and 40 kW of electric power for rotating about its two axes. The smaller RS-11 antenna required only 2 kW of power for rotating about each axis. Special electronic control devices measured and converted antenna angular positions into digital codes and compared them with those sent by the central control computer. Relying on new semiconductor technology, engineers built electronic circuitry without vacuum tubes, using 700 transistors and 650 diodes instead.

Industry produced two types of enclosures, rigid and inflatable radomes, for the 15-m RTN antennas.[22] These Kupol-10 (kupol means a cupola) domes

---

[21] Khodarenok, 2010; Aitkhozhin and Gantsevich, 2010a.
[22] Khodarenok, 2010; Aitkhozhin and Gantsevich, 2010a.

Fig. 7.9 The 49-ft (15-m) diameter antenna of the RTN radar during assembly at the Saryshagan range. Photograph (from Khodarenok, 2010) courtesy of Mikhail Khodarenok.

protected against the elements of the environment with minimal interference to radiowaves (Figs. 7.8, 7.10, and 7.11). The rigid-structure polyhedron radome fit into a sphere with a 115-ft (35-m) diameter. The CIA maps of the Saryshagan range showed them as 110-ft diameter domes (e.g., Fig. 6.13, Chapter 6). They provided protection against strong winds up to 312 ft/s (95 m/s). The shape of the other type, an inflatable air-supported radome, constituted part of a sphere 118 ft (36 m) in diameter and protected against winds up to 100 ft/s (30 m/s). The latter domes required extra pressure of 0.002–0.008 atm (0.03–0.12 psi) above the ambient atmosphere to maintain their shape.

During its operations in active tracking mode, each RTN radar consumed 650 kW of power. Similar to earlier experimental RE radars, the 15-m diameter RS-10 antenna had a 4-horn emitter to achieve the required radiation power. Elevated air pressure of 4–4.5 atm (59–66 psi) in sealed microwave

Fig. 7.10 Top: Highly agile 49-ft (15-m) diameter RS-10 antenna of the RTN radar under the radome. Bottom: A serviceman descends stairs from a platform supporting the antenna. The radome was the height of a 13-story building. From a Soviet documentary film (1961) about the first ballistic missile warhead intercepts by System A; uploaded to YouTube by vpknewsRU, http://www.youtube.com/watch?v=zbR_h__a5ZY&feature=youtu.be (accessed 29 May 2012); frame identification, interpretation, and processing by Mike Gruntman.

waveguides prevented arcing at the operational high-power levels. For the first time designers made many waveguides out of aluminum rather than significantly more expensive copper. Savings of brass exceeded 40 t.

A pencil beam of the 15-m diameter RTN target radar had a 0.7-deg beamwidth.[23] It sent linearly polarized pulses 3 or 0.5 μs long with 30-MW power and repetition rates of 200 or 400 Hz, respectively. The latter shorter and higher-rate pulses tracked target warheads during the precise final phase of the intercept. A single horn emitter fed the 4.6-m diameter RS-11 antenna at a smaller radar that transmitted circularly polarized pulses that were 0.5 μs long with 1-MW power and a repetition rate of 400 Hz to a transponder on the interceptor missile. The RTI institute designed and built powerful transmitters for RTN radar as it had done earlier for the B-200 radar of the S-25.

Aerodynamic drag in the upper atmosphere caused larger deceleration of ballistic missile bodies compared to more compact and denser warheads.

---

[23]Khodarenok, 2010; Aitkhozhin and Gantsevich, 2010b.

Experimental System A

Fig. 7.11 Satellite photograph of large (left) and small (right) radomes of the precise tracking and guidance RTN radar at Site 2 at the Saryshagan range in 1966. A lightning protection mast is next to the larger dome. Original satellite reconnaissance photograph by KH-7 camera (Mission 4032; 20 September 1966) available from the U.S. Geological Survey; photograph identification, interpretation, and processing by Mike Gruntman.

The difference in drag opened the way for identification, selection, and separation of warheads in received radar signals. The computer could not reliably perform such identification in first intercepts. Therefore, RTN operators manually locked on the warhead, which was then tracked. The system automatically guided the missile toward the target during the final 12–14 s of the intercept without operator involvement.

System A accurately measured distances between targets (and interceptors) and RTN radars separated by large distances. In order to obtain positions of warheads and interceptors, the absolute coordinates of all radar systems tracking them had to be precisely determined and accounted for in calculations. In addition, engineers calibrated differences in delays in various electronic circuits transmitting and receiving signals. Each RTN sent measured ranges to the central control computer through microwave relay communication lines as 22-bit binary numbers, with the least significant bit corresponding to an uncertainty of a fraction of 1 m in distances.

Similarly to conventional radars, the narrow pencil-beam RTNs could determine, in principle, not only the target range, but also its direction. Early in 1961, Kisun'ko initiated evaluation of RTN performance in such a mode.[24] After modifications of algorithms and extensive tests, the engineers established accuracies of target tracking by a single radar. This work paved the way for significant simplification of the design of the operational missile defense system, later implemented in the A-35.

The high-altitude nuclear explosions of Operation K had demonstrated the robustness of RTN radar systems in a realistic adverse environment. Kisun'ko later noted that the RTN part of System A, as well as its communications channels and command and control elements, "worked nominally [during nuclear explosions] up to the moment of the target intercept by the V-1000 missile [similarly to] as they had done without nuclear explosions."[25] This robustness resulted, in part, from the radar working at a relatively short wavelength of 15 cm (and a corresponding high frequency of 2 GHz).

In contrast, disruptions of long-range radars, Dunai-2 and TsSO-P, by ionized formations created by nuclear explosions in the atmosphere turned out to be long-lasting and severe. These radar systems operated at significantly longer wavelengths than RTN in the meter range (at lower frequencies, 150–220 MHz). Alexander Kulakov recorded that nuclear explosions blinded Dunai-2 and TsSO-P for dozens of minutes compared to RTNs being affected for only a few seconds.[26] Observed radar responses "confirmed the necessity of switching such [long-range] radars to [shorter] decimeter wavelengths"[27] in operational systems.

Nikolai Ostapenko, who had been responsible for tests of System A in Operation K, wrote that

> [a]fter the detonation of a special [nuclear] warhead the long-range radar Dunai-2 [operating] in the meter wavelength range was blinded by ionized objects [in the atmosphere] for 20 minutes and the TsSO-P radar [was blinded] for much longer time. This fact forced developers of the operational missile defense system to change [its long-range acquisition radar] to [shorter] decimeter wavelength range.[28]

The TsSO-P functioned at a frequency somewhat lower than Dunai-2, which caused the larger effect on the former noted by Ostapenko.

Operation K had thus verified what had been predicted by physics—the dependence of the magnitude of the effects of nuclear explosions on radar frequency. Kisun'ko also pointed out a political fallout from the results of nuclear tests:

---

[24] Aitkhozhin and Gantsevich, 2010b.
[25] Kisun'ko, 1996, p. 430.
[26] Kulakov, 2006, p. 76.
[27] Kamensky, 2003, p. 605.
[28] Belous et al., 2009, p. 187.

A different [from shorter wavelength RTNs] picture [of the effects of nuclear explosions] was observed at the long-range acquisition radar Dunai-2 and especially at the TsSO-P. After a nuclear explosion they were blinded by noise interference from ionized formations created by the nuclear explosion. This confirmed the rationale of our decision to adopt the [shorter] decimeter wavelength range for the [new long-range] search radar Dunai-3 for the [operational] missile defense system [A-35] of Moscow. It was clear from the tests [of Operation K] that switching to decimeter wavelengths was also necessary for the radars of ... [the Ballistic Missile Attack Early Warning System] being built near Murmansk and Riga [and] based on the TsSO-P prototype. Otherwise, they could [also] be blinded by a high-altitude nuclear explosion in space and would not detect an ICBM attack. However, personal ambitions and mutual cover-up by functionaries and operators in the military-industrial complex did not allow them to admit the error [in earlier selection of radar frequencies] and suggest corrective measures to the government.[29]

INTERCEPTOR MISSILE INITIAL GUIDANCE RADAR RSVPR

The small RTN radar with a 4.6-m diameter antenna tracked the interceptor missile in the final phase of the intercept. It could not, however, "see" the missile during its initial trajectory from the launch site. Therefore, Kisun'ko included an additional Antimissile Missile Initial Guidance Radar, RSVPR, to guide the missile after launch toward the predicted intercept point and into RTN radar beams. In addition, RSVPR included a command unit—a Station for Command Transmission (Stantsiya Peredachi Komand, or SPK)—to control the missile. The interceptor carried a transponder to assist in its tracking and a command receiver.

Kisun'ko had assigned development of the RSVPR radar to NII-20 of the Ministry of Defense Industry in the Moscow suburb of Kuntsevo; Samuil P. Rabinovich headed this effort. The radar had three coaligned antennas with a total mass of 1.5–2 t on a common platform: two antennas, each 8.2 ft (2.5 m) in diameter, and one smaller antenna that was 3 ft (0.9 m) in diameter (Fig. 7.12).[30] The latter antenna first captured the missile a few seconds after the launch and then handed it over to a (lower) larger 2.5-m antenna for accurate tracking. The other (upper) large antenna aided in determination of angular coordinates. This antenna also sent controlling SPK commands to the missile. (Kisun'ko built the SPK command unit internally in his design bureau.) The radar transmitted 1-μs-long pulses with 1-MW power and a repetition rate of 880 Hz. The command line sent out a modulated continuous signal with a power of 180 kW; its beamwidth was 14 deg.

---

[29]Kisun'ko, 1996, pp. 430, 431.
[30]Davydov, 2009, p. 115.

Fig. 7.12 The initial guidance RSVPR radar with two antennas 8.2 ft (2.5 m) in diameter and one antenna 3.0 ft (0.9 m) in diameter, mounted on a common pointing platform. The interceptor missile launch site operated two such RSVPR radars for redundancy with a control blockhouse nearby. Photograph (from Khodarenok, 2010) courtesy of Mikhail Khodarenok.

Collocated with the launch pads, the RSVPR radar could lock onto the interceptor missile within ±7 deg from the nominal trajectory and track and guide it at distances up to 37 miles (60 km). The radar measured the range to the missile with accuracy better than 160 ft (50 m) and determined its elevation and azimuth with accuracies better than one-sixth of a degree.[31]

For redundancy, System A included two identical RSVPR radars in the vicinity of missile launch pads at Site 6 (Fig. 7.12). A 26-ft (8-m) layer of concrete protected the control room in an underground blockhouse nearby. A television camera mounted on one antenna allowed operators to visually follow the interceptor missile flight. As with all other System A radar, the RSVPR sent its data in real time to the central control computer.

After the first successful intercepts, Kisun'ko decided to bring the best engineers who had built the RSVPR radar into his design bureau due to the expanding work on the operational A-35 missile defense system. To identify

[31]Khodarenok, 2010.

those who had contributed most to the development, he used a trick. The semiofficial history of NII-20 described Kisun'ko's stratagem:

> After completion of the development work [on the RSVPR radar] ... G.V. Kisun'ko asked NII-20 to prepare a suggestion for awarding state decorations to [those] specialists who had distinguished themselves [in this work]. The prepared list was sent [to Kisun'ko]. Then, there was an order by the minister to transfer a group of specialists, headed by S.P. Rabinovich, from NII-20 to SKB-30 to continue work on System A-35. The group for transfer consisted of those proposed for the awards. As it became clear later, Kisun'ko did not want to bring NII-20 [as an institution] to missile defense work, but he wanted their experienced specialists. If he had requested [unnamed] specialists [through a usual appeal to the ministry], then the institute [NII-20] would have sent him not the very best. Therefore, there was this sly approach, a suggestion of decorations.[32]

The transferred lead RSVPR designer, Samuil Rabinovich, worked with Kisun'ko for a few years and then left for another research institute.

INTERCEPTOR MISSILE V-1000

The design bureau of Petr Grushin, OKB Fakel, built the V-1000 interceptor missile for System A. It drew on its experience with the antiaircraft V-750 missile of the S-75 (SA-2) air defense system. The military constructed two identical launch pads, side by side, at Site 6 of the Saryshagan range and set up a technical position for storage, assembly, and maintenance of the missiles at Site 7 in the vicinity of the main airfield. Two different groups led by Konstantin I. Kozorezov and Alexander V. Voronov[33] developed two types of warheads for the interceptors.

The V-1000 missiles launched from inclined ramps at a fixed 78-deg elevation angle in a desired azimuth direction determined by the central control computer. OKB Fakel designed the missile to intercept a target within 55 s after its launch at a distance of 34 miles (55 km) from the launch pad and at a 27-deg angle with respect to the horizon, which corresponded to a 82,000-ft (25-km) altitude at the maximum distance. System A thus required a 3280-ft/s (1000-m/s) average velocity of the missile, with a maximum velocity reaching 4920 ft/s (1500 m/s) and high maneuverability at altitudes above 66,000 ft (20 km).

The new V-1000 represented a major improvement beyond the V-750's capabilities. The new interceptor came out three time heavier than its predecessor, carried a 3.8 times more massive warhead, and achieved a range that was twice as long. The missile had to be ready for launch only 30 s after

---
[32]Davydov, 2009, p. 115.
[33]Konstantin Isaakovich Kozorezov, b. 1920; Alexander Vasil'evich Voronov.

receiving the command. This requirement precluded the use of vacuum tubes, which took some time to warm up for proper operation. Therefore, engineers built electronics based on the just-appearing new semiconductor components, transistors and first integrated circuits.

The new two-stage interceptor had a solid-propellant booster and a main stage with a liquid-propulsion engine (Fig. 7.13). The main stage was ignited after separation of the burned out booster. The 48-ft (14.5-m) long V-1000 had an initial mass of 19,370 lb (8785 kg) and carried a 1100-lb (500-kg) warhead.

Dominik D. Sevruk began development of a liquid-propellant engine for the V-1000 in NII-88 in 1956, but could not overcome technical difficulties. Then, in 1958, Alexei M. Isaev, also in NII-88, took over the task and produced reliable engines. The liquid-propellant engine of the main stage burned for 36.5–42 s with 10.5 t (23.15 klbf, 103 kN) thrust. The missile acceleration reached 2.5–3 g.[34]

Because of mounting pressures of the new programs, Isaev and his design bureau, KB Khimmash, gradually transferred production and design support for engines for the V-750 and V-1000 missiles to a design bureau of Plant No. 466, later known as Krasnyi Oktaybr', in Leningrad. Another Leningrad factory, Bolshevik, built the launch stand SM-71P for the V-1000. The stand weighed 39 t and inherited its design from a naval artillery gun system. The mounted V-1000 missile stayed in readiness horizontally, then turned, under command from the central control computer, to a desired azimuth, raised to the 78-deg-elevation angle, and fired.[35] The mechanisms turning the missile stand in azimuth and elevation consumed 50 kW and 20 kW of electric power, respectively. Two identically equipped launch pads at Site 6 could automatically fire the second missile in case of launch failure of the first interceptor.

Initial problems with building solid-propellant and liquid-propellant engines as well as the launch stand caused delays in development of the V-1000. So, OKB Fakel improvised a simplified launch structure and clustered solid-propellant motors from another missile to imitate the missile booster. In October 1957, the engineers test-fired the first put-together missile with an inert mass mockup of the main stage. First test launches of completely assembled missiles took place in spring 1958.[36]

By February 1960, engineers had conducted a total of 25 stand-alone missile launches. Then, 23 more launches followed during the period from May to November of 1960. The missiles now interacted with other elements of System A under commands of the central control computer. More than 35 additional launches would be performed after that, leading to the first successful intercepts in 1961.

---

[34] Svetlov, 2008; Belous et al., 2009, pp. 122–124.
[35] Svetlov, 2008; Khodarenok, 2010.
[36] Svetlov, 2008.

EXPERIMENTAL SYSTEM A 183

Fig. 7.13 V-1000 interceptor missile and the control blockhouse at the launch pad. (a) V-1000 interceptor missile on a ramp allowing launch from an inclined position in a desired direction, with the second missile seen at some distance. (b) Solid-propellant booster (first stage) of the V-1000 interceptor missile. (c) A serviceman enters the underground control blockhouse near the launch pad. (d) Armor-plated door in the underground blockhouse. Frames from a Soviet documentary film (1961) about the first ballistic missile warhead intercepts by System A; uploaded to YouTube by vpknewsRU, http://www.youtube.com/watch?v=zbR_h__a5ZY&feature=youtu.be (accessed 29 May 2012); frame identification, interpretation, and processing by Mike Gruntman.

Fig. 7.14 Concept of a missile launch pad at Launch Complex B (Site 6) sketched by U.S. intelligence analysts and based on the U-2 overflight in April 1960. Figure 36 from Central Intelligence Agency, Photographic Intelligence Center, 1961, p. 78.

Saryshagan veteran Zhadeiko recalled that

> [o]ne time an interceptor missile (V-1000) flew [in a wrong direction and landed near a major town] Karaganda [to the north from Saryshagan]. The system of rocket destruction [for mission abort] malfunctioned and [the missile hit and] destroyed railroad tracks, leading to the town. [Representative of Kisun'ko at Saryshagan] Nikolai K. Ostapenko had to go to the Chairman of the Council of Ministers of Kazakhstan D.A. Kunaev[37] with explanations.[38]

The U-2 photographs obtained in April 1960 led U.S. analysts to accurate conclusions about the nature of the Saryshagan installations at Site 6, or Launch Complex B (Fig. 7.14) in the CIA nomenclature. A contemporary report stated that

> [t]he unique design of facilities in the Launch Area differs so radically from that of conventional surface-to-surface and surface-to-air missile site configurations that it suggests the development of a totally different system. In view of this design and the presence of the many long-range electronic facilities located throughout the Sary Shagan Antimissile Complex, Launch Complex B can best be postulated as a developmental

---

[37]Dinmukhamed Akhmetuly (Akhmedovich) Kunaev, 1912–1993.
[38]Zhadeiko, p. 10.

countermissile launch complex. This hypothesis is substantiated not only by the novel configuration of the semicircular probable launch pads ..., but also by their overall size, the apparent capability of trainable launchers, and the geographic location of the launch complex with respect to Kapustin Yar and Tyura Tam.[39]

DATA TRANSMISSION SYSTEM

Real-time computer control of the complex System A depended on highly reliable, high data rate communication channels. The military did not consider it realistic to connect remote sites of the Saryshagan range by underground cables in a short period of time. Therefore, microwave radio relays offered an effective alternative. Consequently, in March 1956, the Central Committee of the CPSU and the USSR Council of Ministers ordered establishment of a new scientific-research institute, NII-129, to develop such a system. Frol P. Lipsman and a group of almost 130 engineers and technicians transferred from NII-244 to form the core of the new organization. An experienced specialist, Lipsman had been working in the field of radio relays since after World War II.

The communications network of System A included 17 retransmission and terminal receiving posts (Figs. 7.1, 7.15, and 7.16), each with a tower usually 160–260 ft (50–80 m) high, a technical building, diesel power generators, and a barrack for soldiers. Five main radio relay lines connected the central control computer at Site 40 with Dunai-2 at Sites 14 and 15; launch pads with interceptor missiles and initial guidance RSVPR radar at Site 6; and three precise guidance radars RTN-1 (Site 1), RTN-2 (Site 2), and RTN-3 (Site 3). These links stretched for 6.2 miles (10 km), 91 miles (146 km), 84 miles (135 km), 184 miles (296 km), and 93 miles (150 km), respectively. The total length of all radio relay lines reached 764 miles (1230 km).[40]

The system transmitted data in a binary code organized as 14-bit words, with pulses typically 0.3–0.6 µs long. Each line had the rate capacity of 100 kpulse/s and sent information through 16 independent channels with a 10^{-7} bit error rate. In addition to data exchange and telephone connections, the communications supported the synchronized signals of the unified time reference system.

CENTRAL COMPUTING STATION

The reliance on a digital computer for intercept control in real time represented a bold enabling feature in the design of System A in 1956. At first engineers installed one of the first Soviet digital computers, Strela (strela

---

[39]Central Intelligence Agency, 1961, p. 74
[40]Khodarenok, 2010.

Fig. 7.15 Microwave relay communication tower. A network of 17 towers connected the central command and control center of System A with the long-range Dunai-2 radar, precise tracking RTN radars, and the launch site of interceptor missiles by high-capacity digital communications links. The lines stretched for 764 miles (1230 km). Photograph (from Khodarenok, 2010) courtesy of Mikhail Khodarenok.

means an arrow), in KB-1 for processing and reduction of experimental data of warhead tracking by experimental radars RE-1 and RE-2 at Saryshagan.[41]

A digital computer played a much bigger role in System A. It determined in real time an absolute position of the warhead from measurements of three distances by RTN radars; integrated equations of warhead motion in the gravitational field of the Earth with aerodynamic drag in the atmosphere and predicted the warhead's trajectory; transformed coordinates among different reference frames; worked out the intercept solution; processed and fused data from various sources of information; and developed commands for launch and guiding of the interceptor missile and finally for detonating its high-explosive charge. The computer controlled the final phase of the intercept, which lasted 12–14 s without operator involvement.

The Institute of Precision Mechanics and Computer Engineering (ITMVT) under Chief Designer Sergei A. Lebedev had completed crash development of

---

[41] Aitkhozhin and Gantsevich, 2010a.

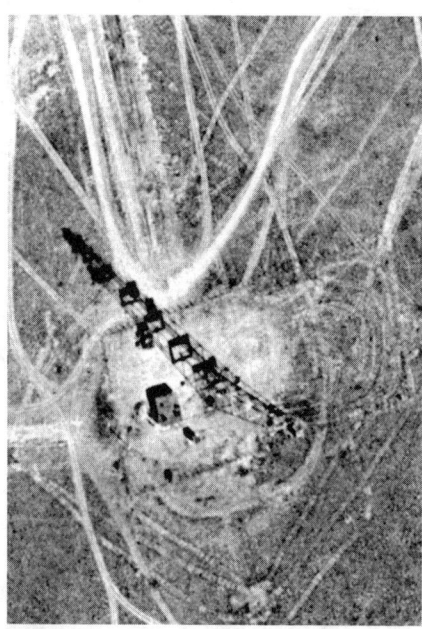

Fig. 7.16 Satellite photograph of a terminal communication tower of System A at Site 2 in 1966. Original satellite reconnaissance photograph by KH-7 camera (Mission 4032; 20 September 1966) available from the U.S. Geological Survey; photograph identification, interpretation, and processing by Mike Gruntman.

computers M-40 and M-50 for System A by the end of 1958. At Saryshagan, the Central Computing Station (Tsentral'naya Vychislitel'naya Stantsiya) housed the computers at Site 40 (Fig. 7.17) near the GNIIP-10 headquarters.

The M-40 computer (Fig. 7.18, top) controlled in real time the intercept. It performed necessary calculations and maintained continuous two-way communications with system elements. Five independent relay lines converged on the central computer, providing data exchange with the total rate close to 1 Mbit/s. The computer relied primarily on vacuum tube technology and occupied a 5400-ft^2 (500-m^2) hall on the second floor of a building at Site 40. It executed 40,000 fixed point operations per second. The random access memory of the computer had 4096 words and its external memory constituted 150 kwords.

The Moscow-based ITMVT also built another computer variant, M-50 (Fig. 7.18, bottom), optimized for operations with floating point numbers. The M-50 had a much larger external memory on magnetic tapes and drums. The engineers and GNIIP-10 officers used it primarily for posttest processing of recorded data.

Development of algorithms and highly complex control codes constituted a prodigious task. The Kisun'ko team first installed electronic circuitry simulating an RTN radar in SKB-30 and connected it by communication lines across

Fig. 7.17 Photographs of Site 40 with the computer center and GNIIP-10 headquarters at the Saryshagan range in 1960 (top) and 1966 (bottom). A cloud partially obscured buildings in the 1960 photograph from a U-2 aircraft; the 1966 satellite photograph clearly showed the built up area. Original photographs from (top) National Archives and Records Administration (U-2 Mission 4155; 9 April 1960) and (bottom) by KH-7 camera (Mission 4027; 21 April 1966) available from the U.S. Geological Survey; photograph identification, interpretation, and processing by Mike Gruntman.

Moscow to an M-40 computer prototype in ITMVT. This allowed initial testing of the information exchange and interfaces between critical elements of System A and the control computer. The programmers could also begin development of operational codes.

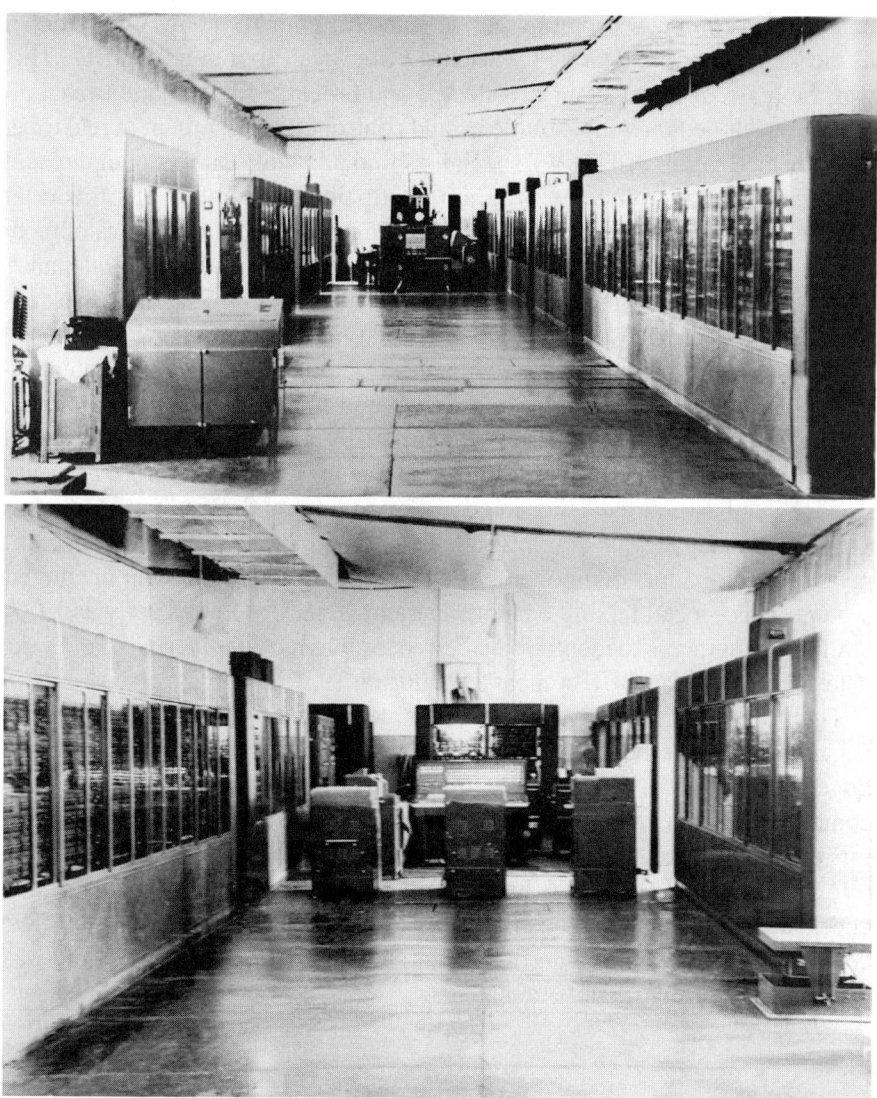

Fig. 7.18 M-40 (top) and M-50 (bottom) computers at the Saryshagan range. The General Combat Code OBP of System A ran on the M-40 computer, controlling in real time information collection and processing, developing intercept solutions, and commanding the interceptor launch and its guidance to the target. The computer occupied a 5400-ft^2 (500-m^2) hall and used operations with fixed point numbers. The M-50 operated with floating point numbers and primarily supported processing missile defense test results and other computational tasks. The obligatory portraits of socialist "saints," Karl Marx (top) and Vladimir I. Lenin (bottom), decorated the far walls of the halls. Photographs (from Khodarenok, 2010) courtesy of Mikhail Khodarenok.

The main computer program that operated the entire System A was known as the General Combat Code (Obshchaya Boevaya Programa, or OBP). This was the brain of the system; it collected and processed information from all radars, developed intercept solutions, and controlled the intercept in real time.

A selected group of ITMVT specialists initiated work on this complex and most challenging computer program. With time, a number of officers at GNIIP-10 learned programming and joined in testing and further development of the codes. In 1962, Kisun'ko transferred the responsibility for maintenance and modifications of the operational OBP combat code of System A to a specialized unit of GNIIP-10 at Site 40 of the Saryshagan range.

Missile defense tests required launches of ballistic missiles and interceptors, operation of multiple radar systems, and measurements by numerous remote instrumentation sites, all supported by extensive communications networks, which made tests exceptionally expensive. Therefore, engineers designed and assembled a dedicated experimental stand, first in Moscow and then at Saryshagan. The stand simulated System A and assisted in development of its individual components and then their functioning as an integrated system. This special facility became operational in Moscow as early as 1957. It allowed engineers to identify and correct problems with interfaces and validate elements of the OBP combat code during 1957 and 1958.

Then the work shifted to Saryshagan, where components of System A began to arrive. There again, the special stand tested interceptor guidance and target tracking and assisted in development and validation of the entire OBP combat code. Numerous trials with computer-simulated targets and interceptors proved particularly useful.

All elements of System A had been assembled at the Saryshagan range, checked, and connected to each other by the end of the summer of 1960. The control computer was running, with the OBP code operational. The time was approaching for missile intercepts.

Chapter 8

INTERCEPTS

AUTONOMOUS AND SYSTEM TESTS

After stand-alone autonomous trials of System A components, the tests progressed to incremental integration with the M-40 central computer that now controlled radar systems and interceptors. Independent launches of V-1000 missiles in complete configurations began on 31 August 1958. Then testing advanced to evaluation of missile response to real-time commands generated by the General Combat Code, OBP. As a first step, missiles "learned" to follow preprogrammed trajectories.

In November 1958, the long-range Dunai-2 radar detected and tracked a ballistic missile and sent its coordinates to the M-40 computer. The control center then automatically performed predictions of the target trajectory and pointed precise guidance RTN radars.

Initial tests of the entire system included two preliminary and complementary steps, with simulated warheads and simulated interceptors, respectively. At first, in May 1960, System A began launching and guiding interceptor missiles against computer-simulated trajectories of approaching warheads instead of real ballistic missiles. Then, the long-range Dunai-2 radar detected real ballistic missiles and the RTNs tracked them. The control computer predicted target trajectories and finally commanded simulated launches of interceptors and their guidance toward real targets.

The Central System Indicator (Display), or Tsentral'nyi Indikator Sistemy (TsIS), represented the focal point for control of System A and real-time display of its status and operations (Fig. 8.1). TsIS occupied an isolated room on the third floor of the building of the main computer center at Site 40 (Fig. 8.2). The rules restricted access to the room "to the duty shift of test range officers and a strictly limited circle of closest assistants of the chief designer

Fig. 8.1　Central control post TsIS of System A. Photograph (from Khodarenok, 2010) courtesy of Mikhail Khodarenok.

Fig. 8.2　Expanding Site 40 with the GNIIP-10 headquarters and the computer center in 1961. The M-40 computer occupied rooms on the second floor of the center. The main control post TsIS of System A on the third floor operated the entire system and controlled warhead intercepts. Frame from a Soviet documentary film (1961) about the first ballistic missile warhead intercepts by System A; uploaded to YouTube by vpknewsRU, http://www.youtube.com/watch?v=zbR_h__a5ZY&feature=youtu.be (accessed 29 May 2012); frame identification, interpretation, and processing by Mike Gruntman.

[Kisun'ko]."[1] Either Kisun'ko himself or his assigned representative was usually present in the control room during System A tests.

The main display of TsIS showed the situation in an area 280 miles (450 km) in radius, including the current positions of the target and the interceptor missile as well as the locations of the interceptor launch site and RTN radars.[2] An additional display plotted in real time the changing altitudes of the target and the interceptor during the final 130-s interval of the engagement. The system could display altitudes up to 140 miles (225 km). The panel also had numerous control switches and special indicators that lit up when Dunai-2 locked on the target as well as individual RTN radars.

In addition to controlling the entire System A, performing computations in real time, and displaying the developing situation during an intercept, TsIS also maintained the system in a standby mode in anticipation of the forthcoming engagement. It could also remotely test individual system elements. The designers referred to the intercept and standby modes as a "combat operation" and a "precombat operation," respectively. All the incoming and outgoing computer information for the M-40 and all its commands were recorded for postoperation processing and analysis. Local control centers supported major components of System A such as Dunai-2, RTNs, and the missile launch site.

The Soviet military apparently did not preserve technical records of first intercepts. Therefore, one has to rely on the recollections of the participants to reconstruct the events. Saryshagan officer Kulakov wrote much later that

> [u]nfortunately, it is impossible now to present a complete description of [System A] combat [test] launches. The archives with the results of [our combat] work are either inaccessible or destroyed. In my time [at Saryshagan] as head of the department of tests of the [General Combat Code] OBP, I maintained the log journal [for recording details] of combat operations. The entries were made in this journal with the results of every combat operation as well as the actions taken to [address and] eliminate uncovered deficiencies. Prior to my departure from the [Saryshagan] test range, I had handed this log journal to the secret department [responsible for storage of classified documents] where it was apparently also destroyed.[3]

In fall 1960 the time had finally come to try the complete, assembled System A against a real target. The nominal missile intercept included about 30–35 min of preparations followed by less than 7 min for detection of the approaching warhead, its tracking, and destruction. In reality, tests typically stretched to many hours because of delays caused by malfunctioning components of the experimental system. Consequently, the preparation process

---
[1]Kisun'ko, 1996, p. 409.
[2]Khodarenok, 2010.
[3]Kulakov, 2006, pp. 67, 68.

frequently halted and the countdown resumed after addressing the identified problems.

Not only technical problems caused delays. The military also tried to minimize intercepts of telemetry and other electromagnetic emissions by the U.S. electronic intelligence. The units of System A usually turned on their radars on the command "30-minute readiness" prior to a test. Emission patterns suggested a forthcoming intercept attempt. Then U.S. aircraft began patrolling along the Soviet southern border, ready to capture electromagnetic signals. Kulakov recalled that

> the headquarters of the Central Asian [Turkestan] air defense district advised [the] Saryshagan [range] about the appearance of such patrolling American airplanes. [The central control post] TsIS then sent an immediate order "2-hour delay" [to stand down] by which all radars suspended their work for several hours until the intelligence gathering aircraft returned to its base. After that the "30-minute readiness" command was announced again and the process repeated. Sometimes, [such suspensions] repeated multiple times, stretching a test for several days....
>
> Many times delays were announced because of such seemingly mundane reasons as passage by a foreign diplomat or some other suspicious foreigner in a train on the railroad crossing the territory of the range [along the lakeshore].[4]

Finally, the experimental System A was ready for intercepts of real ballistic missile warheads. The first intercept attempts engaged relatively short-range R-5 (SS-3) ballistic missiles launched from a field site near the town of Chelkar (Fig. 8.3 and Fig. 6.18, Chapter 6). Later, the military fired larger R-12 (SS-4) missiles as targets. The strategic rocket forces could launch them from bases at the Kapustin Yar test range. In addition, they fired R-12s from a temporary launch site in the vicinity of the railroad station Makat about 75 miles (120 km) from Gur'yev, or Guryev (today's Atyrau), in Kazakhstan (Fig. 8.3). The latter launches provided steeper trajectories of target warheads.

The first attempt at a ballistic missile intercept took place on 5 November 1960. It failed because the R-5 rocket malfunctioned soon after the launch. During the second attempt, the long-range Dunai-2 radar failed.

INTERCEPTOR WARHEAD

Finally, the test on 24 November 1960 partially succeeded. Dunai-2 successfully detected the approaching inert warhead of an R-5 ballistic missile; RTN radars, on cue, found and tracked it; and the V-1000 missile (Figs. 8.4

---

[4]Kulakov, 2006, p. 66.

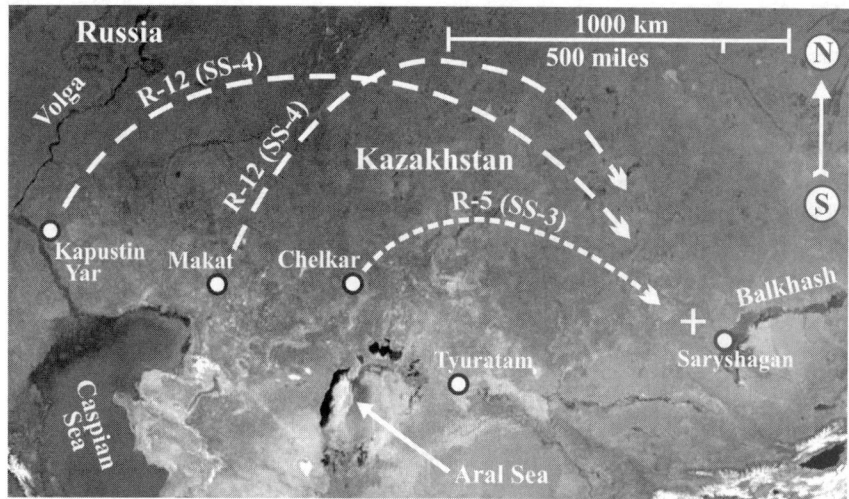

Fig. 8.3 Trajectories of IRBM R-12 (SS-4) fired from the Kapustin Yar missile range and a temporary launch position near the Makat railroad station in Kazakhstan toward the impact point (white cross) at the Saryshagan proving ground in intercept tests of System A in 1960–1962. The strategic rocket forces also launched the shorter range R-5 (SS-3) ballistic missiles from another field position closer to Saryshagan, 31 miles (50 km) north of the small town of Chelkar. Original composite satellite photograph (Landsat, early 2000s) courtesy of NASA WorldWind; image processing, callouts, and markings by Mike Gruntman.

and 8.5) reached the target and detonated its blast fragmentation warhead at the required moment.[5] The next day a search party found the remnants of the target warhead in the desert impact area. The remnants and additional electronic, radar, and optical recordings of the event led to a conclusion that the interceptor had not damaged the target.

Administrative rivalry in the defense establishment resulted in two competing designs of fragmentation warheads for the V-1000 interceptor.[6] First, the Military-Industrial Commission (VPK) assigned Konstantin I. Kozorezov to build the warhead with a 250-ft (75-m) radius destruction field for missile defense applications. At that time, Kozorezov, who was an experienced scientist, worked on warheads for Lavochkin's antiaircraft missiles. His scientific-research institute NII-6 in Moscow specialized in explosives. This was the same institute of the Ministry of Armaments that the government had evicted from its original premises in 1950 to make room for NII-20. The latter, in turn, vacated its buildings and territory for the expanding KB-1 of Kuksenko and Sergo Beria. In 1969, NII-6 changed its name to the Central

---

[5]Golubev et al., 1994, p. 31, 32; Khvatov, 2003, p. 588; Kamensky, 2003, pp. 603, 604; Pervov, 2003, pp. 80, 81; Belous et al., 2009, pp. 171, 172.
[6]Pervov, 2003, pp. 51–56; Belous et al., 2009, pp. 131, 132.

Fig. 8.4 Interceptor missile V-1000 on launch stand SM-71P. Photograph (from Khodarenok, 2010) courtesy of Mikhail Khodarenok.

Fig. 8.5 Two side-by-side launch positions of V-1000 interceptor missiles at Site 6 of the Saryshagan range. The blockhouse for missile checkout and launch is between the launch pads. Farther beyond the pads are initial guidance RSVPR radars, their control blockhouse, and a microwave relay communications tower. Photograph (from Khodarenok, 2010) courtesy of Mikhail Khodarenok.

Scientific-Research Institute of Chemistry and Mechanics. The institute remains active to this day.

As Kozorezov's work progressed, the overseeing ministry, with the support of the institute's director, assigned his rival in the same NII-6, Alexander V. Voronov, to build the fragmentation warhead for System A. However, the VPK wished Kozorezov to continue his development, and therefore transferred him into another special design bureau, SKB-47, of a different ministry (Ministry of Defense Industry). The latter design bureau also focused on production of explosives and ammunition. Today, this organization is known as the Scientific-Industrial Conglomerate Bazalt (the word *bazalt* means basalt). It is one of the major producers and exporters of weapons and ammunition in Russia.

Initially Kozorezov concentrated on a warhead for an antiballistic missile of the Saturn system that had been under development for some time. Vladimir Chelomei actively promoted Saturn. As Kozorezov described it,[7] Chelomei took advantage of his new engineer employee, the recently graduated son of Nikita Khrushchev, to expand his work into new areas. After cancellation of Saturn, Kisun'ko and Grushin took over Kozorezov and his program and assigned him to build an alternative interceptor warhead for their V-1000 missile. Thus, eventually the two teams of Voronov and Kozorezov independently worked on the fragmentation warheads for System A interceptors.

Voronov's warhead included a large number of steel rods 4 in. (10 cm) long and 0.06–0.12 in. (1.5–3 mm) in diameter. Technicians welded a few rods to one end of a base rod, forming a structure in a complex form resembling a broom. The interceptor carried numerous "brooms," as Kisun'ko liked to call the contraptions, and dispersed them to hit the target. Voronov believed that a high-velocity impact by a rod would produce a cumulative jet penetrating through a protective ablation layer and high-strength wall of a target warhead and destroy it. The experimental System A attempted first intercepts by missiles armed with the Voronov warheads.

In the partially successful engagement on 24 November 1960, the V-1000 interceptor carried a Voronov warhead. The records showed a miss distance of only 69 ft (21 m), and the interceptor detonated at the right moment.[8] The failure to damage the warhead mockup pointed to a problem with the Voronov design. Consequently, Kisun'ko made the decision to switch to Kozorezov's interceptor warheads.

Kozorezov's warhead had a mass of 1100 lb (500 kg) and carried 15,000–16,000 spheres with a mass of 0.055 lb (25 g) each and were 0.95 in. (24 mm) in diameter. Each sphere consisted of a 0.4 in. (10-mm) diameter carbide-tungsten-cobalt core surrounded by a high explosive. The interceptor warhead

---

[7]Pervov, 2003, p. 53.
[8]Pervov, 2003, p. 81.

detonated about 0.3–0.45 s before passing the target. At first, special explosive cords fired and removed the container shroud, facilitating dispersion of the spheres. The released spheres created a uniform disk field 165–250 ft (50–75 m) in radius with a high probability of hitting the target. Outward velocities of the spheres at the edge of the disk field reached 560 ft/s (170 m/s).

Initially, Voronov also considered employment of high-explosive spheres. He ultimately chose, however, the alternative "brooms" and cumulative-jet penetration in order to avoid causing danger to the defended territory due to the numerous dispersed parts with explosives.[9]

The target warheads of the R-5 and R-12 ballistic missiles had conical shapes with a length of 11.5–13 ft (3.5–4 m) and a diameter of about 5 ft (1.5 m). Up to 6 in. (15 cm) of ablative protective layer covered a typical warhead with a 0.40-in. (10-mm) thick hardened metal envelope and a nuclear device inside. The R-12 warhead presented a more difficult target because it was designed to enter the atmosphere with a high velocity of 2.5 miles/s (4 km/s); the shorter range R-5 had a smaller 1.9 miles/s (3 km/s) velocity.

A thin steel outer shell contained an explosive charge with a hard core in the center of each sphere dispersed by Kozorezov's interceptor. On high-velocity impact, the outer shell broke through into the protective layer of the target warhead and then the sphere exploded there. It damaged the target regardless of the incident angle and produced destruction much greater than that caused by a hit by an inert metal object of the same mass.

The hard core of the sphere accelerated by the explosion and penetrated further inside the target through the opening created by the initial explosion. It then damaged the nuclear device of the warhead, particularly affecting its trigger mechanisms.[10] In an atomic warhead, all conventional charges surrounding plutonium must be detonated synchronously to achieve the implosion necessary for a nuclear explosion to take place. It was especially hoped that device penetration by the hard core of the sphere would trigger these conventional charges without proper synchronization, thus making the indispensable precisely timed implosion and a subsequent nuclear explosion impossible.

THE SUCCESS ON 4 MARCH 1961

After the unsuccessful intercept on 24 November 1960, six more attempts followed in December against R-5 ballistic missiles. They all failed. First, a short circuit in a vacuum tube in the main computer doomed the engagement on 8 December. Then, the time sequencer in the V-1000 missile malfunctioned during its flight on 10 December. On 17 December, the power units of a receiver in one RTN radar developed problems. An operator made an error at

---

[9]Kamensky, 2003, p. 601.
[10]Golubev et al., 1994, p. 23; Belous et al., 2009, pp. 131, 132; Pervov, 2003, p. 56.

the Dunai-2 radar during an attempt on 22 December. The next day, the main stage liquid-propellant engine of the interceptor missile did not start.

Kisun'ko ordered nonstop simulated tests of the entire system for several days. Then, the final trial of the year took place on 31 December. This time, one RTN failed to transition to the precise tracking mode, which led to a large miss distance.[11] The lethality of Kozorezov's new warheads, now on top of the interceptors, could not be tested yet. The string of December failures increased pressure on Kisun'ko and energized his powerful enemies, rivals, and competitors among government functionaries and chief designers.

The new year began with a new series of unsuccessful intercept attempts. First, on 13 January 1961, a transponder on the V-1000 missile failed 38 s after its launch toward the target. Then the next three engagements included additional complexities of experimenting with new approaches to selecting and separating the warhead from the accompanying ballistic missile body in radar signals. The operator first tried such a selection manually on 14 January; after that, System A attempted automatic signal separation on 18 and 22 February.

These three tests did not succeed in interception, but the entire system otherwise worked. They also pointed to a way to modify radar pulses to achieve automatic warhead selection. Kisun'ko explained that

> fine-tuning of the equipment for [warhead] target selection [and separation from the rocket body] and its delivery to the [Saryshagan] range required some time. We decided to use this time [window] to test the system against an R-12 missile with a maneuver of the rocket body [away from the warhead after its separation].[12]

In this new test on 2 March, the rocket forces fired a larger ballistic missile R-12 (SS-4), built by Mikhail Yangel's Yuzhnoe Design Bureau (Fig. 8.6), toward the impact area at Saryshagan. The range of the new R-12 intermediate-range ballistic missiles (IRBMs; Fig. 8.7) allowed their launch from established positions at Kapustin Yar; however, engineers wanted to engage the warheads in steeper than nominal trajectories in first intercepts. Such trajectories made it easier for System A to detect and track target warheads, but they reduced the effective range of the R-12.[13] Consequently, the military launched the R-12 from a temporary field position near Makat (Fig. 8.8). The U.S. intelligence analysts detected "increased activity and enlargement of the [Makat] site since December 1960."[14]

On 2 March, System A worked flawlessly during the attempted intercept of a warhead with a postseparation maneuver of the rocket body. A human error,

---

[11]Kisun'ko, 1996, pp. 412, 413.
[12]Kisun'ko, 1996, p. 415.
[13]Pervov, 2003, p. 85.
[14]National Photographic Interpretation Center, 1961, p. 1; National Photographic Interpretation Center, Jan. 1962, III-M-1.

Fig. 8.6   Mikhail K. Yangel directed the special design bureau OKB-586, later known as Yuzhnoe Design Bureau, in Dnepropetrovsk, Ukraine, from its inception in 1954 until his death in 1971. He pioneered a family of ballistic missiles with storable, noncryogenic propellants and made the bureau a leading Soviet spacecraft development and production establishment, with more than 400 satellites placed into orbit throughout its history. It was Yuzhnoe's IRBM R-12 (SS-4) that System A successfully intercepted for the first time at Saryshagan. Under Yangel, the bureau also actively advanced the first Soviet penetration aids and other countermeasures against missile defenses. It produced numerous space launchers, including for the operational antisatellite weapon system. By 1990, the enterprise employed 50,000 people. Photograph courtesy of the Russian Academy of Sciences.

however, again resulted in failure. An operator of the RTN-2 radar locked on and tracked the missile body instead of the warhead. As a result, the system guided the V-1000 interceptor to a point between the warhead and the missile body, leading to the failure.

Success came on 4 March 1961. After the usual technical delays, System A was ready. Then, security officers halted the preparations because of a foreigner on a train passing through the area.[15] Finally, a unit of the strategic rocket forces launched another IRBM R-12 towards Saryshagan. The rocket body moved away from the target warhead after its separation. Six minutes before the intercept took place, the long-range Dunai-2 radar detected the approaching target at an altitude of 280 miles (450 km) and a distance of 606 miles (975 km) from the projected impact point. The warhead flew with a velocity of 2.5 miles/s (4 km/s).

---

[15]Kisun'ko, 1996, p. 416.

Fig. 8.7 Single-stage liquid-propellant intermediate-range ballistic missile R-12 (SS-4) developed by Mikhail K. Yangel's Yuzhnoe Design Bureau in Dnepropetrovsk, Ukraine. In the late 1950s and 1960s, the Soviet Union deployed more than 500 of these IRBMs carrying nuclear warheads. Photograph courtesy of Yuzhnoe Design Bureau (Office), Dnepropetrovsk, Ukraine.

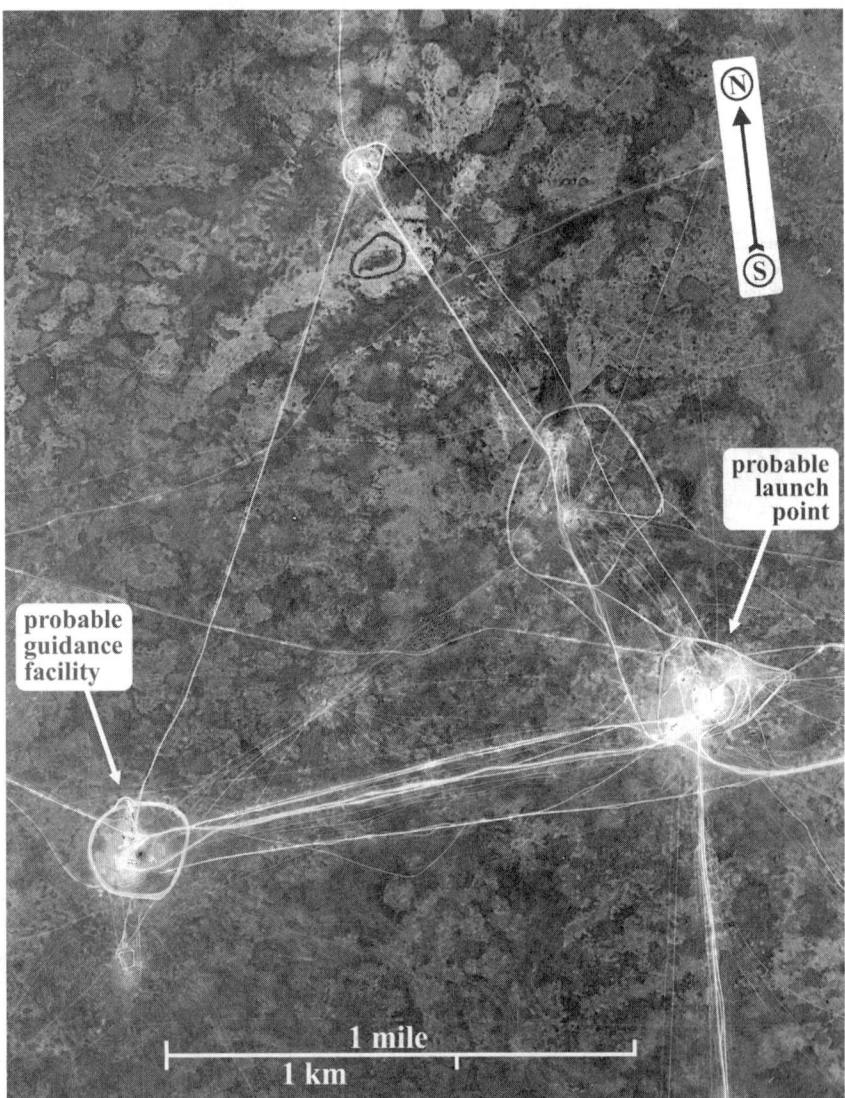

Fig. 8.8 Field launch position 38 miles (60 km) northwest from Makat in 1966. The U.S. intelligence analysts identified this launch area and guidance facility in 1961 and described that "[t]he probable... launch site consists of a probable launch point and two possible bunkers or revetments which are located 330 and 390 feet from the point..." (National Photographic Interpretation Center, 1961, p. 1; National Photographic Interpretation Center, Jan. 1962, III-M-1). Original satellite reconnaissance photograph by KH-7 camera (Mission 4030; 15 July 1966) available from the U.S. Geological Survey; photograph identification, interpretation, and processing by Mike Gruntman.

Dunai-2 locked on the target at a distance of 491 miles (790 km). Then, the RTNs took over warhead tracking. Soon the control computer had to initiate execution of a code calculating the intercept solution. Suddenly, a fateful event almost led to a failure.

An ITMVT engineer, Andrei M. Stepanov, was at the control console of the computer system on that day and ran the operational OBP combat code. Stepanov described later that,

> the M-40 computer... [and computational devices] were built using vacuum tubes and had low reliability. Especially unreliable was the circuit controlling the random access memory. It used powerful vacuum tubes that frequently failed. Not only did they stop working, but usually exploded. There were unused sockets in one electronic rack where spare vacuum tubes were mounted during important [intercept] launches and kept in a warmed-up ready-to-work state. [Fortunately], the operational [OBP] code periodically wrote down data [with the status of the entire system] to a magnetic drum [memory] necessary to restart the code in case of malfunction or [other interruptions]....
>
> [On 4 March] the target was detected by the long-range search [Dunai-2] radar and the RTN radars, based on that information, locked on [and tracked] the target.... The moment was approaching for starting... [execution of a special code] calculating the trajectory of the intercepting missile.... Suddenly, an explosion sounded—one of the vacuum tubes blew out. Engineers rushed to replace the tube, and then I began a procedure to restart the code which involved a sequence of actions at the control console of the computer. [Precise guidance] RTN radars had lost the target.... After the restart of the code, the RTNs again locked on the target,... [and the code calculating the interceptor trajectory] became operational—that meant that we had succeeded to do it in time.[16]

So, at a distance of 298 miles (480 km) from the impact point and 145 s before the intercept, the central control computer restarted execution of the OBP code. It soon began calculations of the projected target trajectory. Precise guidance radars, first RTN-3, then RTN-2, and finally RTN-1 locked again on the target 125 s, 115 s, and 95 s, respectively, before the intercept and resumed its tracking.[17]

Then the computer sent a command for launch of the V-1000 missile 43.7 s before the intercept (Fig. 8.9). The target warhead was 98 miles (158 km) from the projected impact point at this moment in time.

The final fully automatic phase of the engagement lasted 14 s, with RTNs tracking with high precision both the target and the interceptor. At the intercept point, atmospheric drag had reduced the warhead velocity to 1.6 miles/s

---
[16]Kulakov, 2006, p. 69, 70; Belous et al., 2009, pp. 173, 174.
[17]Pervov, 2003, pp. 88–91; Belous et al., 2009, pp. 173–179.

Fig. 8.9 Launch of the V-1000 interceptor missile against the approaching target warhead of the IRBM R-12 (SS-4) on 4 March 1961. The initial guidance RSVPR radar (left) follows the missile and guides it into the beams of precise guidance RTN radars. Frame from a Soviet documentary film (1961) about the first ballistic missile warhead intercepts by System A; uploaded to YouTube by vpknewsRU, http://www.youtube.com/watch?v=zbR_h__a5ZY&feature=youtu.be (accessed 29 May 2012); identification, interpretation, and processing by Mike Gruntman.

(2.5 km/s) as the interceptor missile approached it with a velocity of about 0.6 miles/s (1 km/s). The interception took place 37 miles (60 km) away from the V-1000 launch pad at Site 6. The Kozorezov warhead detonated, on command, at an altitude of 15 miles or 82,000 ft (25 km), approximately 0.4 s before the point of the closest approach to the target. Later analysis established the miss distance as being 104 ft (31.8 m) to the left and 7.2 ft (2.2 m) above the target. For the interceptor missile, it was the 80th launch of the V-1000 at the Saryshagan range and the 38th since September 1960.[18]

A special radio system installed in the inert ballistic missile target warhead stopped sending signals 6 s after the intercept. When the optical and radar records had confirmed the success, chief designer of the experimental System A, Grigorii Kisun'ko, and GNIIP-10 commander Stepan Dorokhov sent a secret cable to Soviet leader Nikita Khrushchev. After reporting the details of the intercept, the cable concluded, "[T]he destruction of a ballistic missile warhead in its flight trajectory by means of missile defense has been demonstrated for the first time in the national and world practice. Tests of System A continue as planned."[19]

It took days for search parties to find the main parts of the shot-down warhead in the desert. They recovered a 1100-lb (500-kg) steel piece simulating the warhead, a ring frame, and a nose section of the cone body (Fig. 8.10).

---

[18]Svetlov, 2008.
[19]Kisun'ko, 1996, pp. 4, 5.

Fig. 8.10 Found in the desert, three major pieces of the intercepted and destroyed IRBM R-12 (SS-4) target warhead by System A on 4 March 1961. Top: A 1100-lb (500-kg) steel mockup of the ballistic missile warhead; middle: a ring frame; bottom: a nose section of the cone body. Frames from a Soviet documentary film (1961) about the first ballistic missile warhead intercepts by System A; uploaded to YouTube by vpknewsRU, http://www.youtube.com/watch?v=zbR_h__a5ZY&feature=youtu.be (accessed 29 May 2012); frame identification, interpretation, and processing by Mike Gruntman.

The parts showed that the interceptor had indeed destroyed the incoming target warhead.[20]

Tense relations among rival groups in the government and industry showed up in the System A success and even increased because of it. In a clear sign of a power struggle, often turning personal, Kisun'ko did not notify his immediate boss, chairman of the State Committee on Radioelectronics Minister Kalmykov, about the successful intercept. However, he did report it to the highest authorities in the Communist Party and to Khrushchev. "Let him [Minister Kalmykov] enjoy a fantasy for some time that the attempt [of the intercept] on March 4 by System A also failed [as he wished it]," wrote Kisun'ko later.[21] Prior to reporting to Khrushchev, however, he did telephone his supporter, Chairman of the Military Industrial Commission Ustinov.

Saryshagan officer Kulakov also noted that not everybody appreciated the first successful intercept. As a result, it took until 1966, 5 years later, for the government to decorate a group of designers, industry managers, military commanders, and test engineers and range officers who had distinguished themselves in this important accomplishment.[22]

Tests Continue

After celebrations of the first intercept, the tests of System A continued. Already on 26 March 1961, Kisun'ko's team repeated the success. This time the V-1000 interceptor dashed towards the R-5 ballistic missile with a standard operational 1100-lb (500-kg) warhead. The intercept caused an explosion of the target warhead charge.

Then, on 9 June, System A intercepted another R-12 warhead. Optical instruments (Fig. 8.11) of the Saryshagan range recorded fragmentation of the target after crossing the disk field of explosive spheres (Figs. 8.12, 8.13, and 8.14). Two large parts detached from the main target body, slowed down, and burned out (Fig. 8.14). Drag forces acted on moving bodies proportional to their ballistic coefficients. The resulting acceleration was thus larger for less massive and less compact bodies. Consequently, smaller detached burning parts slowed down and stayed behind the larger, more massive main body of the damaged target warhead.

Tests of System A continued until 1964, with a total of 16 interceptor missiles fired after 4 March 1961. Leading participants summarized later that, after the first successful intercept, "[d]irect destruction of warheads of ballistic missiles was achieved in two launches of antimissile missiles against warheads of R-5 ballistic missiles and in three launches against warheads of R-12

---

[20]Golubev et al., 1994, pp. 32, 33; Kisun'ko, 1996, p. 5.
[21]Kisun'ko, 1996, p. 420.
[22]Kulakov, 2006, p. 72.

Fig. 8.11 Optical instrument KST-60 records the ballistic missile warhead intercept (Figs. 8.12, 8.13, and 8.14) on 9 June 1961. Frame from a Soviet documentary film (1961) about the first ballistic missile warhead intercepts by System A; uploaded to YouTube by vpknewsRU, http://www.youtube.com/watch?v=zbR_h__a5ZY&feature=youtu.be (accessed 29 May 2012); frame identification, interpretation, and processing by Mike Gruntman.

[missiles]."[23] The V-1000s intercepted a total of 11 warheads. Overall, System A achieved direct destruction of targets in six engagements, which showed, as some specialists noted, a rather low efficiency of the fragmentation interceptor warhead.[24]

Some tests also included evaluation of alternative infrared guidance sensors as well as radio and optical triggers of the interceptors.[25] Many characteristics of System A could only be determined from multiple engagements under varying conditions. Studies of the ability of radar systems to distinguish between warheads and accompanying ballistic missile bodies were of particular importance, especially because the observation geometry of such target pairs substantially differed for three widely separated RTNs.[26]

BATTLE AGAINST PENETRATION AIDS BEGINS

In 1958, Aksel' Berg's NII-108 initiated preliminary studies of penetration aids for ballistic missiles to break through missile defenses. Petr S. Pleshakov[27] replaced Berg as the institute director in that year and continued the development. The Central Committee of the CPSU and the Council of Ministers

---

[23]Golubev et al., 1992, p. 173.
[24]Belous et al., 2009, p. 182.
[25]Kisun'ko, 1996, p. 420.
[26]Kulakov, 2006, p. 57.
[27]Petr Stepanovich Pleshakov, 1922–1987.

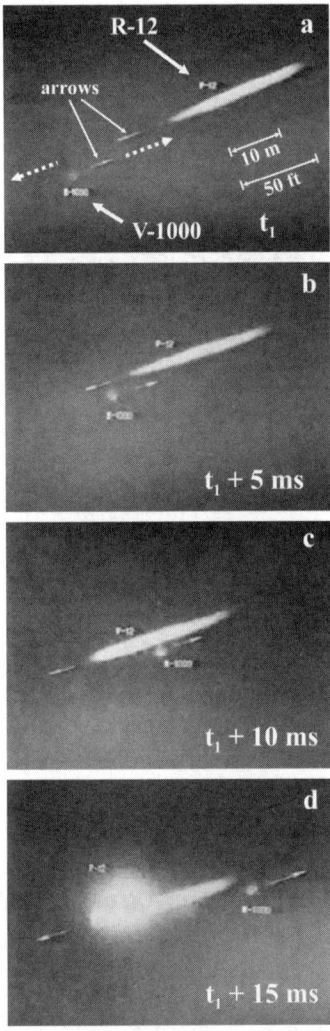

Fig. 8.12 Sequence of frames of optical recordings (by KST-60, Fig. 8.11) of the R-12 intercept on 9 June 1961. Frame sequence continues in Figs. 8.13 and 8.14. (a) Fragments of the V-1000 interceptor on the left (moving to the right) and the target warhead (R-12) on the right (moving to the left) at the moment of time $t_1 = 0.43$ s after the detonation of the interceptor warhead. Two blurred arrows (in each frame) show the directions of motion of the interceptor and target; high velocities of the interceptor (~1 km/s) and target (~2.5 km/s) resulted in their elongated shapes, accumulated during the frame exposure, in the photographs. Spatial scales (10 m and 50 ft) show distances along the intercept trajectory, which was inclined with respect to the image plane. (b) The intercept at the moment of time $t_1 + 5$ ms, that is, 5 ms after frame a. (c) The intercept at $t_1 + 10$ ms. (d) The intercept at $t_1 + 15$ ms, showing disintegration of the target warhead. Frames from a Soviet documentary film (1961) about the first ballistic missile warhead intercepts by System A; uploaded to YouTube by vpknewsRU, http://www.youtube.com/watch?v=zbR_h__a5ZY&feature=youtu.be (accessed 29 May 2012); frame identification, interpretation, processing, and callouts by Mike Gruntman.

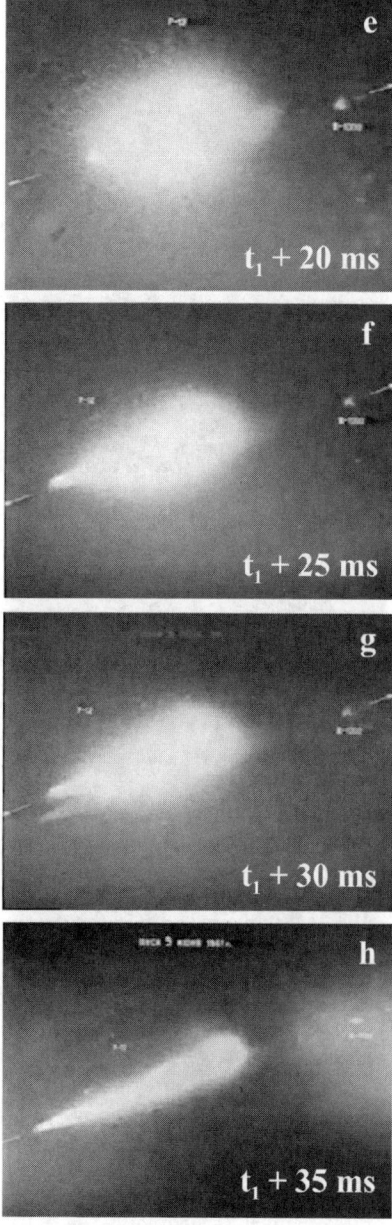

Fig. 8.13 Sequence of frames of optical recordings of the R-12 intercept on 9 June 1961; continuation of the frame sequence in Fig. 8.12, continued in Fig. 8.14. (e) The intercept at $t_1 + 20$ ms; disintegration of the target continues. (f) The intercept at $t_1 + 25$ ms. (g) The intercept at $t_1 + 30$ ms. (h) The intercept at $t_1 + 35$ ms. Frames from a Soviet documentary film (1961) about the first ballistic missile warhead intercepts by System A; uploaded to YouTube by vpknewsRU, http://www.youtube.com/watch?v=zbR_h__a5ZY&feature=youtu.be (accessed 29 May 2012); frame identification, interpretation, processing, and callouts by Mike Gruntman.

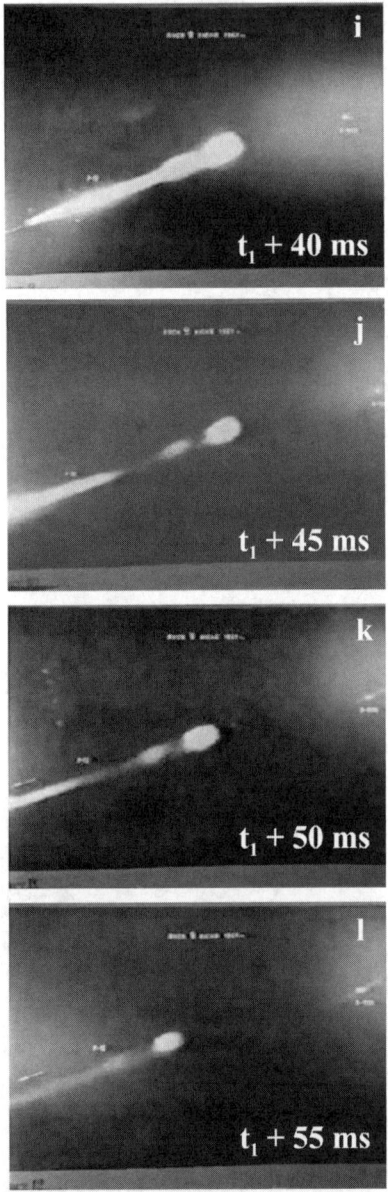

Fig. 8.14 Sequence of frames of optical recordings of the R-12 intercept on 9 June 1961; continuation of the frame sequences in Figs. 8.12 and 8.13. (i) The intercept at $t_1 + 40$ ms; one can see two objects slowing down and separating from and staying behind the main target. (j) The intercept at $t_1 + 45$ ms; the two burning parts separated and stayed behind the main body of the target. (k) The intercept at $t_1 + 50$ ms. (l) The intercept at $t_1 + 55$ ms. Frames from a Soviet documentary film (1961) about the first ballistic missile warhead intercepts by System A; uploaded to YouTube by vpknewsRU, http://www.youtube.com/watch?v=zbR_h__a5ZY&feature=youtu.be (accessed 29 May 2012); frame identification, interpretation, processing, and callouts by Mike Gruntman.

authorized specific focused experimental programs in penetration aids after the successful intercept at Saryshagan.[28]

These programs, Verba, Kaktus, and Krot, tested the performance of System A in the presence of emerging countermeasures. [Verba, kaktus, and krot translate as a willow tree, cactus, and mole (animal), respectively.] Kisun'ko considered these first tests of penetration aids, forced on him by superiors, as continuing attempts by his adversaries and rivals to diminish the accomplishments of System A.

Pleshakov, who oversaw the Verba, Kaktus, and Krot programs, flew to Saryshagan to introduce them to Kisun'ko and organize their trials. Kisun'ko could have stopped this first attempt of testing countermeasures by bureaucratic maneuvers; however, he let them be deployed against System A in order, as he put it, "to break the habit of both, Pleshakov and those [in the government] who had sent him, to poke into our [Kisun'ko's] work with careless interference."[29]

The Verba program deployed numerous inflatable and dipole reflectors. The RTN radar systems found a way to distinguish and filter out signals from inflatable decoys because they were relatively steady, in contrast to the rapid variation in time of signals from the target warheads. The Kaktus technology introduced warhead coatings that absorbed incident radar pulses. High temperatures quickly damaged such coatings as warheads entered the upper atmosphere, making them ineffective. The Krot program employed special radio equipment that emitted noise signals in response to the arriving pulses of RTN radars. To counteract this particular technique, System A added small pulses, preceding main radar pulses, that triggered the Krot circuitry at wrong times. Later, a specially generated sequence of pulses at higher rates completely overwhelmed Krot devices.[30]

System A had thus succeeded in overcoming these first crude countermeasures. However, continuous improvements in penetration aids would present permanent challenges from now on. The battle between missile defense and countermeasures had thus started simultaneously with the first intercepts. In a historical twist, it would be the same Pleshakov who had overseen the first penetration aids in 1961 who, as Minister of Radio Industry, removed Kisun'ko from work on missile defense in 1975.

Whatever the internal politics and power struggle, the first intercepts by Kisun'ko led to a vigorous development program to find the technical means for breaking through possible U.S. missile defenses by the Soviet intercontinental ballistic missile (ICBM) forces. A very important builder of ICBMs, the Yuzhnoe Design Bureau in Dnepropetrovsk, initiated a major program to develop technologies against missile defenses in the early 1960s. Already in

---

[28]Pervov, 2003, p. 102.
[29]Kisun'ko, 1996, p. 423.
[30]Kisun'ko, 1996, pp. 421–423; Khvatov, 2003, p. 590.

1963, the Ministry of Defense had expanded formal technical requirements for the new ICBM R-36 (SS-9) to include additional capabilities to break through the U.S. Nike-Zeus antiballistic missile system.

Naum I. Ur'ev[31] led this effort at Yuzhnoe from the early 1960s into the 1990s. The first Soviet countermeasures, the List system (list means a leaf or a sheet, as in a sheet of paper), included all three components tried against System A: radar-absorbing coatings (combined with thermal protection of the warhead), active electronic units interfering with radar signals, and decoys.[32] In 1965, the first flight tests of penetration aids were conducted at Saryshagan by deploying them on ballistic missiles fired from Kapustin Yar.

It was clear that such trials would be observed by U.S. technical means located outside the Soviet Union. Consequently, the flight tests of decoys could not be concealed. Therefore, engineers conducted these launches with decoys that differed from those to be deployed on operational ICBMs.[33]

The barriers of secrecy between developers of penetration aids such as those at Yuzhnoe and missile defense designers led to confusion. Two different ministries, the Ministry of General Machine Building and the Ministry of Radio Industry, respectively, oversaw these two administratively and technologically separate parts of the military-industrial establishment. Missile defense specialists assumed certain countermeasures deployed against them by attacking ballistic missile forces. At the same time, the military and government bodies considered other, more effective penetration aids resulting from ICBM work. Neither formal requirements to missile defense systems nor decrees from the Central Committee of the CPSU and the USSR Council of Ministers that had authorized the development of missile defenses said "a single word about the characteristics of the targets [warheads] and means [of countermeasures deployed against missile defense]."[34]

The importance to missile defense designers of the "consistent description of [ballistic missile] targets and [possible] penetration aids with their projected advancement for the next 10–15 years" was abundantly clear. The participants in the events noted that military institutes and organizations involved in building ballistic missiles and countermeasures worked together to provide such a description, and

> as a result, the so called "White Book" [with the desired information] was produced in 1972 (Kisun'ko coined the name for the book from the color of its cover).... It became a go-to handbook for developers of missile defenses as well as for designers of the means of early warning of ballistic missile attack. A number of additions to the book appeared [with time].

---

[31]Naum Isaakovich Ur'ev, b. 1926.
[32]Konyukhov, 2009, pp. 159–162.
[33]Konyukhov, 2009, p. 162.
[34]Golubev et al., 1994, p. 64.

In 1985, it was reissued based on improved understanding of expected [in the future] characteristics of ballistic missiles.[35]

The intercept on 4 March 1961 at the Saryshagan range presented a major milestone in development of missile defenses. After the first U.S. intercepts of smaller missiles in 1960 (see Appendix A), the Soviet Union had demonstrated nonnuclear destruction of long-range ballistic missile warheads. The kill technology for direct head-on collisions was not mature yet for operational applications. Nevertheless, successful Soviet and U.S. tests ended the argument about whether it was possible to "hit a bullet with a bullet"—it could be done.

Next generations of missile defense systems would rely on nuclear-tipped interceptors. Thirty years later, however, nonnuclear interception has again become the focus of efforts to protect rather than avenge.

---

[35]Golubev et al., 1994, pp. 64, 65.

Chapter 9

BEYOND EXPERIMENTS

TOWARD OPERATIONAL MISSILE DEFENSE

The success of the experimental System A in the first nonnuclear intercepts of long-range ballistic missile warheads put the Soviet Union on the road toward operational missile defense. Many powerful groups now tried to get a piece of the growing and lavishly funded pie. Grigorii Kisun'ko described a new environment created by the "unexpected by many results of testing of System A" in 1961,

> If there had been an earlier notion that missile defense was the same stupidity as shooting [an artillery] projectile against another projectile, now the skeptics wished "to stake" this prestigious field [and get a piece of the action] for themselves. To achieve this, they had to eliminate the pioneers, who had blazed the trail and who thus came under concerted attack by these new enthusiasts of missile defense. New amateurish projects began to crop up, as mushrooms after a rain, promoted by the [Chairman of the State Committee on Radioelectronics] Minister [Kalmykov, responsible for missile defense] himself and tempting to the military customer.[1]

Saryshagan veteran Kulakov, who had known Kisun'ko for many years, observed that the "essence of the conflict he [Kisun'ko] saw not [in technical disagreements originating] in search of the optimal way for developing missile defense but in power struggle."[2]

Nikolai V. Mikhailov,[3] a leading missile defense manager in the 1980s who had become a top Russian national security official a decade later and then

---

[1] Kisun'ko, 1996, pp. 420, 421.
[2] Kulakov, 2006, p. 12.
[3] Nikolai Vasil'evich Mikhailov, b. 1937.

served as First Deputy of the Minister of Defense from 1997 to 2001, also noted that

> sometimes problems that appeared [in development of missile defense in the USSR] were the result of personal circumstances and ambitions rather than objective conditions. This caused many disappointing losses, undoubtedly diminishing the overall effectiveness of our work.
>
> Unfortunately, none of our major institutions and perhaps none among general [chief] designers escaped this problem.[4]

After the intercepts of 1961, Kisun'ko finally succeeded in separating administratively from KB-1. He faced resistance from the powerful Raspletin and Kalmykov, but enjoyed the support of Ustinov. The ordinance of the Central Committee of the CPSU and the USSR Council of Ministers on 30 December 1961 formally established his new special design bureau, OKB-30.

Kisun'ko's new organization, however, remained within the territory of KB-1 and thus was dependent on it. Only in March 1966 did OKB-30 become an operationally independent Special Design Bureau OKB Vympel, and it finally occupied its own new premises in 1969. (Vympel means a pennant, especially when given as an award.) The government later renamed it the Scientific-Research Institute of Radio Instrument Building (Nauchno-Issledovatel'skii Institut Radiopribiborostroeniya, or NIIRP).

On 8 April 1958 the CPSU Central Committee and the Council of Ministers authorized development of the operational missile defense system of Moscow, A-35, with Kisun'ko as chief designer. The A-35 requirements remained in flux for many years. They changed in response to improving understanding of the evolving missile threats and the first results of tests of System A at Saryshagan.

The A-35 built on the success of System A but became significantly more complex. Importantly, the military required it to defend the Soviet capital against eight approaching pairs of warheads with the accompanying ballistic missile bodies, compared to a single pair dealt with by System A. Relying on the method of three distances would have resulted in 24 RTN-type radars operating in the same limited space. Such a design appeared exceedingly complex and impractical, and thus a new approach was needed. Missile defense veteran Oleg Golubev observed that

> a fruitful idea [of the method of three distances] for an experimental complex [System A] of missile defense showed no promise for a combat [operational] complex. ... This was one of the profound reasons of the following delay in development of the national missile defense system

---

[4]Mikhailov, 2005, p. 42.

and unfavorable developments personally for Kisun'ko. One arrives at the thought that a common saying that "our flaws are an extension of our virtues" could be applied not only to individuals but also to technical systems.[5]

Subsequently, for the operational missile defense system, Kisun'ko decided to switch to a traditional radar mode in which radar determined not only the distance (range) to a target but also its two angular coordinates, elevation and azimuth. The resulting decrease in tracking accuracy forced the selection of nuclear interceptor warheads instead of the nonnuclear high-explosive fragmentation warheads of System A. Golubev gave the full credit to Kisun'ko for this concept and called him "the father" of nuclear-armed interceptors (see Chapter 5).[6]

The final concept of the A-35 emerged in the early 1960s. Now the system had to defend against 16 pairs of warheads with the accompanying ballistic missile bodies. The plan included eight sites around Moscow with two firing complexes at each site. Each such complex, in turn, was to be armed with eight antiballistic missile launchers. The system would fire two interceptor missiles against each approaching target and could deliver two such salvos.

Vladimir Sosul'nikov designed the new continuous-wave long-range radar Dunai-3 (Fig. 9.1) for the A-35. He built it based on his earlier experience with the Dunai-2 radar of System A. The new radar operated at a higher frequency of about 400 MHz (wavelength 0.75 m), used a 40-MHz band for modulation, and covered a 45-deg × 45-deg sector. CIA analysts later nicknamed Dunai-3 the Dog House. Petr Grushin's Design Bureau Fakel developed the new interceptor, A-350, known as Galosh in the West. Two command centers completed the envisioned system architecture.

Powerful rival organizations and groups in industry stepped up their efforts to introduce new concepts, with varying degrees of maturity and merit. The government authorized work on a number of proposals, including building experimental testbeds for their evaluation at Saryshagan. The proving ground started work on such systems as Aldan for Kisun'ko's A-35 and Azov for Raspletin's S-225.

In the mid-1960s, the military also initiated construction of the new Argun' testbed (Fig. 9.2) for contemplated modernization of the A-35 system, to be known as the A-35M. This program included building and evaluating two advanced experimental radar systems, a target radar RKTs-35TA and an interceptor missile radar RKI-35TA at Site 38 of the range. RKTs and RKI stood for Radiolokator Kanala Tseli (Target Channel Radar) and Radiolokator Kanala Izdeliya (Article Channel Radar, where an article means a missile in this case).

---

[5]Golubev et al., 2003, p. 577.
[6]Golubev et al., 2003, p. 577.

Fig. 9.1 (Top) Transmitting (T) and receiving (R) sections (separated by almost 3 km, or slightly less than 2 miles) of the continuous-wave long-range radar Dunai-3 (Dog House) of the A-35 missile defense system in 1967. Dense forests surrounded this site 45 miles (70 km) west-southwest from downtown Moscow. The face of the phased-array receiver (bottom) was 330 × 330 ft (100 × 100 m) and the structure was 420 ft (128 m) tall; the adjacent control building measured 560 × 270 ft (170 × 83 m). The bottom photograph is flipped vertically. Original satellite reconnaissance photograph by KH-7 camera (Mission 4038; 11 June 1967) available from the U.S. Geological Survey; photograph identification, interpretation, processing by Mike Gruntman.

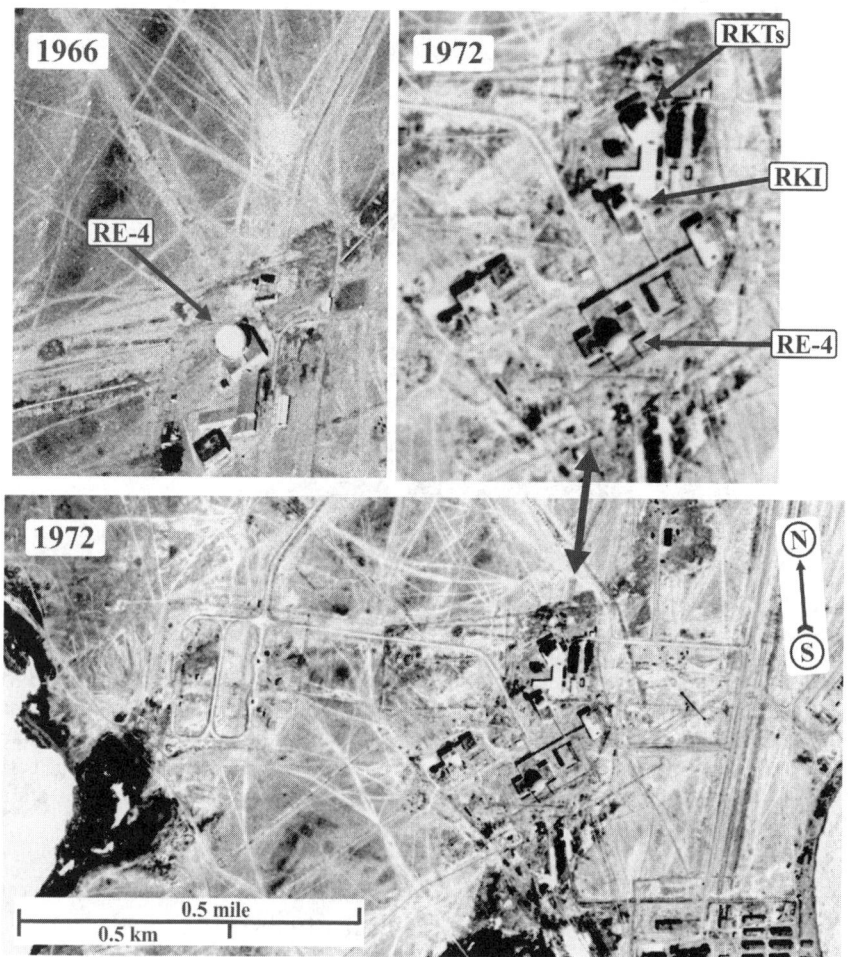

Fig. 9.2 Expansion of Site 38 from 1966 to 1972 at the Saryshagan range. Top left: experimental radar RE-4 in 1966. Top right: the same area in 1972 with two new prototype radars, the target radar RKTs-35TA Istra and the interceptor missile radar RKI-35TA, of the experimental testbed Argun' to assess new technologies for the modernization of the A-35 missile defense system. Two rectangular areas with rounded corners 0.6 mile (1 km) to the west from the radars in the bottom figure are launch fields of interceptor missiles. From the 1970s, Site 38 also engaged in evaluating prototype laser weapons such as Terra. Original satellite reconnaissance photographs (top left) by KH-7 camera (Mission 4027; 21 April 1966) and (top right and bottom) by KH-4 camera (Mission 1117-2; 29 May 1972) available from the U.S. Geological Survey; photograph identification, interpretation, and processing by Mike Gruntman.

The RKTs-35TA radar had to demonstrate the new approach, advanced by Kisun'ko, for identifying warheads among multiple accompanying decoys by detailed analysis of amplitude and phase characteristics of reflected radar pulses. Finally, in the mid-1970s, work began at the Saryshagan range on the

Fig. 9.3 Vladimir N. Chelomei (1914–1984) became the major developer of Soviet ICBMs in the 1960s. The Soviet Union had deployed 930 of his UR-100 (SS-11) ICBMs by 1970. Chelomei's design and production establishment in Moscow also substantially contributed to building space launchers (e.g., Proton), space stations, and various military space systems. He proposed the global national missile defense system Taran in the early 1960s and initiated the antisatellite weapon program that would reach operational status in the late 1970s. Photo courtesy of the Russian Academy of Sciences.

testbed Amur-P for the next-generation missile defense system of Moscow, A-135, which would become operational in the 1990s.

Changing requirements, technical difficulties, and power struggles caused delays in the development of the A-35. In particular, the government cut the program funding in 1963 as Vladimir Chelomei (Fig. 9.3) injected into competition his global national missile defense system, Taran. The Central Committee of the CPSU and the USSR Council of Ministers authorized development of Taran in 1962. The system relied on Chelomei's new "universal" rocket UR-100 (8K84), being developed at that time for the strategic rocket forces. This silo-based two-stage ballistic missile would become the most numerous Soviet intercontinental ballistic missile (ICBM) SS-11, with 930 operationally deployed by 1970.[7]

A leading specialist in missile defense, Leonid Khvatov, pointed to S-225 and Taran as examples of "fantastic or scientifically unfounded projects of ... missile defense concepts [of Raspletin and Chelomei, respectively]." He attributed promotion of these and some other programs to the desire of leaders

---

[7]Gruntman, 2004, pp. 288–290.

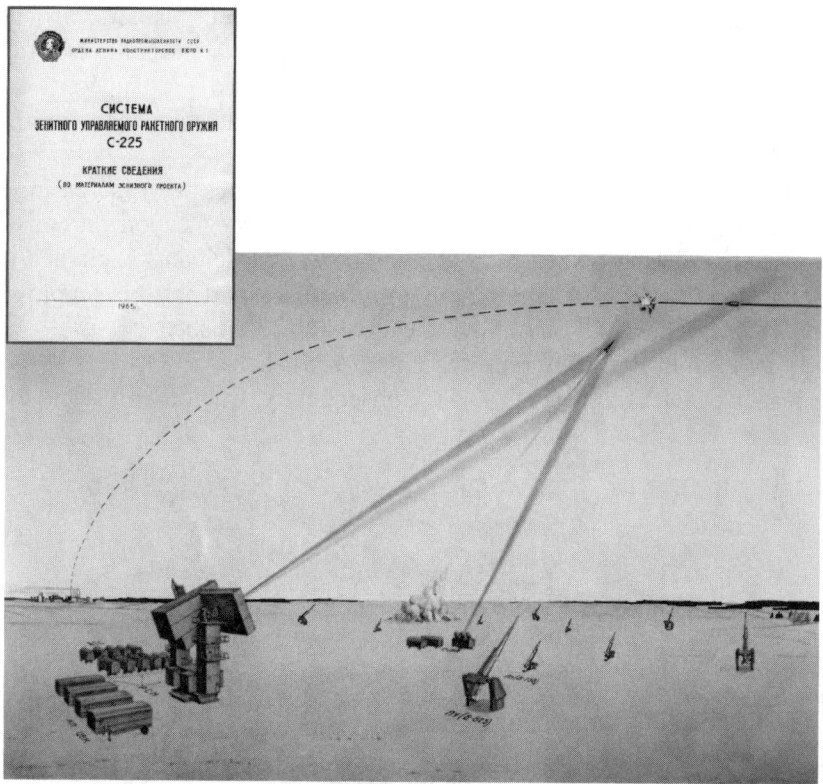

Fig. 9.4 Cover page of the "brief description" of the preliminary design and schematic of the S-225 complex for defense "against ballistic missiles, global missiles, and future aerodynamic means of attack" advanced by Alexander Raspletin's KB-1 in 1965. The proposal rationalized that the desire of the "potential adversary," that is the United States, to inflict maximum damage would result in an attack on a large number of targets at the same time. The latter would necessarily lead to a "relatively small number of ballistic missiles simultaneously attacking a particular [defended] object." Therefore, the proposed system focused on defense against a limited number of warheads and their intercept at relatively low altitudes. The latter feature opened "ways for a solution to one of the most complex problems of missile defense that consisted of the necessity of identifying the ballistic missile warhead in the presence of accompanying decoys." The proposed S-225 also promised to defend objects near the borders and seashores, where the time available for interception was shorter than for important areas and centers deep inside the territory of the country (Ministry of Radio Industry, 1965, pp. 5–9). Figures from Ministry of Radio Industry, 1965, cover page and Fig. 1 on p. 4.

of various organizations to "capture the area" because of its rapid growth in importance.[8]

Raspletin proposed his S-225 (Fig. 9.4) as a "universal system" for defending various important areas and centers near the borders and deep inside the

---

[8]Khvatov, 2003, p. 592.

country as well as operationally deployed troops from "ballistic missiles, global missiles, and future aerodynamic means of attack."[9] Chelomei's Taran relied on his universal UR-100 ballistic missiles armed with superpowerful nuclear charges that were to engage targets at large distances from the defended area.[10]

In the early 1960s, Vladimir Chelomei had been aggressively increasing his share of the ballistic missile, space system, antisatellite weapon, and missile defense effort in the USSR. Many contemporary participants explained this successful expansion of Chelomei's programs, in part, by his employing Sergei Khrushchev, the son of Soviet leader Nikita Khrushchev.[11]

Then, Leonid Brezhnev replaced Nikita Khrushchev in October 1964 as the leader of the country (Fig. 9.5). Consequently, Chelomei lost his most

Fig. 9.5 Commander (1963–1972) of the Strategic Rocket Forces Marshal Nikolai I. Krylov greets Soviet leader Nikita S. Khrushchev at a Tyuratam (Baikonur) airfield on 23 September 1964. Behind Khrushchev are (left to right) Dmitrii F. Ustinov (overseeing the defense establishment) and very senior Communist Party officials Leonid I. Brezhnev and Frol A. Kozlov. Less than three weeks later Khrushchev was ousted from power in a "palace coup" in the Kremlin and Brezhnev became the new leader of the country. Photograph from collection of a Baikonur veteran, Colonel Yu. V. Bonchkovsky; courtesy of Alexander Yu. Bonchkovsky.

---

[9]Ministry of Radio Industry, 1965, p. 9.
[10]Zavaly, 2003, p. 742.
[11]e.g., Golubev et al., 1993, p. 192; Golubev et al., 1994, p. 55; Kuchma, 2003, p. 354; Pervov, 2003, p. 53.

important backers. As a result, the government cancelled his Taran program, which many specialists had unsuccessfully criticized earlier on technical grounds. Elimination of this program brought new life to and focus on fielding Kisun'ko's A-35 missile defense system. The government resumed funding of the latter in 1965. The work on the A-35 experimental testbed Aldan at Saryshagan also restarted at the end of that year. By 1967, many elements of Aldan had been constructed.[12]

Colonel N.D. Drozdov served from 1962 to 1978 in the scientific research institute NII-2 of the Ministry of Defense and specialized in missile defenses. Drozdov observed that "talented" and "overambitious" chief designers

> [had] appealed to the political leadership of the country and apparatchiks [functionaries] with the arguments about the importance of their proposals ... [and thus] helped in creating confidence among incompetent [and] sometimes illiterate individuals in their right to make decisions about any subject and to be arbiters in scientific arguments.
>
> This brought dependence of scientific-technical decisions on personal links to individuals from the nomenklatura [highly positioned functionaries and administrators whose appointments required approval by the top-level Communist Party bodies]. This was the way the decisions on missile defense were also made. When Khrushchev was the leader, then Chelomei was right with his [global national missile defense system] Taran. Khrushchev left [the leadership position, ousted in 1964], but [responsible for the defense industry in the CPSU] Ustinov remained, so Kisun'ko was now right with his deep modernization of the A-35, and volumes of the Taran proposal [and associated studies] went [figuratively] to burn in the stove. Other competitions [in ballistic missiles, space exploration, and space applications] were handled in a similar way, for example, [the competition] between S.P. Korolev and V.P. Glushko about [the directions of] the lunar program. ...
>
> Some General Designers liked the situation, while others took it as a given [and went along]. In any case, nobody showed signs of self-criticism, including Kisun'ko.[13]

Drozdov's criticism particularly singled out selection of the A-350 interceptor missiles of Grushin's Fakel design bureau for the A-35 as a glaring example of pulling political strings to overcome unfavorable technical reviews in a competition.[14]

The operational configuration of the fielded A-35 significantly differed and was smaller compared to the original plans. It consisted of only one control

---
[12]Belous et al., 2009, p. 237.
[13]Drozdov, 1998, p. 41.
[14]Drozdov, 1998, pp. 7, 8.

center instead of two, one long-range radar Dunai-3, and four sites, instead of the planned eight, with two firing complexes each armed with interceptor missiles and radars.

Each firing complex included one RKTs-35 radar to track targets. It had the 18-m primary antenna protected by a radome and could separately track the warhead and the accompanying rocket body at distances up to 930 miles (1500 km). The complex also had two smaller RKI-35 radars for tracking and controlling interceptors, which allowed the complex to engage one target with two missiles simultaneously or two different targets at the same time. Contemporary CIA reports identified these groups of three radars as "the terminal target tracking and missile guidance radar installations" and called them "Triads."[15]

The designers developed the target radar RKTs-35 in two configurations, named Enisei and Tobol, differing in supporting computers and algorithms. They completed development of Enisei first. The government accepted the first phase of the partially completed A-35 missile defense system of Moscow in 1971. In addition to the control center and Dunai-3 radar, it had only three firing complexes activated at three bases near the towns of Klin, Zagorsk, and Naro-Fominsk, all with the Enisei radar (Fig. 9.6). After state tests, the military declared the system "experimentally operational" in 1972. The A-35 went on combat duty in July 1973.

Figures 9.7 and 9.8 show firing complexes at the system pilot site near Klin (E-33) in 1966 and at a site near Zagorsk (E-05) in 1972, respectively, for the first phase of the A-35. Each base had one operational firing complex with a target RKTs-35 Enisei radar, two interceptor missile RKI-35 radars, and eight launch positions of the interceptors.

Then, in the second phase of the A-35 development, in December 1974, the government accepted five additional firing complexes, all equipped with the new Tobol radar. One such complex was added to each of the three existing sites (E-05, E-24, and E-33 in Fig. 9.6). In addition, two complexes with Tobol radar were activated at the new fourth site (E-31) near Nudol'. Finally, the A-35 missile defense system consisting of the four bases with two firing complexes at each was thus complete.

The map in Fig. 9.6 also shows the location of two additional long-range Dunai-3U radars, deployed later at a location near the town of Chekhov for the modernized A-35M. Alexander N. Musatov[16] led the design of Dunai-3U, nicknamed by U.S. intelligence analysts as the Cat House, at the NIIDAR institute. Musatov based these new radar systems, which used the 388–429 MHz range, on semiconductor technology, in contrast to the earlier Dunai-3 that had relied on vacuum tubes. Sosul'nikov had also upgraded his

---

[15]Central Intelligence Agency, Dec. 1965, p. 12; Central Intelligence Agency, 1967, pp. 17, 18.
[16]Alexander Nikolaevich Musatov, 1925–2008.

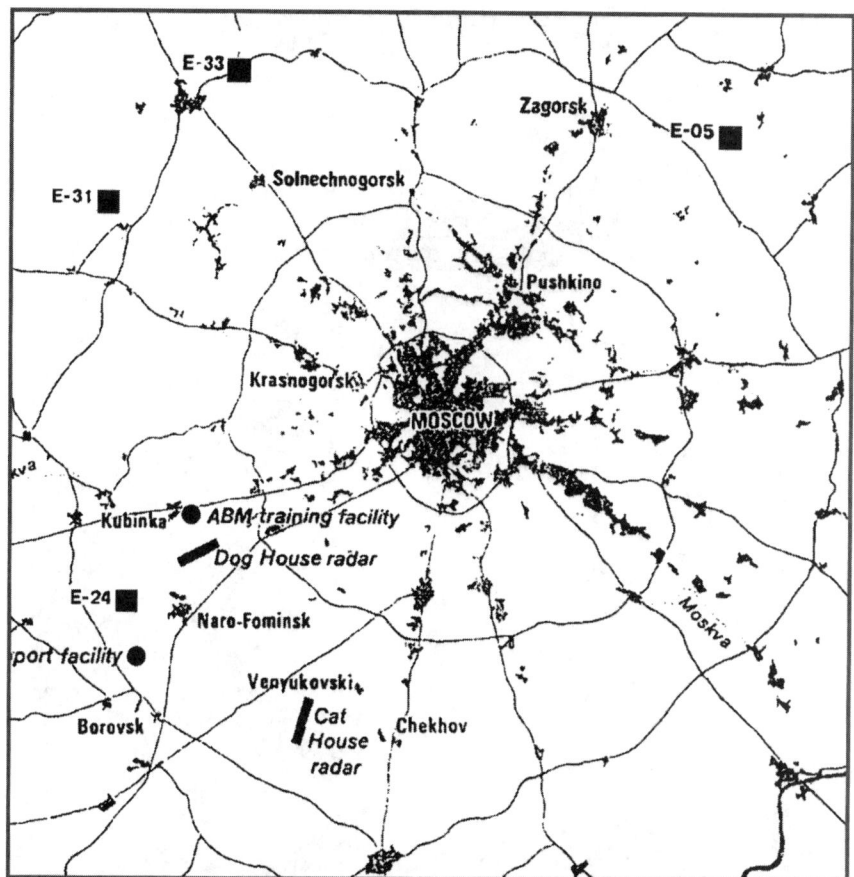

Fig. 9.6 Map of elements of the A-35 (and its upgrade the A-35M) missile defense system of Moscow from the CIA's National Intelligence Estimate, 1982. The map shows Dunai-3 (later Dunai-3M) as the Dog House radar. The modernized system A-35M also included another long-range radar, Dunai-3U, near Chekhov, marked as the Cat House radar. The black circle near Borovsk shows the ABM support facility. The interceptors with fire control RKTs-35 Enisei radars of the first phase of the A-35 were deployed at the sites (black squares) E-05 (near Zagorsk), E-33 (Klin), and E-24 (Naro-Fominsk). The second phase of the A-35 augmented these three sites with additional firing complexes with newer RKTs-35 Tobol radars and also activated an additional site E-31 near Nudol' with two firing complexes, each equipped with the Tobol radar. All four bases with two firing complexes at each were deployed at converted sites of the outer ring of the S-25 (SA-1) air defense system. Both the outer and inner ring roads built for the S-25 prominently stand out on the map. From Central Intelligence Agency, 1982, Figure 1, p. 11.

Dunai-3 in 1977, and it was then designated Dunai-3M. All Dunai-3U and Dunai-3M radar systems became operational in 1978. Dunai-3U continues to function to this day. In 1989, a fire destroyed Dunai-3M and it has not been rebuilt.

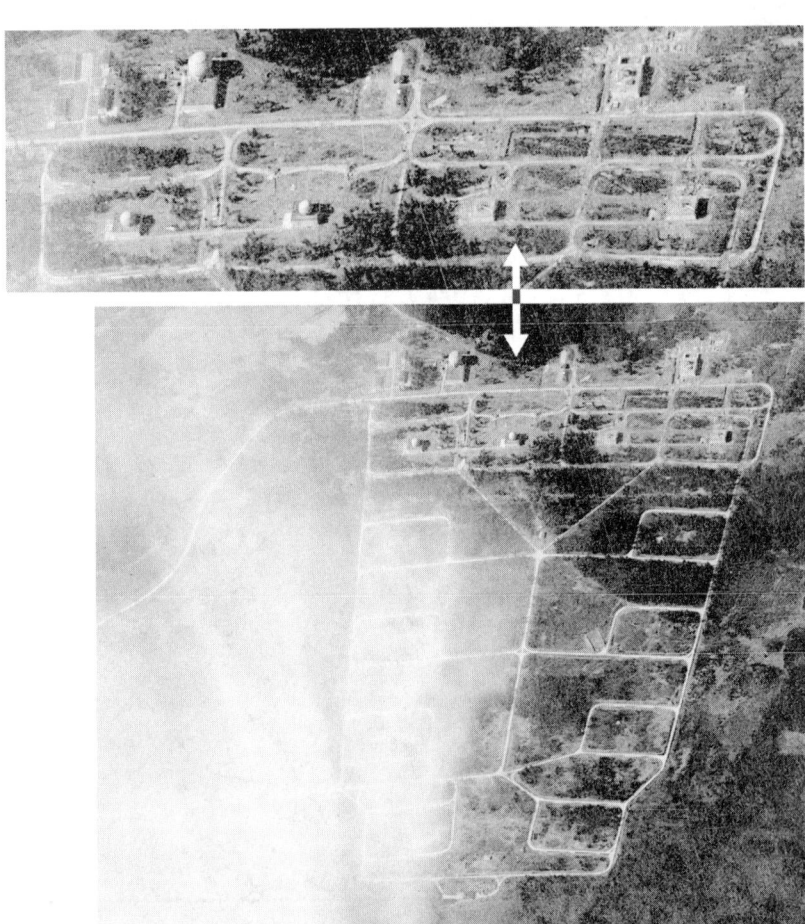

Fig. 9.7 Pilot site of the first operational missile defense system, A-35, near the town of Klin (E-33 in Fig. 9.6) northwest of Moscow in 1966. The missile firing positions and radars (top) were deployed at the converted outer ring site of the S-25 (SA-1) antiaircraft system, taking advantage of the existing infrastructure and supporting facilities. The military completed construction of this position in 1965 and began assembling equipment in 1966. Evolution of the nature of the targets caused by the introduction of ballistic missiles with multiple warheads and numerous decoys resulted in significant changes in the missile defense system requirements and major delays in achieving its operational deployment. Original satellite reconnaissance photograph by KH-7 camera (Mission 4027; 24 April 1966) available from the U.S. Geological Survey; photograph identification, interpretation, and processing by Mike Gruntman.

As the military accepted the A-35 as operational, Kisun'ko remained its chief designer. He was also in charge of the system modernization, which would lead to the A-35M, although without him. The power had shifted from Kisun'ko in the missile defense establishment by this time, and he would soon be fired.

Fig. 9.8 Operational site of the A-35 missile defense system near the town of Zagorsk (E-05 in Fig. 9.6) northwest of Moscow in 1972. The antiballistic missile positions and radars are deployed at the converted outer ring site of the S-25 (SA-1) antiaircraft system (top). The bottom figure shows the side-by-side areas of two firing complexes, with one larger dome of the target RKTs-35 radar and two smaller domes of the interceptor missile RKI-35 radars at each. The four white and four black small arrows point at pads with launch support equipment of eight A-350 interceptor missiles at one firing complex. Original satellite reconnaissance photograph by KH-4 camera (Mission 1116-2; 5 May 1972) available from the U.S. Geological Survey; photograph identification, interpretation, and processing by Mike Gruntman.

CRISIS IN MISSILE DEFENSE

In 1967 General Yuri Votintsev assumed command of the newly created Missile and Space Defense Forces. He then led a comprehensive government review of principal proposals to deal with evolving ballistic missile threats.

The Votintsev committee included prominent weapons scientists, general and chief designers, and top officials of the Ministry of Defense and defense industry. They rejected three major proposals, including that of Kisun'ko, who had been advancing the concept of a national missile defense system, Avrora (aurora in English), for protection of the European part of the Soviet Union. The committee also declined Mints's proposal for a new multifunctional fire control radar for missile defense, Don-N, and a proposal by Yuri G. Burlakov[17] from NII-244 for a wideband radar based on Luneburg lenses. Votintsev wrote later that the proposals had not addressed the key challenges of missile defense: (1) identifying and selecting target warheads from decoys in the environment of active countermeasures and nuclear explosions, (2) achieving computational capabilities of hundreds of millions of operations per second, and (3) developing effective interceptors for different parts of trajectories of attacking warheads based on various physical principles.[18]

The follow-on committee deliberations recommended focusing on advancing technology for distinguishing target warheads from accompanying decoys and establishing for this purpose two new experimental radar testbeds, Neman and Argun', at Saryshagan.[19] Following the established practice, the military named these testbeds after rivers. (River names were also commonly assigned to operational radar.)

The Neman radar would test the design proposed by Burlakov. It promised to diminish the effectiveness of decoys by enabling the reception of radar signals at different frequencies and from various directions. The Argun' experimental radar would evaluate Kisun'ko's concept for distinguishing targets from decoys by detailed mathematical analysis of the amplitude and phase characteristics of reflected radar pulses. The designers later incorporated the results of the latter work into the modernized missile defense system, A-35M.

The rejection of the proposals for major new systems by the Votintsev committee and the uncertainty about future directions of the national program increased tensions in the defense establishment. Some leading participants called the situation "critical" and even a "crisis" in Soviet missile defense.[20]

By the late 1960s, the nature of ballistic missile targets had fundamentally changed from the original pairs, each consisting of a warhead and the accompanying body of a ballistic missile. As Vladimir Markov described it, the target now

> ... represented, instead of a warhead and a [body of a] ballistic missile, a "sausage" 20–40 km [12.5–25 miles] in diameter and 200–400 km

---

[17]Yuri Grigorievich Burlakov, 1928–2002.
[18]Votintsev, 1993, Part 2, pp. 35, 36.
[19]Votintsev, 2003, pp. 31–33.
[20]Votintsev, 1993, part 2, p. 35; Golubev et al., 1994, pp. 55–58; Markov, 2003, pp. 403, 405; Markov, 2005, p. 31; Repin, 2005, p. 58.

[125–250 miles] long packed, in addition to the warhead and the [ballistic missile] body, with hundreds of passive reflectors and with systems of active electronic countermeasures.[21]

There was also growing acceptance of a view that a realistic national missile defense based on a missile-against-a-warhead approach would not be able to protect against a massive attack with multiple sophisticated decoys and other penetration aids. Instead, perhaps it could and should focus on defense against a limited ballistic missile attack. Missile defense specialists and witnesses of those events described such a moderated objective as "defense of the capital [Moscow] from single, accidental, [or] provocative strikes by . . . [ballistic missiles] or against a limited group of . . . [ballistic missiles launched] from territories of third [i.e. non-U.S.] countries or from a single out-of-control [by the national authority, rogue] submarine."[22]

In addition, some believed that high priorities should include development of capable systems for ballistic missile and space situational awareness; completion and operational deployment of the A-35, which was being constructed at that time; and concluding a treaty with the United States limiting missile defense systems.[23] The latter thrust would ultimately lead, as U.S. analysts described, to

> a two-track approach [by the Soviet Union simultaneously] calling for arms control and a Soviet military buildup. During 1970s, the USSR achieved limits on the number of US delivery vehicles through the SALT [Strategic Arms Limitation Talks] process, constrained US defenses through the ABM Treaty, and gave priority to building up its own offensive forces. This two-track approach worked well [for the USSR] in the 1970s. . . .
>
> . . . [T]he Soviets apparently see the [ABM] treaty as having slowed US ABM research and development, while they moved ahead with their own.[24]

Vladislav G. Repin,[25] who had led development of Soviet early warning systems for ballistic missile attack from 1970 to 1987, observed that

> [u]nfortunately, at this time [in the late 1960s and early 1970s] G.V. Kisun'ko did not show the so characteristic for him technical courage. As I can judge based on numerous discussions with him about features and consequences of fundamental changes of the target environment

---

[21] Markov, 2003, p. 403.
[22] Golubev et al., 1994, p. 64.
[23] Golubev et al., 1994, pp. 63, 64; Pervov, 2003, p. 213; Repin, 2003, p. 445; Repin, 2005, pp. 59–60.
[24] Central Intelligence Agency, 1982, p. 9.
[25] Vladislav Georgievich Repin, 1934–2011.

[evolving] from simple to complex targets, he understood them very well and realized that neither the A-35 missile defense system built by him for the Moscow region nor the Avrora proposal for territory defense [of the European part of the USSR] provided an adequate response to these new targets in both its principle of approach and the [implemented] technical means.[26]

Interestingly, similarly to Kisun'ko, the critical Repin was also eased from his position in the 1980s when he began to appear "too independent and frequently contradicted superior authorities on technical issues viewed [by them] as beneficial for some reasons to [their] part of the industrial establishment," as the outspoken Kisun'ko had done in the 1970s. Even the same particular functionary who pushed Repin out had earlier conducted a "similar [administrative] operation against Kisun'ko."[27]

To some degree, the designers responded to new target challenges by adding new features and increasing the capabilities of the A-35 missile defense system. A detonation of a nuclear interceptor warhead, for example, could affect the performance of other interceptors. Consequently, the interceptors now switched to autonomous guidance before the detonation of nearby nuclear interceptors took place. An interceptor carried a special radiation-hardened electronic unit that stored in its memory the latest information about the projected time of its own detonation.[28] Such a design would be particularly important for planned future missile defense engagements where interceptors operate in a hostile environment caused by multiple nuclear explosions, including those of other interceptors, and at large distances from the defended area.

Nuclear explosions also created ionized formations in the atmosphere that interfered with radar, as Operation K had demonstrated at Saryshagan in the early 1960s. An explosion in a denser atmosphere produced a weak formation that nevertheless attenuated radar signals, causing angular errors as well as errors in target velocities obtained from the measured frequency Doppler shifts in received reflected pulses. The effective range of the fire control radar and its accuracy thus decreased.

A more distant nuclear intercept at higher altitudes created global ionized formations that effectively shielded targets. Dealing with such phenomena required new generations of radar operating at various frequencies, tracking multiple targets with high accuracy, and employing sophisticated algorithms. Atmospheric drag remained the best means of separating warheads from decoys. Relying on the atmosphere, however, dramatically decreased distances to intercepts, brought them closer to the defended object, and shrank the time interval for engagement and destruction of the targets.

---

[26]Repin, 2003, p. 443.
[27]Kuriksha, 2005, p. 110.
[28]Golubev, 2003, p. 578.

SCIENTIFIC-INDUSTRIAL ASSOCIATION VYMPEL

Many Soviet officials saw that lack of coordination and institutional and personal rivalry had been delaying advancements in missile defense. Consequently, on 15 January 1970 the government reorganized the national effort by putting several leading missile defense design bureaus and manufacturing plants under one administrative roof in the new Central Scientific-Industrial Association (Tsentral'noe Nauchno-Proizvodstvennoe Ob"edinenie, or TsNPO) Vympel.[29] Vladimir Markov became general director of the association with Kisun'ko as his deputy for science. At the same time, Markov retained his position of deputy minister in the overseeing Ministry of Radio Industry. The association used the same name, Vympel, as Kisun'ko's design bureau, and the latter was then renamed the Scientific-Research Institute of Radio Instrument Building (NIIRP).

The new TsNPO Vympel included Grigorii Kisun'ko's OKB Vympel (NIIRP); NII-37 (NIIDAR), which was responsible for the development of long-range Dunai-2, -3, -3M, and -3U radars as well as for the over-the-horizon radar Duga and later the space object identification system Krona; Alexander Mints's RTI institute; and several other development organizations and major industrial plants. In addition, the government strengthened the new association by transferring 1000 engineers and technicians from other institutions and design bureaus.[30]

Markov formed three special design bureaus, SKBs, in the new Scientific-Programmatic Center (Nauchno-Tematicheskii Tsentr, or NTTs) created within TsNPO Vympel:[31]

SKB-1 for development of systems for early warning of ballistic missile attack and control of space (space situational awareness) under Vladislav G. Repin
SKB-2 with a focus on the next-generation missile defense system (to be known as the A-135) under Anatolii G. Basistov[32]
SKB-3 with responsibility for the A-35 under Kisun'ko

A new rising leader in the field, Basistov, became chief designer of the A-135 in 1973. Its development included the experimental testbed Amur-P at Saryshagan. After the firing of Kisun'ko in 1975, Basistov assumed the responsibility of chief designer of national missile defense. After 20 years of development, the A-135 became operational in the mid-1990s and stands on combat duty today.

---

[29] Anisimov et al., 2005.
[30] Golubev et al., 1994, p. 58.
[31] Pervov, 2003, p. 218; Anisimov et al., 2005.
[32] Anatolii Georgievich Basistov, 1920–1998.

A leading missile defense specialist in interceptor warheads for Systems A, A-35, and A-135, Yuri A. Kamensky, saw administrative games as a reason for establishing TsNPO Vympel. "[I]n order to deal with Kisun'ko, our KB [design bureau] was merged into a [new] scientific-production association [Vympel]. After some administrative maneuvers, Basistov was appointed the chief designer of missile defense, while Kisun'ko was left out."[33]

In the late 1970s the government gave major decorations and awards to the developers of the first Soviet operational missile defense system, A-35. The bitterness and vindictiveness of the establishment clearly showed in the fact that Kisun'ko, who had designed, built, and fielded the A-35 system and laid the foundation and was the driving force for its modernization to the A-35M, received none.[34]

Alexander Mints objected to putting his RTI institute under control of the new organization. According to Repin, Mints told him that he [Mints] "particularly did not want to work under direction of V.I. Markov" and that he "did not like Kisun'ko very much as well as all his team."[35] Repin wryly observed that Kisun'ko "responded to [this attitude of] Mints with full reciprocity." He also noted that "personal relations between G.V. Kisun'ko and A.L. Mints [were so bad that they] even excluded possibility of any serious discussion of interaction between developers of missile defense [led by Kisun'ko] and means of [ballistic missile] early warning [led by Mints]."[36]

Subsequently, Mints wrote a letter to Minister Kalmykov, stating his negative position on the reorganization of missile defense and the creation of TsNPO Vympel. The minister responded by thanking Mints for his service and asked him to retire. Mints was 75 years old at that time, and he died 5 years later. Actually, RTI fared well as part of Vympel by developing and fielding a number of early warning and missile defense radars such as Dnepr, Daryal, Daugava, and Don-2.

Establishment of Vympel reorganized national work on missile defense. The new administrative structure lasted into the 1990s, until the dissolution of the Soviet Union and fielding of the A-135. It was a long road that started with the first intercepts by the experimental System A at the Saryshagan range and led to the operational missile defense system A-35; then to its upgrade, A-35M; and finally to the A-135.

FIRING OF KISUN'KO

By 1975 the new requirements called for modernization of the A-35 to defend Moscow against either eight pairs of simple targets or a single complex

---

[33] Kamensky, 2002.
[34] Golubev et al., 1994, p. 61; Kamensky, 2002.
[35] Repin, 2003, p. 451.
[36] Repin, 2003, p. 453.

multicomponent target consisting of up to 10 warheads accompanied by numerous light and heavy decoys.[37] Kisun'ko advanced the technology for identification of target warheads by detailed real-time analysis of amplitude and phase properties of reflected radar pulses, which would be implemented in the system upgrade. The modernized A-35M partially met these new requirements by providing some required capabilities.

As work on A-35 improvements progressed, there was no agreement among leading designers and officials on further development of missile defense. Participants at the events noted that already in the early 1970s, Deputy Minister of Radio Industry Vladimir Markov had "been purposefully diminishing the role of [Kisun'ko]." This led to resistance by the latter and his associates to the directions of future programs approved by the Ministry. Power struggles thus continued.

> General Director of [Scientific-Industrial] Association [Vympel] V.I. Markov took on himself the solution of the problem [of disagreements] and he succeeded in having Kisun'ko relieved from his position and fired from the Association. The hope of Kisun'ko for decisive support from omnipotent at that time Minister of Defense D.F. Ustinov was not realized.[38]

In June 1975, the new Minister of Radio Industry, Petr Pleshakov, personally removed Kisun'ko from being chief designer of the A-35 and appointed Kisun'ko's first deputy, Ivan D. Omel'chenko,[39] in his stead. Omel'chenko completed the system upgrade, and the government accepted the A-35M on 28 December 1977. It became operational and went on combat duty on 15 May 1978.

The commander of the Missile and Space Defense Forces, General Votintsev, observed that "as a result of intrigues in the Ministry of Radio Industry, an extraordinary and gifted designer [Kisun'ko] was eliminated [from a leadership position] literally on his ascent [to new accomplishments]."[40] Looking back on the role of Kisun'ko, Oleg Golubev recalled that from the 1950s until 1975, "Grigorii Vasil'evich [Kisun'ko] was remembered by all of us as an energetic, life-loving man with good will, wide views, and broad interests. His authority as a scientist and a designer was without challenge."[41] Kisun'ko did not take his firing lightly and, in the words of Golubev, felt many years later a "grudge against everybody and everything."

Among the reasons for firing Kisun'ko, Golubev listed "a ten-year delay in completing the A-35, rejection by the [military] customer of the proposed [by him] system of territorial [national missile] defense Avrora, and Kisun'ko's

---

[37]e.g., Golubev, 2003, p. 579; Belous et al., 2009, p. 249.
[38]Golubev et al., 1994, pp. 65, 66.
[39]Ivan Dmitrievich Omel'chenko, 1919–2000.
[40]Votintsev, 1993, Part 2, p. 33.
[41]Golubev, 2003, pp. 577–578.

disagreement with the concept of modernizing the system as well as other subjective reasons."[42] The latter "subjective reasons" meant interpersonal relations and power struggle. Golubev emphasized, however, that circumstances external to Kisun'ko and his team caused, in part, the rejection of Avrora and the A-35 delays. Markov himself confirmed that Kisun'ko's objections to the approved directions of development of national missile defense and active criticism resulted in his "transfer to another job," a euphemism for being fired.[43]

A GIGANTIC ENTERPRISE

The growing Soviet missile defense establishment included numerous research institutes, design bureaus, industrial plants, and various military units. The national missile defense that emerged (Protivoraketnaya Oborona, or PRO) evolved through a sequence of the operational systems A-35, A-35M, and present-day A-135 protecting Moscow. In addition, the country deployed the Space Control System (Sistema Kontrolya Kosmicheskogo Prostranstva, or SKKP) and the Ballistic Missile Attack Early Warning System (Sistema Preduprezhdeniya o Raketnom Napadenii, or SPRN). The early warning SPRN grew to include ground-based and space-based layers, with radar systems on the ground and optical and infrared sensors on satellites, respectively. Another major program, started around 1960, strived to develop the over-the-horizon radar for ballistic missile early warning and remained highly controversial for many years.[44] Antisatellite weapons also expanded into an important undertaking and reached operational status.

TsNPO Vympel coordinated the national effort in missile and space defenses by providing direction to 150 leading organizations in the country and contracting more than 600 various other scientific, design, and industrial institutes, bureaus, and plants. At its peak, Vympel directly employed 80,000 people, including more than 700 with PhD degrees in science and engineering.[45]

Huge resources went into the gigantic enterprise. In 1965, a U.S. National Intelligence Estimate described the emphasis on strategic air and missile defenses in the USSR:

> Soviet expenditures for strategic defense have grown steadily since 1950. In recent years, these expenditures have roughly equaled those for strategic attack, when the major buildup of strategic missile forces was in process. The USSR devoted a much larger share of its military expenditures to strategic defense during the 1961–1964 period than did the US. Manpower allocated to the strategic defense mission has also increased

---

[42]Golubev, 2003, p. 579.
[43]Markov, 2003, p. 416.
[44]Markov, 2003, pp. 417–426; Repin, 2003, pp. 444, 467–469; Babakin, 2008; Belous et al., 2009, pp. 204, 347–351.
[45]Litvinov, 2005, p. 19.

markedly—from about 200,000 in 1950 to almost 500,000 men at present. This increase occurred during a period of large scale reductions in military manpower.[46]

A recent study put Soviet expenditures on strategic air and missile defenses at 50–100% of the funds spent on development of strategic offense capabilities in the years 1955–1972.[47] Total military production constituted 20% of the *gross social product* in the 1980s, with defense programs accounting for 80% of the national research and development effort.[48] (The USSR commonly used a quantitative measure of gross social product, or gross output of industries, for statistical purposes.)

The communist state pored enormous resources into weapon systems while basic items of life were in short supply. With a low standard of living, national pride and the demonstration of ideological superiority and military prowess served as substitutes for bread on the table or shoes on the feet. Government rewards to the first Soviet cosmonauts in the early 1960s starkly illustrated conditions in the country.

A secret decree of the USSR Council of Ministers awarded the first cosmonaut, Yuri Gagarin, not only a significant monetary bonus and a car, a vacuum cleaner, and floor rugs among other life necessities, but also such small items as six pairs of socks and six sets of underwear shorts and singlets.[49] The socialist government also calibrated these rewards to achieve social justice by giving, for example, three blankets, three bedspreads, and six pillows to another cosmonaut who was a married man, whereas a single woman, Valentina Tereshkova,[50] received only two blankets, two bedspreads, and four pillows.

General Yuri Votintsev recalled that

> in the 1970s more than 100,000 construction troops toiled on several missile and space defense sites. During the final phase of construction . . . at [just] one such site the number of military construction troops was more than 30 thousand, with an additional two to three thousand specialists from the Ministry of Assembly and Special Construction, and [additional] three to four thousand people representing the industry [contractors].[51]

Missile defense troops became a highly technical and educated branch of the armed forces, with 60% of servicemen being officers and noncommissioned officers.[52] Typically, a town with 10,000–12,000 inhabitants attached

---

[46]Central Intelligence Agency, Nov. 1965, p. 4.
[47]U.S. Army, 2009, Vol. 2, p. 121.
[48]Kuchma, 2003, p. 336.
[49]Gruntman, 2011.
[50]Valentina Vladimirovna Tereshkova, b. 1937.
[51]Votintsev, 1993, Part 4, p. 22.
[52]Votintsev, 1993, Part 4, p. 25.

to and supported operations of each major radar site of missile defense, ballistic missile early warning, or space control systems.[53]

Development of early missile defenses in the 1950s and 1960s with steadily increasing government support and funding produced a crop of young, ambitious, and already accomplished scientists, engineers, and managers. Golubev wrote that "[t]he youngest missile defense system in our country—System A—was developed and built also by young people. The age of the most "old" was 30–40 years. And the most senior among us—G.V. Kisun'ko—by the time of completing [first intercept] tests was only 43 years old."[54] This new generation joined the aggressive push to further expand the field.

ANTISATELLITE WEAPONS

A modified version of the long-range TsSO-P radar, Dnestr, built by Alexander Mints's RTI institute, first found its application in antisatellite weapons. Vladimir Chelomei initiated development of antisatellite capabilities in 1959. As early as 23 June 1960, the Central Committee of the CPSU and the USSR Council of Ministers authorized this major program. The work soon pointed to the particular importance of monitoring and cataloging orbiting space objects. Consequently, on 15 November 1962, special government decrees formalized development of the national *system of control of space*, better described as space situational awareness. These decrees also started large-scale programs in early warning of ballistic missile attack, including over-the-horizon radars.

Chelomei's antisatellite system evolved to include two major facilities to track target satellites and guide interceptors, known as the Satellite Finder nodes (uzel Obnaruzheniya Sputnikov, or OS; uzel means a node) and equipped with Dnestr radar. The designers placed one such node, OS-1 (Fig. 9.9), about 60 miles (100 km) northwest of the town of Irkutsk in Siberia and the other node, OS-2 (Fig. 9.10), at the Saryshagan range 50 miles (80 km) northeast of Priozersk.

The original system operated as follows. The OS-1 site in Siberia first detected a U.S. satellite and determined its orbit parameters. The accuracy of orbit determination was insufficient, however, for immediate launch of an interceptor and destruction of the target. After one orbit around the Earth, the satellite again flew over the Soviet Union, with its ground track now shifted to the west, due to the Earth's rotation, by about 1550 miles (2500 km). Then the other node, OS-2, at Saryshagan took measurements. The shift of groundtracks for a satellite in a typical low-Earth orbit determined the distance between the two OS sites.

---

[53]Kosmachev, 2005, p. 121.
[54]Golubev, 2003, p. 576.

Fig. 9.9 Satellite photograph (bottom) of the OS-1 node of the antisatellite system 60 miles (100 km) northwest of Irkutsk in Siberia in 1966. The bend of the river Belaya (top), a tributary of the Angara, confines the military base with the Dnestr radars. Together with the OS-2 node near Saryshagan and the interceptor launch site at Tyuratam, this Siberian base comprised the operational antisatellite weapon system. The installation also served as an element of the early warning system of ballistic missile attack. Original satellite reconnaissance photograph by KH-7 camera (Mission 4032; 22 September 1966) available from the U.S. Geological Survey; photograph identification, interpretation, and processing by Mike Gruntman.

Soon after the first detection and orbit measurement of the target satellite by OS-1, the military made an intercept decision and initiated the launch sequence of a Satellite Destroyer interceptor (Istrebitel' Sputnikov, or IS). Site 90 at Tyuratam launched IS interceptors, sometimes called satellite killers or killer

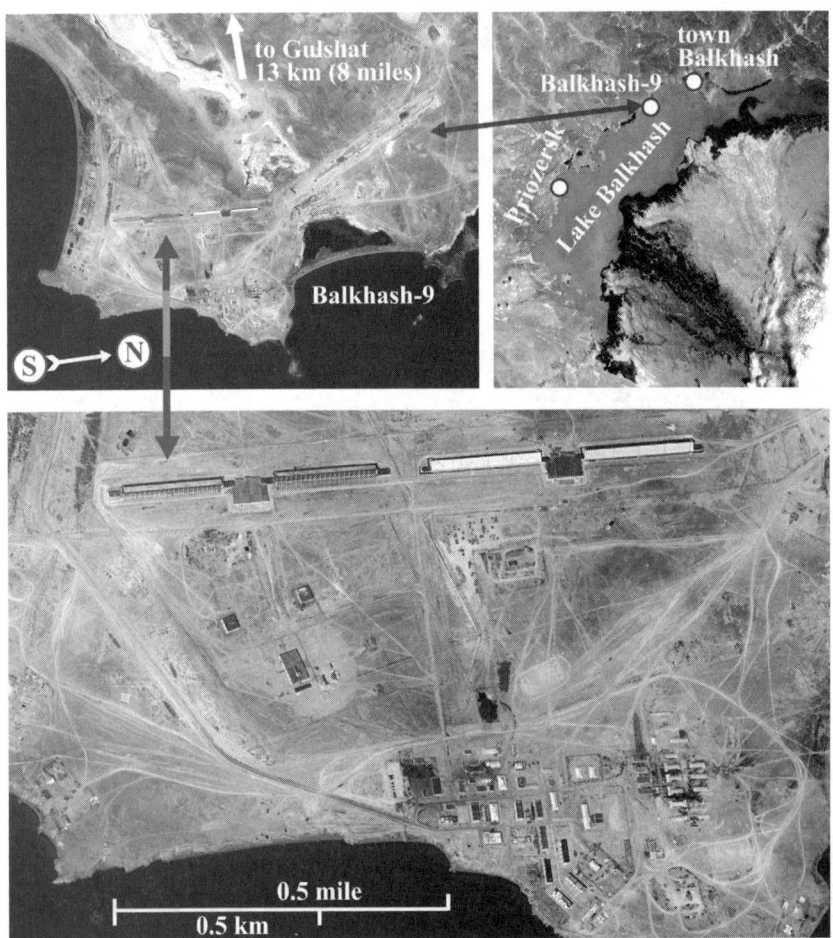

Fig. 9.10 Satellite photograph (bottom) of the OS-2 node of the antisatellite weapon system with the Dnestr radars that also tracked orbiting objects in space and provided early warning of ballistic missile attack. These radars evolved from the original work of Mints's RTI institute on the long-range TsSO-P radar at Saryshagan. The military built four such radars (top left) from 1964 to 1968 at a peninsula 50 miles (80 km) northeast of Priozersk (top right). Today, the Russian military operates an early warning radar at this site, known as Balkhash-9, in Kazakhstan. Original satellite reconnaissance photographs by (top left and bottom) KH-7 camera (Mission 4027; 21 April 1966) and (top right) KH-5 Argon mapping camera (Mission 9066A; 21 August 1964) available from the U.S. Geological Survey; photograph identification, interpretation, and processing by Mike Gruntman.

satellites, into orbit. The U.S. intelligence listed Tyuratam's Site 90 as Launch Complex G in its reports.[55] It was located about 70 miles (110 km) northwest of the main residential area of the Tyuratam (Baikonur) missile range.

---

[55]e.g., Central Intelligence Agency, Dec. 1965.

The military timed the moment of launch of the satellite interceptor in such a way as to place it into an orbit approximately coplanar with the target satellite when the latter appeared in the range of the radar at the Saryshagan OS-2 node. It was essential to achieve a nearly coplanar orbit because inclination changes of the interceptor orbital plane would have required unacceptable amounts of onboard propellant; such changes thus had to be minimized.

The second measurement by OS-2 improved the knowledge of the target satellite orbit and enabled the engagement. The ground control first guided the interceptor, which then locked onto the target, autonomously maneuvered, approached it, and finally detonated its high-explosive fragmentation warhead, destroying the target satellite.

Initially, Vladimir Chelomei led the entire antisatellite program. He planned to use UR-200 space launchers and satellite interceptor space vehicles built by his design bureau. Already on 1 November 1963, Chelomei's team had placed into orbit the first prototype of the IS satellite destroyer, Polet-1 (polet means flight), capable of maneuvering in orbit. The modified ICBM R-7 launched the satellite. Deployment of the second satellite, Polet-2, followed on 12 April 1964.

Early in development, Chelomei brought in the special design bureau SKB-41 of KB-1 to help him with radio control systems. By that time, Anatolii Savin had replaced (in 1960) Andrei Kolosov as head of SKB-41. The inclusion of a unit of the powerful KB-1 turned out to be consequential for the program.

The ousting of Khrushchev from power in 1964 changed all the arrangements, because Chelomei had lost his most important supporter in the Kremlin. The government then took the program away from Chelomei and made Savin and his SKB-41 the lead organization. Now, Yuzhnoe Design Bureau in Dnepropetrovsk in Ukraine built the space launcher. Chelomei's organization retained only the satellite destroyer, or killer, spacecraft. Ultimately, Savin's design bureau became the independent Central Scientific-Research Institute Kometa[56] in 1973 (Fig. 5.8 in Chapter 5).

Mikhail Yangel's Yuzhnoe Design Bureau planned to build a new space launcher, Tsyklon-2 (11K69), for this program based on its successful ICBM R-16 (8K64, SS-7). It first produced and used a simplified version of the launcher with limited control capabilities, Tsyklon-2A (11K67). A young Yuzhnoe engineer, Leonid D. Kuchma,[57] who would later become head of the Dnepropetrovsk rocket and space enterprise and then the president of independent Ukraine from 1994 to 2005, spent some time at Tyuratam in the early 1960s supporting tests of Tsyklon space launchers.

---

[56]Vlasko-Vlasov, 2002.
[57]Leonid Danilovich Kuchma, b. 1938.

The intercept of a satellite required quick determination of orbital parameters of the deployed IS interceptor vehicle with high precision. A special control post, conceptually similar to testbed installations seen as cross-type structures in satellite photographs at Saryshagan (e.g., Fig. 6.15 in Chapter 6), was developed for this purpose. The post consisted of a central receiver and four additional receivers displaced by approximately 1 km in four directions to the sides. The receivers operated as an interferometer and achieved angular accuracy of about 1′ of arc.[58] It roughly corresponded to a 330 ft (100 m) spatial uncertainty in a measured position of a satellite at a distance of 210 miles (340 km) in low-Earth orbit.

The antisatellite system successfully intercepted and destroyed a target satellite for the first time on 1 November 1968. The completed new space launcher Tsyklon-2, known in the West as SL-11, soon replaced the prototype Tsyklon-2A rockets.

The requirement to launch a satellite only 1 hour after receiving the command represented a very demanding challenge for the antisatellite weapon system. This time interval resulted from orbital mechanics. A typical orbital period of a satellite in low-Earth orbit is 90–100 minutes, and it takes about 10–20 minutes for a space vehicle to reach orbit after the rocket is fired on a launch pad. Correspondingly, the intercept command issued after the first detection of a target satellite by the OS-1 node required the launch of a killer satellite in about 1 hour in order to place it in a coplanar orbit in the vicinity of the U.S. target satellite when it appeared in the range of the OS-2 node.

To meet the 1-hour launch requirement, engineers built a futuristic automatic launch system. One of the leading system designers, Konstantin A. Vlasko-Vlasov,[59] described its operation:

> Exactly at [receiving the] one-hour readiness [command], the gates of the storage depot automatically opened and an electric locomotive rolled out with a transport-erection mechanism and a space launcher [on a platform car] along the railroad track towards the launch pad. No personnel accompanied this movement [of platforms]. The electric locomotive went [automatically] beyond a switch and stopped. The switch automatically changed its position. Then, the locomotive moved in reverse and pushed [the platform cars with] the erection mechanism and the rocket to the launch pad. After reaching the contacts near the pad opening, it stopped, disconnected [from the platforms], and moved to a dead-end siding.
>
> A special pin captured the erection mechanism [platform] and pulled it towards the pad. Then, a special plate connected [to the erection mechanism] with 50 electric connectors, four fuel funnels and two air pipes, the latter used for maintaining required air temperature and humidity of the

---

[58]Vlasko-Vlasov, 2003, p. 512.
[59]Konstantin Alexandrovich Vlasko-Vlasov, 1920–2009.

spacecraft. After a green sign lit up on a control board showing the proper connections [of the plate], the launcher [with the rocket] began to rise until standing up on the pad on its supporting heels.[60]

The official history of the Yuzhnoe Design Bureau notes that the system conducted preparation operations automatically, without the presence of a launch crew on the pad.[61] First, it took 2 min to automatically connect the pipes supplying high-pressure nitrogen and air, liquid and air systems for thermal control, and electric connections. The latter included more than 5000 individual contacts leading to the rocket and the spacecraft. Then, the system automatically attached propellant lines.

As the mechanisms raised the rocket and installed it on the launch pad, the ground equipment tested the launcher and the spacecraft. Finally, the control center uploaded launch commands and targeted the rocket. After that, pressurized gas expelled propellants from storage tanks and loaded them into the rocket, completing the automatic preparation process. The system now stood ready for launch.

The space interceptor vehicle, IS, had substantial capabilities for orbital maneuvering. The onboard liquid-propulsion engine with a gas-pressure propellant feed system allowed multiple restarts and provided spacecraft velocity changes (the delta-V capability) up to 3940 ft/s (1200 m/s). The initial mass of the IS satellite destroyer deployed in orbit was 5400 lb (2450 kg), including its fragmentation warhead.[62]

Tests of the antisatellite system continued after the first success of 1968. The military accepted the satellite destroyer for experimental deployment in 1972, and it became fully operational in 1978. The antisatellite system launched a space vehicle for the last time in 1982. From 1969 to 1982, the Tsyklon-2 space launchers placed into orbit 3 target spacecraft and 18 interceptor satellites as part of this program.[63] Following Soviet–American treaties, the USSR deactivated the operational system in 1983. Its capabilities remained, however. A leading U.S. specialist in and historian of national security space noted that after the formal program termination

> the Soviets continued to keep five antisatellites housed in Tyuratam should the need arise. Every time the Soviets went through the launch process for the SL-11 [Tsyklon-2] booster, they were exercising the antisatellite process flow. The Soviets could, with very little advance warning, roll out, erect and launch an antisatellite aboard the SL-11 and within an hour of launch, intercept and destroy a U.S. satellite.[64]

---

[60]Vlasko-Vlasov, 2003, p. 514.
[61]Konyukhov, 2009, pp. 168, 169.
[62]Efremov, 2005, p. 95.
[63]Konyukhov, 2009, p. 171.
[64]Temple, 2005, p. 526.

Tsyklon-2 emerged as a reliable space launcher in the Soviet arsenal. Kuchma wrote that it

> placed into orbit reconnaissance satellites, target satellites, and satellite destroyers. The military unit v/ch 46180 launched them from Baikonur. Once we conducted three launches during three days.... During a quarter of a century of its use (from August 1969), more than one hundred launches were conducted and none of them failed![65]

BALLISTIC MISSILE EARLY WARNING

Improved Dnestr radars of the RTI institute became a key element of the SPRN ballistic missile early warning system. In October 1976, the operational SPRN went on combat duty, exchanging information with the TsKKP space control center and the PRO missile defense system. In addition to two original sites at Saryshagan and near Irkutsk in Siberia, it included at that time two more above-the-horizon radars in Latvia and near Murmansk in the north. The system soon expanded with the activation of new radars near Sevastopol and Mukachevo in Ukraine and deployment of the new receiver near Murmansk. The military added a few more operational sites in the 1980s.

Around 1960, the NII-37 institute (future NIIDAR) began a major and expensive program to develop over-the-horizon radar systems for early warning of ballistic missile attack. Such radars relied on their pulses bouncing between the ionosphere and the ground, which dramatically increased radar reach. The first large radar prototype with its two transmitting and receiving antennas measuring 660 and 1000 ft (200 and 300 meters) in length and 360 and 460 ft (110 and 140 meters) in height, respectively, was erected near the town of Nikolaev (Mykolaiv) in Ukraine (Fig. 9.11). It began experiments in the early 1970s.

Then, the military built two over-the-horizon operational radars near Chernobyl' in Ukraine and near Komsomolsk-on-Amur in the Soviet Far East. The radars operated at frequencies of 5–28 MHz (wavelengths 11–60 m). Their ability to detect ballistic missile launches at large distances, especially in directions over polar regions, remained highly controversial and uncertain during the program's existence.[66] Vladislav Repin, who had directed development of national early warning systems, praised one-time NII-37 and Vympel director Vladimir Markov for his "enormous in its scale" and "positive" contributions to space and missile defenses, but noted that

> [p]erhaps his [Markov's] only major technical mistake was persistent lobbying for super long distance over-the-horizon radars for detecting

---

[65]Kuchma, 2003, p. 544, 545.
[66]Babakin, 2008.

[attacking ballistic] missiles. Here he remained "deaf" to any most persuasive criticism and sped up work on radars without sufficient scientific and experimental foundation as well as ignored negative results.[67]

Fig. 9.11 Receiving antenna of the prototype over-the-horizon 5N77 radar located 6 miles (10 km) east of the town of Nikolaev (Mykolaiv) in Ukraine. It began operations in 1972. White arrows point at a metal structure 1000 ft (300 m) long and 460 ft (140 m) high. The structure resembled a linear array of masts with 330 oscillators mounted on them. It casts a sharp shadow in the photograph. The total of 26 transmitters of 50 kW each sent synchronized signals from another radar antenna that was 660 ft (200 m) long and 360 ft (110 m) high, located 12 miles (20 km) to the south of the receiver. Original satellite reconnaissance photograph by KH-4 camera (Mission 1114-2; 6 April 1971) available from the U.S. Geological Survey; photograph identification, interpretation, and processing by Mike Gruntman.

---

[67]Repin, 2003, p. 444.

Another complementary element of the early warning SPRN system included deployment of satellite-based optical and infrared sensors for detection of ballistic missile launches during their powered ascent.

Soviet Princelings

The importance of missile defense rapidly grew in the Soviet establishment. In a clear sign of its prominence, the field attracted a number of children of top Communist Party and government officials, the Soviet princelings, to make their careers. Their life paths did not differ much from the rise of the young Sergo Beria as chief designer at SB-1 (later KB-1) in 1947. Professional qualification and merit were not deciding factors in their promotion to leadership positions. As Colonel Drozdov explained,

> since the work in the defense industry was prestigious and well paid, relatives of highly placed officials, without adequate qualifications, took over [many] leading positions in the military-industrial complex instead of a [previous] generation of talented specialists of the postwar [World War II] era. Bootlickers, sycophants, and political operators also successfully advanced [their careers]. A "leader-organizer" became typical for [heading] scientific-research and design organizations.[68]

An old joke in the former Soviet Union was that the answer to the question of whether a son of a colonel could also become a colonel was affirmative. But could he become a general? The answer to that was negative because generals had their own sons.

Nikolai Ustinov,[69] the son of a very powerful defense industry official, Dmitrii Ustinov, appeared in Kisun'ko's design bureau (the future OKB-30 and NIIRP) in the late 1950s. After a short trip to Saryshagan "to familiarize with the concepts of instrumentation and the current status [of the test site]," he began work on emerging laser and semiconductor technologies. Eventually, the government created a special independent organization for the young Ustinov as director, the Central Design Bureau Luch, or TsKB Luch. (The word *luch* means a ray, as in a ray of the sun or a laser ray.) The design bureau later evolved into the Scientific-Production Association Astrofizika, or NPO Astrofizika.[70] (Astrofizika means astrophysics.) Nikolai Ustinov became a corresponding member of the prestigious USSR Academy of Sciences in 1981. NPO Astrofizika played a leading role in the development of Soviet laser weapons, with many tests conducted at Site 38 at Saryshagan.

Yuri Kamensky, who had been working with Kisun'ko since the mid-1950s, recalled that young

---
[68]Drozdov, 1998, p. 37.
[69]Nikolai Dmitrievich Ustinov, 1931–1992.
[70]Khvatov, 2003, p. 592.

Kolya[71] [Ustinov] was a great guy and selected for his [new] organization a picturesque place on the bank of the Moscow river. It was convenient for him for water skiing. After the death of his powerful father [Politburo Member and Minister of Defense D.F. Ustinov], the government sent the young Ustinov to direct the Institute of History of Natural Sciences and Technology [of the USSR Academy of Sciences].[72]

Many viewed this directorship as a sinecure appointment with good pay and few responsibilities.

A leading Soviet specialist in missile defense phased-array antennas, Dmitrii B. Zimin,[73] wryly observed that "an administrative structure of any Soviet organization was never determined by considerations of economic purpose and from the modern [post-Soviet] point of view was inexplicable."[74] Zimin is a most credible commentator—he became a highly successful businessman after dissolution of the Soviet Union and was president of the first Russian publicly traded company listed on the New York Stock Exchange, VympelKom.

Another Soviet princeling, Revolii M. Suslov, was the son of a Politburo member and the feared ideological chief and watchdog of the Communist Party, Mikhail A. Suslov.[75] The young Suslov held the rank of captain when he joined Kisun'ko's NIIRP. Kamensky saw the importance of such appointments for the field, "Revolii worked somewhere in the state security, dealing with radio technology. His appearance [in our institute] meant that the very top leadership [of the country] considered missile defense highly promising."[76]

The government also quickly created a new institute for the young Suslov. Kamensky noted that

> in personal contacts he [Revolii Suslov] did not produce an impression of a very bright person. However, smart guys were selected [to support him] as deputies who knew "how to put a harness on a horse." [The government] built for his institute a good building with winter gardens and a swimming pool on Prospekt Mira [Peace Avenue in a prestigious and desirable part of Moscow]. He has been director there for many years.[77]

A mission defense veteran, Khvatov, noted that the young Suslov had the rank of general and that Yu. L. Smirnova, the daughter of the powerful chairman of the Military-Industrial Commission (VPK), Leonid V. Smirnov, also worked for a time at NIIRP. Khvatov wrote that he "had an opportunity to

---

[71]Kolya is a diminutive of the full first name Nikolai.
[72]Kamensky, 2002.
[73]Dmitrii Borisovich Zimin, b. 1933.
[74]Zimin, 2003, p. 637.
[75]Revolii Mikhailovich Suslov, b. 1929; Mikhail Andreevich Suslov, 1902–1982.
[76]Kamensky, 2002.
[77]Kamensky, 2002.

closely work with N.D. Ustinov and R.M. Suslov and they produced a good impression on me by their managerial and human qualities."[78]

Employing the relatives of powerful leaders benefitted chief designers. Kisun'ko and the man overseeing his design bureau, Minister of Radio Industry Kalmykov, had tense relations. However, the minister could not "touch" Kisun'ko for a long time, being "afraid of the reaction [to firing him]"[79] by Secretaries of the Central Committee of the CPSU Suslov and Ustinov and VPK Chairman Smirnov because their children worked in Kisun'ko's organization. By the time Kisun'ko was fired in 1975, the units led by the new crop of Soviet nobility, Nikolai D. Ustinov and Revolii M. Suslov, had separated from his organization and formed independent TsKB Luch (NPO Astrofizika) and All-Union Scientific-Research Institute of Radioelectronic Systems, respectively.

The story of the son of Soviet leader Nikita Khrushchev particularly illustrates the consequential nature of employing a princeling. In 1958, the graduating young engineer Sergei N. Khrushchev showed interest in joining the missile defense program directed by Kisun'ko. Not knowing about it, Kisun'ko sent one of his deputies, Elizarenkov, to the overseeing ministry to select new specialists for his design bureau. Elizarenkov did not mark the name of Sergei Khrushchev in the list of available graduates that was provided to him. The deputy minister responsible for personnel asked him his reasons for not selecting Khrushchev's son. Then Elizarenkov, as described by Kisun'ko,[80] responded that "we had already had once one Sergei [Sergo Beria]. He was there and now gone."

Kisun'ko fumed later after discovering what had happened,

> Only many years later did I learn about this [unauthorized] comment [by Elizarenkov] and became infuriated.
>
> Are you an idiot or a provocateur?—I asked Elizarenkov.—Don't you understand that your stunt became known to Nikita Sergeevich [Khrushchev] as originating not from you personally but from me, the Chief Designer? Don't you understand that the suggestion [of hiring Sergei Khrushchev] could have not been done without approval of the senior Khrushchev? Finally, don't you understand that all turbulence suffered by our organization [including ferocious administrative fights with competing Chief Designers] are the direct consequence of your idiotic stunt?[81]

Subsequently in March 1958, 23-year-old Sergei Khrushchev joined the design bureau of Vladimir N. Chelomei. After only one year of work, Sergei

---

[78]Khvatov, 2003, p. 592
[79]Pervov, 2003, p. 258
[80]Kisun'ko, 1996, p. 384
[81]Kisun'ko, 1996, p. 384.

received, together with a few of Chelomei's leading specialists, the country's top science and monetary award, the Lenin Prize. Then, four years later, he became a Hero of Socialist Labor.[82] At the same time, Commander of the Tyuratam Missile Range (1961–1965), General Alexander G. Zakharov,[83] noted the personal modesty of Sergei Khrushchev.[84]

Employing the son of Nikita Khrushchev gave Chelomei, a newcomer to ballistic missiles, missile defense, and space systems and vehicles, an important political advantage.[85] The official history of Chelomei's design bureau noted "a special role" played by the young Khrushchev in its growth:

> Being a patriot of the organization and rooting for our common cause, he [Sergei N. Khrushchev] tried to prepare [his father, Soviet leader] N.S. Khrushchev to understand the new ideas and proposals which were presented by the General Designer [Chelomei] to the government. [Such preparation] often helped in gaining approval of our projects by the leadership of the country as well as later to implement them into new . . . [rockets and spacecraft]. In the environment of tough competition among leading defense KBs [design bureaus] and negative attitude toward V.N. Chelomei by the leadership of the defense-industrial complex, this played a major role.[86]

Leonid Khvatov, a collaborator of Kisun'ko's since 1955, thought that the fact that Sergei Khrushchev had ended up in the design bureau of Chelomei

> played not the last role in pushing through the [national missile defense] Taran project [of Chelomei], creation of a formidable competitor to the missile defense work of Kisun'ko, and in forming negative assessments of the results of the latter. I cannot restrain myself from asking a rhetorical question that remains unanswered—what would have been the path of development of missile defense if . . . [Sergei] Khrushchev had joined SKB-30 [of Kisun'ko in 1958].[87]

Other missile defense specialists also pointed to the delays in development and deployment of the A-35 caused by the aggressive promotion of Taran by Chelomei, which was facilitated to a "not-small degree" by employing Sergei Khrushchev. They also noted that these circumstances helped Chelomei "to easily win the struggle [against]" and "push back the [share of strategic ballistic missile] production by the old [established] rocket organizations of Korolev and Yangel."[88]

---

[82]Efrémov et al., 2009, p. 34, 56.
[83]Alexander Grigorievich Zakharov, 1921–2010.
[84]Zakharov, 1996, p. 97.
[85]Gruntman, 2004, p. 208.
[86]Efremov et al., 2009, p. 52.
[87]Khvatov, 2003, pp. 592, 593.
[88]Golubev et al., 1994, p. 55.

The demise of Nikita Khrushchev in 1964 led to the loss of some important programs by Chelomei and reduction of his role in the others. The times had changed in the Soviet Union, so the new authorities did not banish Sergei Khrushchev from Moscow, unlike what had happened to Sergo Beria. The young Khrushchev had to change his place of work, however, and built up his technical and managerial career in a different research institute. In the 1990s he moved to the United States and became a naturalized U.S. citizen.

WEAPONS IN SPACE

In the 1970s, the Soviet missile defense and space establishments initiated a major effort in space-based weapons. It culminated in an attempt to orbit a partially complete prototype of a laser space battle station, Polyus (Skif-DM). (Polyus means a pole, as in the north pole.) The gigantic Polyus measured 120 ft (37 m) in length and 13 ft (4 m) in diameter with a mass, including its propulsion system, of about 80 t.[89]

The government had embarked on a new program that eventually led to Polyus in 1976 by

> a special decree of the Central Committee of the CPSU and the USSR Council of Ministers "On Feasibility Study of Development of Weapons Operating in Space and from Space." This decree initiated in the USSR work of the type that would be supported by the American president only seven years later. This development began in spite of the [limitations imposed by] the [US–USSR] missile defense [ABM] treaty, signed in Moscow on May 26, 1972. According to the treaty, its participants, including the Soviet Union, agreed "not to develop, test, and deploy sea-based, air-based, space-based, or mobile ground-based missile defense systems and their components." Many of those same individuals, who had prepared and signed this treaty [in 1972], only four years later prepared and arranged signing of the [new] government decree [in 1976] fully contradicting [and violating] the USSR obligations from [the earlier] 1972 [treaty].[90]

The first flight test of the largest Soviet space launcher Energia took place on 15 May 1987. Instead of the originally planned dummy payload, the launcher carried a prototype of the laser space battle station into orbit (Fig. 9.12). The Energia booster performed nominally and Polyus separated from it. The latter then fired its own propulsion system to provide the final velocity increment of 200 ft/s (60 m/s) to reach the desired low-Earth orbit. The latter maneuver required burning 1.5–2.0 t of the propellant.

Polyus separated from Energia with its thrusters pointing in the direction of the velocity vector. The space vehicle thus first had to turn and change its pitch

---

[89]Kornilov, 1992; Gubanov, 1998, Chapter 36; Lantratov, 2005; Lukashevich, n.d.
[90]Lantratov, 2005.

Fig. 9.12 Prototype laser space battle station Polyus (black-color body) with 80 t wet mass and attached to the side of the largest Soviet space launcher, Energia, being installed at the launch site at Tyuratam (Baikonur) in May 1987. Photo from Gubanov, 1998, Chapter 35; http://www.buran.ru/htm/gubanov3.htm; photo URL: http://www.buran.ru/images/jpg/pole2.jpg; accessed 27 March 2011.

angle by 180 deg and the roll angle by 90 deg, and then start its engines for the final push into orbit. After the required turn of Polyus, its engines began firing but the vehicle rotation did not stop, as planned, at that time and continued.[91] Consequently, Polyus failed to reach orbit because of the error during the maneuver. The next—and it would become the last—launch of the Energia booster successfully deployed a Soviet space shuttle, Buran, into orbit without a crew on 15 November 1988. After two revolutions around the Earth, Buran automatically landed at Tyuratam as a plane.

Space-Based Weapons and NPO Energia

In the late 1960s and early 1970s, the USA began work on feasibility studies of conducting military operations in space and from space. A series of USSR government decrees (first in 1976) assigned the

---

[91] Kornilov, 1992; Semenov, 1996, pp. 369, 370; Gubanov, 1998, Chapter 36.

responsibility for work [in the Soviet Union] in this area to a team of organizations led by the Scientific-Production Association (NPO) Energia. (Note: NPO Energia, established by Sergei Korolev, is located in the town of Korolev, also known as Podlipki and Kaliningrad, near Moscow; e.g., Gruntman, 2004, pp. 279–282.)

In the 1970s–1980s, system studies were conducted to determine possible ways of development of space systems capable of solving the problems of destroying military spacecraft, ballistic missiles in flight, as well as especially important air, sea, and ground targets. . . . Two combat spacecraft, based on a common design but utilizing different types of weapons, have been developed for destroying military space vehicles. One system was based on laser weapons and the other on missiles. . . .

It was planned to deploy these systems in orbit from a cargo bay of the reusable space vehicle Buran (and the Proton space launcher during tests). Orbit refueling was also considered with the help of systems deployed by the reusable vehicle Buran. There was a possibility of visiting these space systems by two-man crews for up to seven days to assure long-term deployment in orbit and maintaining a high degree of combat readiness.

The weapon system based on missiles had a smaller mass than the one based on laser weapons and thus allowed carrying more propellant. Consequently, it was desirable to deploy a constellation of space vehicles, some armed with missiles and others with lasers. The former type of vehicles would target low-earth orbit spacecraft while the latter deal with space assets in middle altitude and geostationary orbits. . . .

NPO Energia developed a design of a space-based interceptor missile for destruction of ballistic missiles during the launch and their warheads in flight. It was the smallest but most energetic rocket developed by NPO Energia. Suffice to say that for the initial mass of a few dozen kilograms, the interceptor missile had a velocity increment capability comparable to that of rockets deploying spacecraft into orbit. . . .

In the beginning of the 1990s, development of combat space systems ceased in NPO Energia due to changes in the military-political environment.

All departments of the leading design bureau [in NPO Energia] and broad cooperation of specialized development organizations of the national military-industrial complex as well as leading research establishments of the Ministry of Defense and Academy of Sciences were engaged in this work.

(Semenov, 1996, pp. 419, 420)

Soviet leader Mikhail S. Gorbachev[92] visited the launch base at Tyuratam, the cosmodrome Baikonur, in May 1987. After some hesitation, he authorized the launch of the space battle station Polyus. Chief designer of the Energia launch vehicle, Boris I. Gubanov,[93] recalled that he had unsuccessfully urged General Secretary Gorbachev to attend the launch. A top Communist Party official, Lev N. Zaikov,[94] who accompanied Gorbachev, got irritated and explained to Gubanov,

> Can't you understand? If Mikhail Sergeevich [Gorbachev] attends the launch and an accident occurs, then all the world would talk that even the General Secretary [Gorbachev] could not help. If everything is nominal [and the prototype laser space battle station Polyus successfully deployed], then they would say that the General Secretary steps up the arms race.[95]

If successful, the deployment of this gigantic space weapon system prototype could have indeed profoundly changed the dynamics of the arms race and relations between the superpowers.

Nikolai V. Mikhailov led TsNPO Vympel and its successor organization from 1987 to 1996 and then served as Deputy Minister of Defense from 1997 to 2001. He noted the role that the U.S. Strategic Defense Initiative and the Soviet effort in space weapons played in historical developments,

> The name of U.S. President Reagan is linked with "the beginning of the end" of our country, the USSR. Perhaps, there is a reason for this, which has many aspects. One of them is the [U.S.] Strategic Defense Initiative and the aspiration of the Soviet political leadership to counteract it by our "asymmetric response."
>
> From today's point of view, the efforts of then political leadership which permitted to draw itself into senseless competition with the USA look especially dubious. The [two] leading ministries of Radio Industry [responsible for missile defense] and General Machine Building [MOM, responsible for ballistic missiles and space] prepared and went through all stages of [Communist Party and government] approval for two exceptionally ambitious programs which, in essence, were not "asymmetric" but, to the contrary, just repeated in most areas main features of the American programs. Inertia of thinking then did not leave any hope for our political leaders to critically reevaluate the situation. "The process has started," as people used to say in those days. With the backdrop of empty shelves in the stores at that time, this seemed to become the weighty [last] drop that overfilled the sea of otherwise accumulated problems in the society.[96]

---

[92]Mikhail Sergeevich Gorbachev, b. 1931.
[93]Boris Ivanovich Gubanov, 1930–1999.
[94]Lev Nikolaevich Zaikov, 1923–2002.
[95]Gubanov, 1998, Chapter 34.
[96]Mikhailov, 2005, pp. 44–45.

The Soviet development effort and attempted deployment in orbit of a prototype of the laser space battle station remains rarely mentioned in the United States, even in specialized publications. This selectivity reflects the ideologically charged partisan nature of policy debates on arms race and missile defense.

Post-USSR Era

Dismantling of the Soviet Union in the early 1990s led to major turbulence and changes in the military-industrial complex of the country. Nevertheless, Russia deployed the new operational missile defense system of Moscow, the A-135, in those uncertain times.

Reorganization and consolidation of the industry first resulted in the formation of two major organizations in air and missile defenses, associations Antei[97] and Almaz, led by NIEMI (former NII-20) and the successor of KB-1, respectively. Then, in 2010, they combined into one government-controlled association Almaz-Antei that placed the most important former leading research institutions, design bureaus, and production plants engaged in air and missile defenses under one corporate roof.

After the breakup of the Soviet Union, the Saryshagan test range ended up in the independent Kazakhstan. Russia leased parts of the proving ground for weapons development while other abandoned installations decayed (Figs. 9.13 and 9.14). The number of personnel serving at GNIIP-10 and its successor organizations had dropped precipitously by a factor of 10 by 2001 (see Table 9.1).[98]

Russia's Ministry of Defense transferred the Saryshagan range to the Strategic Rocket Forces on 11 April 1998.[99] Then on 5 November of the same year, the government formally put the test site under control of the State Central Interservice Test Range with headquarters at Kapustin Yar,[100] the latter remaining on territory of the Russian Federation. The Russian military personnel from a smaller tactical air defense test site in Emba, Kazakhstan, transferred to Kapustin Yar in 1999.

Grigorii V. Kisun'ko died on 11 October 1998. The pioneer builder of the Saryshagan range, retired General Alexander Gubenko, buried his old friend. Even in death and with the Soviet Union no more, old bitterness, tensions, and frictions in the defense officialdom were alive and on display. Gubenko wrote that

> the fate had its way that I was the one to bury him [Kisun'ko when he had passed away], to send him on the last journey from where there was no return. . . .

---

[97]Davydov, 2009, p. 442.
[98]Belous et al., 2009, pp. 220, 221.
[99]Kulakov, 2006, p. 180.
[100]Kavel'kina et al., 2006.

Fig. 9.13 Remnants of the precise tracking and guidance radar RTN of System A at Saryshagan's Site 2 in 2011. Photograph by user badger-16 (Google's Panoramio); http://www.panoramio.com/photo/56705205; accessed 17 October 2011.

Fig. 9.14 Deserted garrison clubhouse with the obligatory statue of Vladimir I. Lenin in front at Saryshagan's Site 35, which had specialized in testing of air defense missiles. Photograph (2006) by user anleon (Google's Panoramio); http://www.panoramio.com/photo/4918089; accessed 16 October 2011.

Table 9.1 Personnel at GNIIP-10

Year	Number of People
1983	15,288
1992	10,848
1995	5659
1998	2491
2001	1551

Now I cannot express [enough] my indignation at how present authorities handled his untimely death. Did not this outstanding son of his country deserve the funeral at least at the [prestigious] Novodevich'e cemetery,[101] where his comrade-in-arms, the first commander of the [Saryshagan] test range, General S.D. Dorokhov, had been buried? Could not then [First] Deputy Minister of Defense N.V. Mikhailov intervene into this issue and help? Did not Grigorii Vasil'evich [Kisun'ko] deserve a decent tombstone at his burial place, erected by the [Russian] State [as frequently done for individuals with major national accomplishments], rather than we together with Rear-Admiral M.A. Arkharov collected kopecks [from missile defense veterans and friends] for a very modest tombstone?

It is a point of honor for the [Russian] State to correct this error and give the General Designer [of missile defense] his due![102]

More than 50 years ago, the pioneers blazed the trail in search of technical means for defense against deadly ballistic missiles. As life goes on and new threats emerge, the eternal competition between the sword and the shield continues.

---

[101] Novodevich'e cemetery is the most prestigious elite cemetery in Moscow, where burial requires approval of high-level authorities.
[102] Gubenko, 2003, p. 669.

Appendix A

FIRST U.S. MISSILE INTERCEPTS

The prospect of a long-range ballistic missile carrying an atomic bomb, the first of which was detonated in July 1945, had alarmed leading U.S. officers and scientists, who began considering approaches to counteract this new dangerous threat.

As early as fall 1945, physicist Edward Teller outlined his thoughts about missile defense in a report for the U.S. Navy. Then, in December 1945, the Air Force's Special Advisory Group, led by Theodore von Kármán,[1] issued its highly influential report "Toward New Horizons,"[2] which also discussed the missile threat. In 1946, the War Department Equipment Board produced the "Stilwell report," named after Chairman of the Board General Joseph W. Stilwell.[3] The document considered defenses against ballistic missiles armed with atomic bombs.

After World War II, the Army, the Navy, and the Air Force (which in 1947 had been reorganized into an equal and independent service) competed in development of guided missiles, including for air defense. Performance characteristics of advanced antiaircraft missiles pointed to a logical extension of their capabilities to interception of emerging ballistic missiles. Subsequently, all three services established partially overlapping exploratory programs and tried to add missile defense to their roles and missions.

The U.S. Army prevailed in this bitter rivalry. In 1958, Secretary of Defense Neil McElroy[4] assigned to the Army the primary responsibility for ballistic missile defense, which it maintains to this day.[5] The growth of the Army's air defenses led to the first missile intercepts and then to theater and strategic missile defense systems.

---

[1] Theodore von Kármán, 1881–1963.
[2] Air Force System Command, 1992.
[3] Joseph W. Stilwell, 1883–1946.
[4] Neil H. McElroy, 1904–1972.
[5] *Investigation . . .* , 1958, pp. 4196–4197.

In 1949 the Army initiated Project Plato to develop theater defenses to protect deployed units against short-range ballistic missiles. The Field Army Ballistic Missile Defense System (FABMDS) succeeded Plato in 1959, only to be cancelled three years later.[6] These early development efforts contributed to today's antimissile capabilities of the Patriot (Phased Array Tracking Radar Intercept on Target).

Another Army program contracted with Bell Telephone Laboratories in February 1945 to develop an antiaircraft guided missile system. It would become known as the Nike.[7] The first antiaircraft units armed with new guided missiles reached operational status in 1954. The Nike-Ajax missiles, as they were called since 1956, could engage targets at a 25-mile (40-km) range and up to a 12-mile (19-km) altitude. The Army air defense completed transition to guided missiles from antiaircraft guns in 1958. In the same year, a new antiaircraft missile of this family, the more maneuverable and powerful Nike-Hercules, began replacing the Nike-Ajax.[8] Further advancement of technology soon led to the Nike-Zeus missile and simulated intercepts of warheads of intercontinental ballistic missiles (ICBM) in 1962.

The first missile intercepts by the U.S. Army took place at the White Sands Missile Range (WSMR) in New Mexico. The range is the largest land-area military reservation in the United States. The government established it in 1945, and WSMR plays a major role in the development of various weapons to this day.[9]

In the first successful intercept, a Hawk (Homing-All-the-Way Killer) antiaircraft missile destroyed an unguided rocket, Honest John (Fig. A.1),

Fig. A.1 Short-range unguided rocket Honest John on a launch platform (left) and during launch (right) at a test site. The Douglas Aircraft Corporation built these rockets, which could deliver 1500-lb (680-kg) nuclear warheads. Photos courtesy of U.S. Army Aviation and Missile Command.

---

[6]Bullard, 1963.
[7]Gruntman, 2004, p. 202.
[8]Cagle, 1959, 1973.
[9]Gruntman, 2004, p. 198.

Fig. A.2 Launch of a Hawk antiaircraft missile at the White Sands Missile Range. The highly mobile Hawk targeted aircraft at low and medium altitudes and complemented capabilities of the high-altitude Nike-Ajax and later Nike-Hercules. Frames from the U.S. Army's motion picture *The Big Picture. Army Digest Number 5*. Original film courtesy of National Archives and Records Administration; frame identification, interpretation, and processing by Mike Gruntman.

demonstrating the possibilities for theater missile defense. In the 1950s, Raytheon developed the Hawk missile (Fig. A.2) for low- and medium-altitude air defense, complementary to the higher-altitude Nike. The Hawk air defense units became operational in 1959.

After the Hawk success, modified Nike-Hercules missiles scored three intercepts, first destroying two tactical Corporal ballistic missiles, followed by an intercept of another Nike-Hercules that was serving as a target. The Nike-Hercules intercepts contributed to technology advancement that ultimately led to strategic missile defense against ICBMs.

All of these first four intercepts had taken place at White Sands in 1960, as shown in Table A.1, before the Soviet Union intercepted the R-12 (SS-4) IRBM at the Saryshagan range on 4 March 1961.

Table A.1 First U.S. Army Intercepts in 1960

Date	Target	Interceptor
25 January 1960	Honest John	Hawk
3 June 1960	Corporal	Nike-Hercules
15 August 1960	Corporal	Nike-Hercules
14 September 1960	Nike-Hercules	Nike-Hercules

The unguided rocket Honest John played the role of a target in the first successful missile intercept by the Hawk on 25 January 1960 (Fig. A.3). The Army described that on that day "the interception had occurred within fifteen seconds four miles from the Hawk launching site and ten miles from the Honest John launching site."[10]

Then, the modified Nike-Hercules surface-to-air missiles (Fig. A.4) twice intercepted and destroyed Corporal ballistic missiles (Fig. A.4) on 3 June 1960, and again on 15 August of the same year (Fig. A.5). The supersonic vertically launched liquid-propellant Corporal reached distances of 75 miles (120 km).[11] In 1950, it became the first U.S. tactical missile approved as an atomic weapon carrier. The intercept on 3 June demonstrated the first-ever ballistic missile kill by another missile, if one disregards the intercept of an unguided rocket, Honest John.

The highly maneuverable and high-velocity Nike-Hercules proved to be a convenient and useful target for missile defense tests. In the last intercept of 1960 at the White Sands Missile Range, a Nike-Hercules missile intercepted and destroyed another Nike-Hercules, serving as a target.

Almost one year had passed before the new and more powerful Nike-Zeus missile (Fig. A.6) became available for interception tests. The Army began Nike-Zeus trials with the first launch in August 1959. One year earlier, in 1958, the Atomic Energy Commission had initiated development of a special warhead, W-50, with a high-kiloton yield for this missile.[12]

The test program of the new interceptor at White Sands involved "Zeus missile firings and installation of major components of the Zeus system—large spherical acquisition radar, the associated missile and track radars, and the ground guidance computer—to prove in the designs."[13] The system introduced a number of important innovations; for example, the acquisition radar was "the first track-while-scan radar system to successfully cover the

---

[10]Raymond, 1960.
[11]Gruntman, 2004, pp. 202–204.
[12]Hansen, 1988, pp. 107, 109.
[13]Bell Laboratories, 1975, p. I-21.

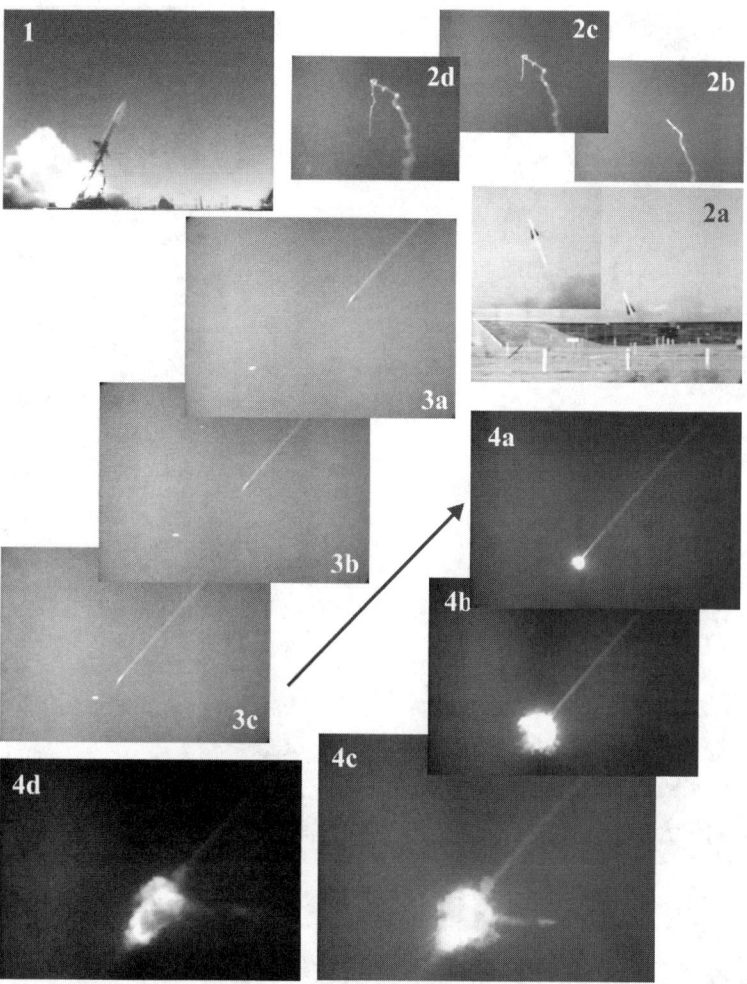

Fig. A.3 A Hawk missile successfully intercepts the unguided rocket Honest John on 25 January 1960 at White Sands Missile Range: (1) launch of Honest John; (2a,b,c,d) launch and contrails of Hawk; (3a,b,c) final approach to the target; (4a,b) intercept; (4c,d) fragments of Honest John continue on its initial trajectory. Frames from the U.S. Army's motion picture *The Big Picture. Tularosa Frontier*; original film courtesy of National Archives and Records Administration and White Sands Missile Range Archives. Frames identification, interpretation, and processing by Mike Gruntman.

entire hemisphere surrounding a radar position, detect the objects in that space, remember their past positions, and predict where the objects would next be in three dimensions—all automatically, beginning with initial detection."[14]

---

[14]Bell Laboratories, 1975, p. I-24.

Fig. A.4 Ballistic missile Corporal (left) and antiaircraft missile Nike-Hercules (center) on display at the U.S. Space and Rocket Center in Huntsville, Alabama, in 2003 (photos courtesy of Mike Gruntman). The Nike-Hercules (right) rapidly rises from a launch position (photo courtesy of U.S. Army Aviation and Missile Command).

Building these new antiballistic missile capabilities involved leading U.S. companies. As Secretary of the Army Wilber N. Brucker[15] described in testimony to Congress in 1958,

> Under the supervision of the Army's Redstone Arsenal, the Western Electric Co. acts as a single manager for the Nike-Zeus program. The Bell Telephone Laboratories is the research and development agency of the

---

[15]Wilber M. Brucker, 1894–1968.

Western Electric Co., and the Douglas Aircraft Co. is the subcontractor for the missile. The effort is backed by some 30 industrial and research establishments such as Radio Corp. of America, Goodyear Aircraft Co., Sperry Gyroscope Co., the Stanford Research Institute, Cornell Aeronautical Laboratories, and others. Together they constitute the best qualified team in the free world for surface-to-air missile development.[16]

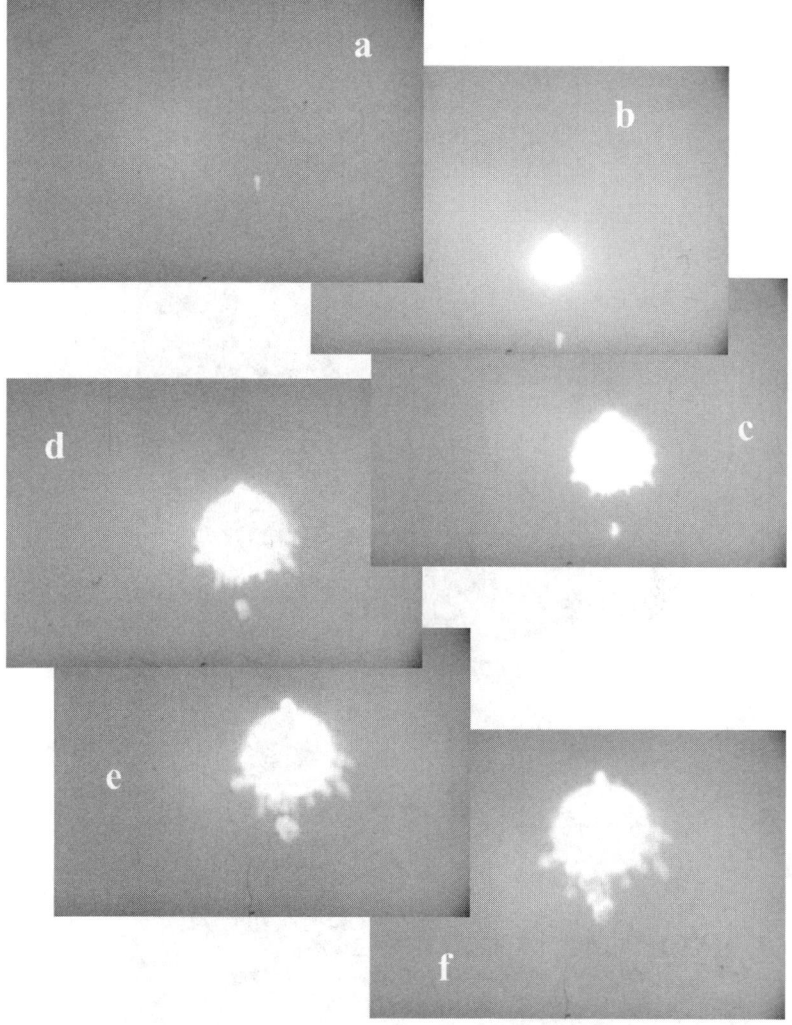

Fig. A.5 Sequence of frames (a-b-c-d-e-f) of the Corporal ballistic missile intercept by the Nike-Hercules over White Sands Missile Range. Still frames from the U.S. Army's motion picture *The Big Picture. Army Digest Number 5*. Original film from the National Archives and Records Administration; frame identification, interpretation, and processing by Mike Gruntman.

[16]*Investigation . . .* , 1958, pp. 4216.

Fig. A.6 Nike-Zeus missile on display at the U.S. Space and Rocket Center in Huntsville, Alabama. Photo (2003) courtesy of Mike Gruntman.

During the time of Nike-Zeus final development, the Soviet System A succeeded in intermediate-range ballistic missile (IRBM) warhead intercepts at the Saryshagan range, the first time on 4 March 1961. In December 1961 and March 1962, the new Nike-Zeus successfully intercepted Nike-Hercules target missiles in trials at White Sands. Then, the U.S. tests advanced to the next phase of targeting ICBM warheads. The Nike-Zeus interceptor would now rise from a base at Kwajalein in the Pacific Ocean to intercept approaching Atlas ICBMs launched from the Vandenberg Air Force Base in California.

The U.S. Army determined that nuclear warheads on interceptors represented the only practical and reliable way of destroying hardened nuclear warheads of strategic ballistic missiles. The Soviet Union also adopted this approach in the late 1950s for the operational missile defense system of Moscow.

Edward Teller explained later that

> [d]efense is easier and more reliable if the antimissile missile is armed with a small nuclear explosive. A well-designed explosive, equivalent to no more than 10 tons of TNT, could destroy the incoming missile from a distance of a few hundred feet, particularly if the explosive is designed to aim its energy in the proper direction. If used within the atmosphere, such explosives must be limited in power, not only to avoid damage on the ground but also so as to cause minimal confusion in the defensive system. If the offensive missile is stopped at least a mile above the earth's surface, the defended population could probably sleep through an effective defense, even if small nuclear explosives are used.[17]

First intercept attempts of Atlas ICBMs on 26 June and 19 July 1962 met with partial success. Then on 12 December 1962, the Nike-Zeus successfully intercepted an ICBM warhead, passing within the 200 m required for destruction of an incoming warhead by a nuclear explosion.[18] Ten days later, another Nike-Zeus missile again successfully intercepted an Atlas warhead. Nine more tests followed in 1963, with Nike-Zeus missiles intercepting Atlas and Titan I ICBMs.[19]

---

[17]Teller, 1987, p. 28.
[18]Walker et al., 1995, p. 19; Walker et al., 2005, p. 46.
[19]Bell Laboratories, 1975, p. I-26.

Appendix B

ACRONYMS AND ABBREVIATIONS

ABM	—	antiballistic missile
AIAA	—	American Institute of Aeronautics and Astronautics
CC	—	Central Committee
CEP	—	circular error probable
CPSU	—	Communist Party of the Soviet Union
CTBT	—	Comprehensive Test Ban Treaty
CW	—	continuous wave
ELINT	—	electronic intelligence
GNIIP	—	Gosudarstvennyi Nauchno-Issledovatel'skii Ispytatel'nyi Poligon (State Scientific-Research Test Range)
GRAB	—	Galactic Radiation Background (satellite)
HE	—	high explosive
HOE	—	Homing Overlay Experiment
ICBM	—	intercontinental ballistic missile
IRBM	—	intermediate-range ballistic missile
KB	—	Konstruktorskoe Byuro (Design Bureau)
KKV	—	kinetic kill vehicle
KS	—	krylatyi snaryad (winged projectile)
MOM	—	Ministerstvo Obshchego Mashinostroeniya (Ministry of General Machine Building)
NIE	—	National Intelligence Estimate
NIEMI	—	Nauchno-Issledovatel'skii Elektromekhanicheskii Institut (Scientific-Research Electromechanical Institute)
NII	—	Nauchno-Issledovatel'skii Institut (Scientific-Research Institute)
NIIDAR	—	Nauchno-Issledovatel'skii Institut Dal'nei Radiosvyazi (Scientific-Research Institute of Long-Range Radio Communications)

NIIRP	—	Nauchno-Issledovatel'skii Institut Radiopriborostroeniya (Scientific-Research Institute of Radio Instrument Building)
NIRTI	—	Nauchno-Issledovatel'skii Radiotekhnicheskii Institut (Scientific-Research Radiotechnical Institute)
NPIC	—	National Photographic Interpretation Center
NPO	—	Nauchno-Proizvodstvennoe Ob"edinenie (Scientific-Production Association)
OBP	—	obshchaya boevaya programa (general combat code)
OKB	—	Osoboe Konstruktorskoe Byuro (Special Design Bureau)
OM	—	obshchego mashinostroeniya (general machine building)
OS	—	[uzel] obnaruzheniya sputnikov ([node of] finding [discovery of] satellites)
PKO	—	Protivokosmicheskaya Oborona (Space Defense)
PRC	—	People's Republic of China
PRO	—	Protivoraketnaya Oborona (Missile Defense)
RALAN	—	Radiotekhnicheskaya Laboratoriya Akademii Nauk (Radiotechnical Laboratory of the Academy of Sciences)
RCS	—	radar cross-section
RE	—	radiolokator eksperimental'nyi (experimental radar)
ROC	—	Republic of China (Taiwan)
RSVPR	—	radiolokatsionnaya stantsiya vizirovaniya protivorakety (anti-missile missile guidance radar)
RTI	—	Radiotekhnicheskii Institut (Radiotechnical Institute)
RTN	—	radiolokator tochnogo navedeniya (precise guidance radar)
SAG	—	Science Advisory Group
SAM	—	surface-to-air missile
SB	—	Spetsial'noe Byuro (Special Bureau)
SDI	—	Strategic Defense Initiative
SIE	—	Special Intelligence Estimate
SKB	—	Spetsial'noe Konstruktorskoe Byuro (Special Design Bureau)
SKKP	—	Sistema Kontrolya Kosmicheskogo Prostranstva (System for Control of Space)
SPK	—	stantsiya peredachi komand (station for command transmission)
SPRN	—	Sistema Preduprezhdeniya o Raketnom Napadenii ([Early] Warning System of [Ballistic] Missile Attack)
SSA	—	space situational awareness
SSSR	—	Soyuz Sovetskikh Sotsialisticheskikh Respublik (The Union of Soviet Socialist Republics, USSR)
TGU	—	Tret'e Glavnoe Upravlenie (Third Chief Directorate)
TNT	—	trinitrotoluene
TsKB	—	Tsentral'noe Konstruktorskoe Byuro (Central Design Bureau)
TsKKP	—	Tsentr Kontrolya Kosmicheskogo Prostranstva (Space Control Center)

TsNPO	—	Tsentral'noe Nauchno-Proizvodstvennoe Ob"edinenie (Central Scientific-Production Association)
TsNII	—	Tsentral'nyi Nauchno-Issledovatel'skii Institut (Central Scientific-Research Institute)
TsNIIMash	—	Tsentral'nyi Nauchno-Issledovatel'skii Institut Mashinostroeniya (Central Scientific-Research Institute of Machine Building)
TsSO-P	—	Tsentral'naya Stantsiya Obnaruzheniya—Predvaritel'naya (Central Finding [Radar] Station—Preliminary)
TsUP	—	Tsentr Upravleniya Poletami (Flight Control Center)
UIR	—	Upravlenie Inzhenernykh Rabot (Directorate of Engineering Works)
USSR	—	Union of Soviet Socialist Republics
v/ch	—	voinskaya chast' (military unit)
VNIIRT	—	Vserossiiskii Nauchno-Issledovatel'skii Institut Radiotekhniki (All-Russian Scientific-Research Institute of Radio Technology)
VPK	—	Voenno-Promyshlennaya Komissiya (Military-Industrial Commission)
WSMR	—	White Sands Missile Range
WWII	—	World War II
ZUR	—	zenitnaya upravlyaemaya raketa (antiaircraft guided missile)

Appendix C

PRONUNCIATION GUIDE

Russian words not only are written in Cyrillic, but also exhibit irregular stress patterns. This guide provides a breakdown of words into syllables and shows stress (by capitalization of the stressed syllable), which should help with proper pronunciation. Most words are personal names, patronymics, and organizational names.

A

akademii	—	a-ka-DE-mi-i
Aksel'	—	AK-sel'
Aldan	—	al-DAN
Alexander	—	a-lek-SANDR
Alexandrovich	—	a-lek-SAN-dro-vich
Alexeevich	—	a-lek-SE-ye-vich
Alexei	—	a-lek-SEY
Almaz	—	al-MAZ
Al'perovich	—	al'-pe-RO-vich
Anatolii	—	a-na-TO-liy
Andreevich	—	an-DRE-ye-vich
Andrei	—	an-DREY
Andropov	—	an-DRO-pov
Anton	—	an-TON
Aral'sk	—	a-RAL'SK
Argun'	—	ar-GUN'
Arkharov	—	ar-KHA-rov
Aronovich	—	a-RO-no-vich
Artem	—	ar-TYOM
Arzamas	—	ar-za-MAS

Astrofizika — as-tro-FI-zi-ka
Azov — a-ZOV

B

Baidukov — bai-du-KOV
Baikonur — bai-ko-NUR
Balkhash — bal-KHASH
Baranovichi — ba-RA-no-vi-chi
Barmin — bar-MIN
Basistov — ba-SIS-tov
Beria — BE-ri-ya
berkut — BER-kut
beton — be-TON
betonka — be-TON-ka
Betpak Dala — bet-PAK da-LA *or* bet-PAK DA-la
bezopasnosti — be-zo-PAS-nos-ti
boevaya — bo-ye-VA-ya
bolshevik — bol'-she-VIK
Bolshevo — BOL-she-vo
Boris — bo-RIS
Borisovich — bo-RI-so-vich
Breitbart — BREIT-bart
Brezhnev — BREZH-nev
Budenny — bu-DEN-nyi
Bunkin — BUN-kin
Buran — bu-RAN
Burlakov — bur-la-KOV
Burshtein — bur-SHTEYN
Burtsev — BUR-tsev
burya — BU-rya
byuro — byu-RO

C

Chekhov — CHE-khov
Chelomei — che-lo-MEY
Chelyabinsk — che-LYA-binsk
Chertok — cher-TOK
Chkalov — CHKA-lov

D

dal'nei — DAL'-ney
Danilovich — da-NI-lo-vich
Daryal — dar'-YAL

Dementiev	—	de-MENT'-ev
divizion	—	di-vi-zi-ON
Dmitrievich	—	DMI-tri-ye-vich
Dmitrii	—	DMI-triy
Dnepr	—	DNEPR
Dnepropetrovk	—	dne-pro-pet-ROVSK
Dnestr	—	DNESTR
doktor	—	DOK-tor
Dolgoprudny	—	dol-go-PRUD-nyi
Dom	—	DOM
Dominik	—	do-mi-NIK
Dominikovich	—	do-mi-NI-ko-vich
Dorokhov	—	DO-ro-khov
Drozdov	—	droz-DOV
duga	—	du-GA
Dunai	—	du-NAY

E

electromekhanicheskii	—	e-LEK-tro-me-kha-NI-ches-kiy
Elizarenkov	—	ye-li-ZA-ren-kov
Emba	—	EM-ba
Energia	—	e-NER-gi-ya
Enisei	—	ye-ni-SEY
Evgenii	—	yev-GE-niy
Evseevich	—	yev-SE-ye-vich
experimental'nyi	—	eks-pe-ri-men-TAL'-nyi

F

fakel	—	FA-kel
Fedor	—	FYO-dor
Fedorovich	—	FYO-do-ro-vich
Fili	—	fi-LI
Filippovich	—	fi-LIP-po-vich

G

Gagarin	—	ga-GA-rin
Gegechkori	—	ge-gech-KO-ri
Georgievich	—	ge-OR-gi-ye-vich
Georgii	—	ge-OR-giy
glavnoe	—	GLAV-no-ye
Glavspetsmash	—	glav-spets-MASH
Galvspetzmontazh	—	glav-spetz-mon-TAZH
Glushko	—	glush-KO

GNIIP	—	gni-IP
Golubev	—	GO-lu-bev
Gonor	—	GO-nor
Gorbachev	—	gor-ba-CHYOV
Gorky	—	GOR'-kiy
gosudarstvennoi	—	go-su-DARST-ven-noy
gosudarstvennyi	—	go-su-DARST-ven-nyi
Grigorievich	—	gri-GOR'-ye-vich
Grigorii	—	gri-GO-riy
Grushin	—	GRU-shin
Gubenko	—	gu-BEN-ko
Gurevich	—	gu-RE-vich
Gur'yev	—	GUR'-yev

I

Ignatievich	—	ig-NAT'-ye-vich
Iliya	—	il'-YA
Ilyich	—	il'-YICH
indikator	—	in-di-KA-tor
INEUM	—	i-NE-um
institut	—	in-sti-TUT
inzhenernykh	—	in-zhe-NER-nykh
Ioganovich	—	i-o-GA-no-vich
Iosif	—	i-O-sif
IS	—	i-ES
Isaak	—	i-sa-AK
Isaev	—	i-SA-yev
ispytatel'nyi	—	is-py-TA-tel'-nyi
istok	—	is-TOK
Istra	—	IST-ra, or is-TRA
istrebitel'	—	is-tre-BI-tel'
ITMVT	—	i-te-em-ve-TE
Ivan	—	i-VAN
Ivanovich	—	i-VA-no-vich
izdelie	—	iz-DE-li-ye

K

kaktus	—	KAK-tus
Kaliningrad	—	ka-LI-nin-grad
Kalmykov	—	kal-my-KOV
Kama	—	KA-ma
Kamensky	—	ka-MEN-skiy
kandidat	—	kan-di-DAT

Kap Yar	—	kap YAR
Kapustin Yar	—	ka-PUS-tin yar
Karaganda	—	ka-ra-gan-DA
Katyusha	—	ka-TYU-sha
Kavkaz	—	kav-KAZ
KB	—	ka-BE
Keldysh	—	KEL-dysh
Khimki	—	KHIM-ki
Khimmash	—	khim-MASH
Khrushchev	—	khru-SHCHYOV
Khvatov	—	KHVA-tov
Kisun'ko	—	ki-sun'-KO
Kobalt	—	KO-bal't
Kolosov	—	KO-lo-sov
Kolya	—	KO-lya
komand	—	ko-MAND
kometa	—	ko-ME-ta
komissiya	—	ko-MIS-si-ya
komitet	—	ko-mi-TET
Konev	—	KO-nev
Konstantin	—	kon-stan-TIN
Konstantinovich	—	kon-stan-TI-no-vich
konstruktorskoe	—	kon-STRUK-tor-sko-ye
kontrolya	—	kon-TRO-lya
Korenev	—	KO-re-nev
Korolev	—	ko-ro-LYOV
Koshlyakov	—	kosh-lya-KOV
kosmicheskogo	—	kos-MI-ches-ko-go
Kosygin	—	ko-SY-gin
krasnyi	—	KRAS-nyi
krona	—	KRO-na
krylatyi	—	kry-LA-tyi
Krylov	—	kry-LOV
Kubinka	—	KU-bin-ka
Kuksenko	—	kuk-SEN-ko
Kuleshov	—	ku-le-SHOV
Kuntsevo	—	KUN-tse-vo
kupol	—	KU-pol
Kuz'mich	—	kuz'-MICH

L

Lavochkin	—	LA-voch-kin
Lavrentievich	—	lav-REN-t'ye-vich

Lavrentii	—	lav-REN-tiy
Lebedev	—	LE-be-dev
Lenin	—	LE-nin
Leningrad	—	le-nin-GRAD
Leningradskoe	—	le-nin-GRAD-sko-ye
Leonid	—	le-o-NID
Leonov	—	le-O-nov
Lev	—	LEV
Levon	—	le-VON
Lipsman	—	LIPS-man
Livshits	—	LIV-shits
Lomonosov	—	lo-mo-NO-sov
Lukin	—	lu-KIN
Lukyanovich	—	luk-YA-no-vich
Luzhniki	—	luzh-ni-KI
Lvovich	—	L'VO-vich
Lyudvigovich	—	LYUD-vi-go-vich

M

Malenkov	—	ma-len-KOV
Malinovsky	—	ma-li-NOV-skiy
Markov	—	MAR-kov
mashin	—	ma-SHIN
mashinostroeniya	—	ma-SHI-no-stro-YE-ni-ya
Maximilianovich	—	mak-si-mil'-YA-no-vich
mekhaniki	—	me-KHA-ni-ki
Michurisnk	—	mi-CHU-rinsk
Mikhail	—	mi-kha-IL
Mikhailov	—	mi-KHAI-lov
Mikhailovich	—	mi-KHAI-lo-vich
Mikoyan	—	mi-ko-YAN
ministerstvo	—	mi-ni-STER-stvo
Minsredmash	—	min-sred-MASH
Mironovich	—	mi-RO-no-vich
Mitrofan	—	mi-tro-FAN
MOM	—	MOM
Mosenergo	—	mos-e-NER-go
Moskovskii	—	mos-KOV-skiy
Mozharovsky	—	mo-zha-ROV-skiy
Mstislav	—	msti-SLAV
Murav'ev	—	mu-rav'-YOV
Musatov	—	mu-SA-tov
Myasishchev	—	MYA-si-shchev

Mymrin	—	MYM-rin
Mytishchi	—	my-TI-shchi

N

Nakhim	—	na-KHIM
Naro-Fominsk	—	na-ro-fo-MINSK
nauchno-issledovatel'skii	—	na-UCH-no-is-SLE-do-va-tel'-skiy
nauchno-tekhnicheskii	—	na-UCH-no-tech-NI-ches-kiy
nauk	—	na-UK
navedeniya	—	na-ve-DE-ni-ya
navodki	—	na-VOD-ki
Nedelin	—	ne-DE-lin
Neman	—	NE-man
NIEMI	—	ni-e-MI
NII	—	ni-I
NIIDAR	—	ni-i-DAR
Nikita	—	ni-KI-ta
Nikitich	—	ni-KI-tich
Nikolaevich	—	ni-ko-LA-ye-vich
Nikolai	—	ni-ko-LAY
Nilovsky	—	ni-LOV-skiy
Nina	—	NI-na
NPO	—	en-pe-O
NTTs	—	en-te-TSE
Nudol'	—	NU-dol'

O

ob"ekt	—	ob-EKT
obnaruzheniya	—	ob-na-ru-ZHE-ni-ya
OBP	—	o-be-PE
obshchaya	—	OB-shcha-ya
obshchego	—	OB-shche-go
ofitserov	—	o-fi-TSE-rov
Oganov	—	o-GA-nov
OKB	—	o-ka-BE
Oleg	—	o-LEG
OM	—	o-EM
Omel'chenko	—	o-MEL'-chen-ko
Orenburg	—	o-ren-BURG
orudiinoi	—	o-ru-DIY-noy
OS	—	o-ES
Osipovich	—	O-si-po-vich
osoboe	—	o-SO-bo-ye

Ostapenko	—	os-TA-pen-ko
otechestvennaya	—	o-TE-chest-ven-na-ya

P

Panteleimonovich	—	pan-te-lei-MO-no-vich
Pavel	—	PA-vel
Pavlovich	—	PAV-lo-vich
peredachi	—	pe-re-DA-chi
Petr	—	PYOTR
Petrovich	—	pet-RO-vich
Pivovarov	—	pi-vo-VA-rov
ploshchadka	—	plo-SHCHAD-ka
Pluton	—	plu-TON
pochtovyi yashchik	—	poch-TO-vyi YA-shchik
podgotovki	—	pod-go-TOV-ki
Podlipki	—	pod-LIP-ki
polet	—	po-LYOT
poletami	—	po-LYO-ta-mi
poligon	—	po-li-GON
poluostrov	—	po-lu-OST-rov
Pomaznev	—	po-MAZ-nev
postoyannay	—	pos-to-YAN-na-ya
PPR	—	pe-pe-ER
predvaritel'naya	—	pred-va-RI-tel'-na-ya
programma	—	pro-GRAM-ma
proizvodstvennoe	—	pro-iz-VOD-stven-no-ye
promyshlennaya	—	pro-MYSH-len-na-ya
prostranstva	—	pro-STRAN-stva
protivorakety	—	pro-ti-vo-ra-KE-ty
punkt	—	PUNKT

R

Rabinovich	—	ra-bi-NO-vich
rabot	—	ra-BOT
radiolokator	—	RA-di-o-lo-KA-tor
radiolokatsionnaya	—	RA-di-o-lo-ka-tsi-ON-na-ya
radiolokatsiya	—	RA-di-o-lo-KA-tsi-ya
radioobnaruzhenie	—	RA-di-o-ob-na-ru-ZHE-ni-ye
radiosvyazi	—	RA-di-o-SVYA-zi
radiotekhnicheskii	—	RA-di-o-tekh-NI-ches-kiy
radiotekhniki	—	RA-di-o-TEKH-ni-ki
raket	—	ra-KET
raketa	—	ra-KE-ta

RALAN	—	ra-LAN
Raspletin	—	ras-PLE-tin
RE	—	er-YE
Repin	—	RE-pin
Revolii	—	re-VO-liy
Rodion	—	ro-di-ON
RSVPR	—	er-es-ve-pe-ER
RTI	—	er-te-I
RTN	—	er-te-EN
Ruvimovich	—	ru-VI-mo-vich
Ryabikov	—	RYA-bi-kov

S

Samuil	—	sa-mu-IL
Samuilovich	—	sa-mu-I-lo-vich
Sarov	—	sa-ROV
Saryshagan	—	sa-ry-sha-GAN
Savin	—	SA-vin
SB	—	es-BE
Semen	—	se-MYON
Semenovich	—	se-MYO-no-vich
Semipalatinsk	—	se-mi-pa-LA-tinsk
Sena	—	SE-na
Sergeevich	—	ser-GE-ye-vich
Sergei	—	ser-GEY
Sergo	—	ser-GO
Sevruk	—	sev-RUK
sharashka	—	sha-RASH-ka
Shchukin	—	SHCHU-kin
Shokin	—	SHO-kin
shosse	—	shos-SE
sistema	—	sis-TE-ma
sistemy	—	sis-TE-my
SKB	—	es-ka-BE
SKKP	—	es-ka-ka-PE
Sliozberg	—	sli-OZ-berg
Smirnov	—	smir-NOV
Smirnova	—	smir-NO-va
Smolensk	—	smo-LENSK
snaryad	—	sna-RYAD
Sokol	—	SO-kol
Sokolovsky	—	so-ko-LOV-skiy
Sosul'nikov	—	so-SUL'-ni-kov

spetsial'noe	—	spe-tsi-AL'-no-ye
Spetsmash	—	spets-MASH
SPK	—	es-pe-KA
SPRN	—	es-pe-er-EN
sputnik	—	SPUT-nik
sputnikov	—	SPUT-ni-kov
srednego	—	SRED-ne-go
SSSR	—	es-es-es-ER
Stalin	—	STA-lin
Stalingrad	—	sta-lin-GRAD
stantsiya	—	STAN-tsi-ya
Stepanovich	—	ste-PA-no-vich
strela	—	stre-LA
Subbotin	—	sub-BO-tin
Sukhoi	—	su-KHOI
Sumbatovich	—	sum-BA-to-vich
Suslov	—	SUS-lov
Sverdlovsk	—	sverd-LOVSK

T

taran	—	ta-RAN
Tashkent	—	tash-KENT
tekhnicheskii	—	tekh-NI-ches-kiy
tekhniki	—	TEKH-ni-ki
tematicheskii	—	te-ma-TI-ches-kiy
Temnikov	—	TEM-ni-kov
Tereshkova	—	te-resh-KO-va
TGU	—	te-ge-U
Tobol	—	to-BOL
tochnogo	—	TOCH'-no-go
tochnoi	—	TOCH'-noi
Tomashevich	—	to-ma-SHE-vich
tret'e	—	TRET'-ye
Trofimchuk	—	tro-fim-CHUK
tsentr	—	TSENTR
tsentral'naya	—	tsen-TRAL'-na-ya
tsentral'nyi	—	tsen-TRAL'-nyi
TsIS	—	TSIS
TsKKP	—	tse-ka-ka-PE
TsNII	—	tsni-I
TsNIIMash	—	tsni-i-MASH
TsSO-P	—	tse-es-o-PE
TsUP	—	TSUP

Tsyklon — tsy-KLON
Tupolev — TU-po-lev
Tushino — TU-shi-no
Tyuratam — tyu-ra-TAM

U
UIR — u-IR
upravlenie — up-rav-LE-ni-ye
upravlyaemaya — u-prav-LYA-e-ma-ya
upravlyayushchikh — u-prav-LYA-yu-shchikh
Ur'ev — UR'-yev
Ustinov — us-TI-nov
uzel — U-zel

V
Valentin — va-len-TIN
Valentina — va-len-TI-na
Valerii — va-LE-riy
Vasil'evich — va-SIL'-ye-vich
Vasilevsky — va-si-LEV-skiy
Vasilii — va-SI-liy
Veisbein — VEIS-bein
venediktov — ve-ne-DIK-tov
verba — VER-ba
Vershynin — ver-SHY-nin
Vetoshkin — ve-TOSH-kin
Viktorovich — VIK-to-ro-vich
Vissarionovich — vis-sa-ri-O-no-vich
vizirovaniya — vi-ZI-ro-va-ni-ya
Vladimir — vla-DI-mir
Vladimirovich — vla-DI-mi-ro-vich
Vladimirovna — vla-DI-mi-rov-na
Vladislav — vla-di-SLAV
voenno-promyshlennaya — vo-EN-no-pro-MYSH-len-na-ya
voinskaya (chast') — VO-in-ska-ya
Volgograd — vol-go-GRAD
Volokolamsk — vo-lo-ko-LAMSK
Volokolamskoe — vo-lo-ko-LAM-sko-ye
Vorob'evy — vo-rob'-YO-vy
Votintsev — vo-TIN-tsev
Voznyuk — voz-NYUK
VPK — ve-pe-KA
vremennaya — VRE-men-na-ya

vserossiiskii	—	vse-ros-SIY-skiy
Vsevolod	—	VSE-vo-lod
Vsevolodovich	—	VSE-vo-lo-do-vich
vychislitel'na-ya	—	vy-chis-LI-tel'-na-ya
vychislitel'noi	—	vy-chis-LI-tel'-noy
vympel	—	VYM-pel

Y

Yakovlev	—	YA-kov-lev
Yakovlevich	—	YA-kov-le-vich
Yangel	—	YAN-gel'
Yekaterinburg	—	ye-ka-te-rin-BURG
Yulievich	—	YUL'-ye-vich
Yuri	—	YU-riy
Yuzhnoe	—	YUZH-no-ye

Z

Zagorsk	—	za-GORSK
Zaikov	—	zai-KOV
Zakharov	—	za-KHA-rov
Zakson	—	ZAK-son
Zelenograd	—	ze-le-no-GRAD
zenitnaya	—	ze-NIT-na-ya
Zhadeiko	—	zha-DEI-ko
Zhukov	—	ZHU-kov
Zhukovsky	—	zhu-KOV-skiy
Zimin	—	zi-MIN
ZUR	—	ZUR

Appendix D

LIST OF FIGURES

Fig. 1.1	Fragments of Scud missile
Fig. 2.1	Joseph V. Stalin
Fig. 2.2	Sergo Beria
Fig. 2.3	Pavel N. Kuksenko
Fig. 2.4	Krug (SA-4)
Fig. 2.5	Alexander S. Yakovlev, Andrei N. Tupolev, Semen A. Lavochkin, and Artem I. Mikoyan
Fig. 2.6	Kometa KS-1 (AS-1)
Fig. 3.1	Joseph V. Stalin and Lavrentii P. Beria
Fig. 3.2	Alexander A. Raspletin
Fig. 3.3	Aksel' I. Berg
Fig. 3.4	KB-1 building
Fig. 3.5	KB-1 building
Fig. 3.6	Antenna aperture
Fig. 3.7	B-200 radar
Fig. 3.8	B-200 radar
Fig. 3.9	Semen A. Lavochkin
Fig. 3.10	ZUR-205 (V-300, SA-1)
Fig. 3.11	Alexei M. Isaev
Fig. 3.12	Vladimir P. Barmin
Fig. 3.13	Vasilii I. Voznyuk, Sergei I. Vetoshkin, and Sergei P. Korolev
Fig. 3.14	Sergo Beria
Fig. 3.15	S-75 (SA-2) missile
Fig. 3.16	Petr D. Grushin
Fig. 4.1	A-100 and Kama radars
Fig. 4.2	A-100 map
Fig. 4.3	A-100 site
Fig. 4.4	S-25 (SA-1) map

Fig. 4.5	S-25 (SA-1) site
Fig. 4.6	S-25 (SA-1) site
Fig. 4.7	V-300 missile
Fig. 4.8	Moscow
Fig. 4.9	Moscow
Fig. 4.10	Moscow
Fig. 4.11	S-75 (SA-2) site
Fig. 4.12	S-75 (SA-2) missile
Fig. 4.13	Grigorii V. Kisun'ko
Fig. 5.1	A-4 (V-2)
Fig. 5.2	A-4 (V-2) hits Antwerp
Fig. 5.3	Vasilii D. Sokolovsky and Dwight D. Eisenhower
Fig. 5.4	Nikita S. Khrushchev
Fig. 5.5	Alexander L. Mints
Fig. 5.6	Grigorii V. Kisun'ko
Fig. 5.7	Timeline of early air and missile defenses
Fig. 5.8	Evolution of KB-1
Fig. 5.9	System A concept
Fig. 5.10	Petr D. Grushin tombstone
Fig. 5.11	Organizations in early air and missile defenses
Fig. 6.1	Major Soviet weapons development test sites
Fig. 6.2	Saryshagan range
Fig. 6.3	Indigenous people
Fig. 6.4	Lake Balkhash
Fig. 6.5	Construction troops badge
Fig. 6.6	Saryshagan railroad station and Priozersk
Fig. 6.7	Saryshagan railroad station
Fig. 6.8	High-voltage power lines at Saryshagan
Fig. 6.9	Main airfield at Saryshagan
Fig. 6.10	Map of Saryshagan range
Fig. 6.11	Map of support base at Saryshagan range
Fig. 6.12	S-75 (SA-2) battery at Saryshagan
Fig. 6.13	Map of Saryshagan range
Fig. 6.14	TsSO-P (Hen House) radar, Saryshagan range
Fig. 6.15	Site 2, Saryshagan range
Fig. 6.16	Fenced-off area, Site 2, Saryshagan range
Fig. 6.17	Airstrip, Site 2, Saryshagan range
Fig. 6.18	R-5 field launch position near Chelkar
Fig. 6.19	RE-4 radar, Site 38, Saryshagan range
Fig. 6.20	Priozersk
Fig. 6.21	Priozersk
Fig. 6.22	Television tower, Priozersk
Fig. 6.23	Eastern part of Saryshagan range

LIST OF FIGURES

Fig. 7.1   System A at Saryshagan range
Fig. 7.2   Dunai-2 (Sites 14 and 15) and TsSO-P (Site 8) area at Saryshagan range
Fig. 7.3   Dunai-2 (Hen Roost North), Site 14
Fig. 7.4   Dunai-2 (Hen Roost)
Fig. 7.5   Dunai-2 (Hen Roost North)
Fig. 7.6   Dunai-2 (Hen Roost North)
Fig. 7.7   TsSO-P (Hen House)
Fig. 7.8   RTN radar
Fig. 7.9   RTN radar
Fig. 7.10 RTN radar
Fig. 7.11 RTN radar at Site 2
Fig. 7.12 Initial guidance RSVPR radar
Fig. 7.13 V-1000 interceptor missile
Fig. 7.14 Missile launch pad
Fig. 7.15 Microwave relay communication tower
Fig. 7.16 Terminal communication tower at Site 2
Fig. 7.17 Site 40, Saryshagan range
Fig. 7.18 M-40 and M-50 computers, Saryshagan range
Fig. 8.1   Central control post TsIS
Fig. 8.2   Site 40, Saryshagan range
Fig. 8.3   Trajectories of ballistic missiles in first intercepts
Fig. 8.4   V-1000 interceptor missile
Fig. 8.5   Launch positions of V-1000 missiles
Fig. 8.6   Mikhail K. Yangel
Fig. 8.7   IRBM R-12 (SS-4)
Fig. 8.8   R-12 field launch position near Makat
Fig. 8.9   Launch of V-1000
Fig. 8.10 Remnants of the warhead shot down on 4 March 1961
Fig. 8.11 Optical instrument KST-60
Fig. 8.12 Sequence of frames of R-12 intercept on 9 June 1961
Fig. 8.13 Sequence of frames of R-12 intercept on 9 June 1961
Fig. 8.14 Sequence of frames of R-12 intercept on 9 June 1961
Fig. 9.1   Dunai-3 (Dog House) radar
Fig. 9.2   RE-4, RKTs, and RKI, Site 38, Saryshagan range
Fig. 9.3   Vladimir N. Chelomei
Fig. 9.4   S-225 concept
Fig. 9.5   Nikolai I. Krylov, Nikita S. Khrushchev, Dmitrii F. Ustinov, Leonid I. Brezhnev, and Frol A. Kozlov
Fig. 9.6   A-35 and A-35M map
Fig. 9.7   A-35 site near Klin
Fig. 9.8   A-35 site near Zagorsk
Fig. 9.9   OS-1 node near Irkutsk

Fig. 9.10   OS-2 node, Saryshagan range
Fig. 9.11   Over-the-horizon radar near Nikolaev
Fig. 9.12   Laser space battlestation Polyus and Energia booster
Fig. 9.13   Remnants of RTN, Site 2, Saryshagan
Fig. 9.14   Remnants of garrison clubhouse, Site 35, Saryshagan range
Fig. A.1   Honest John
Fig. A.2   Hawk
Fig. A.3   Honest John intercept by Hawk
Fig. A.4   Corporal and Nike-Hercules
Fig. A.5   Corporal intercept by Nike-Hercules
Fig. A.6   Nike-Zeus

Appendix E

Selected Bibliography

C.R. Ahern, "The Yo-Yo Story: An Electronic Analysis Case History," *Studies in Intelligence*, vol. 5, pp. 11–23, Winter 1961.

N.A. Aitkhozhin and M.M. Gantsevich, "Radiolokatory Navedeniya Sistemy PRO 'A'. Chast' 1 (Guidance Radars of Missile Defense System A. Part 1)," *Vozdushno-Kosmicheskaya Oborona* (*Air and Missile Defense*), no. 51, issue 2, pp. 62–69, 2010a (in Russian).

N.A. Aitkhozhin and M.M. Gantsevich, "Radiolokatory Navedeniya Sistemy PRO 'A'. Chast' 2 (Guidance Radars of Missile Defense System A. Part 2)," *Vozdushno-Kosmicheskaya Oborona* (*Air and Missile Defense*), no. 52, issue 3, pp. 60–67, 2010b (in Russian).

W. Albring and H. Vinke (editors), *Gorodomlia. Deutsche Raketenforscher in Russland* (*Gorodomlya. German Rocketeers in Russia*), Hamburg, 1991 (in German).

K.S. Al'perovich, *Tak Rozhdalos' Novoe Oruzhie. Zapiski o Zenitnykh Raketnykh Kompleksakh i Ikh Sozdatelyakh*, (*New Weapon Was Born This Way. Notes on Air Defense Missile Complexes and Their Creators*), OAO TsKB "Almaz," Moscow, 1999 (in Russian).

K.S. Al'perovich, *Gody Raboty nad Sistemoi PVO Moskvy—1950–1955. Zametki Inzhenera*, (*Years of Work on the Air Defense System of Moscow—1950–1955. Notes of an Engineer*), NPO "Almaz" im. Akademika A.A. Raspletina, Moscow, 2003 (in Russian).

V.D. Anisimov, G.S. Batyr', V.V. Dementiev, A.D. Luk'yanova, A.V. Menshikov, V.G. Morozov, A.A. Pavlovchev, S.A. Sukhanov, and V.D. Shilin (editors), *Korporatsiya "Vympel." Sistemy Raketno-Kosmicheskoi Oborony* (*Corporation "Vympel." Systems of Missile and Space Defense*), Oruzhie i Tekhnologii, Moscow, 2005 (in Russian).

Army Space and Missile Defense Command, SMDC/ARSTRAT History Office, "The First 'Hit-to-Kill' Kinetic Energy Interceptor Missile," *The Eagle*, Army Space and Missile Defense Command, Huntsville, AL, p. 11, June–July 2007.

L.S. Arzanov, Schast'e ispolnennogo dolga (Happiness of Fulfilled Duty), in N.G. Zavaly, *Rubezhi Oborony—v Kosmose i na Zemle. Ocherki Istorii Raketno-Kosmicheskoi Oborony* (*Defensive Lines—in Space and on Earth. Stories of the History of Missile and Space Defense*), pp. 693–703, Veche, Moscow, 2003 (in Russian).

A. Babakin, *Bitva v Ionosfere. Zhurnalistskoe Issledovanie Tragedii i Triumfa Otechestvennoi Zagorizontnoi Radiolokatsionnoi Boevoi Sistemy* (*Battle in the Ionosphere. Journalistic Investigation of the Tragedy and Triumph of the National Over-the-Horizon Combat Radar System*), Tseikhgauz, Moscow, 2008 (in Russian).

J.A. Baclawski, "The Best Map of Moscow," *Studies in Intelligence*, semiannual edition, no. 1, pp. 111–114, 1997.

D. Ball, *Pine Gap. Australia and the US Geostationary Signals Intelligence Satellite Programs*, Allen and Unwin, Sydney, Australia, 1988.

Yu.M. Baturin, *Sovetskaya Kosmicheskaya Initsiativa v Gosudarstvennykh Dokumentakh, 1946–1964 (Soviet Space Initiative in State Documents, 1946–1964)*, RTSoft, Moscow, 2008 (in Russian).

D.R. Baucom, *The Origins of SDI, 1944–1983*, University Press of Kansas, 1992.

Bell Laboratories, *ABM Research and Development at Bell Laboratories*, Whippany, NJ, Bell Laboratories for U.S. Army Ballistic Missile Defense Systems Command, 1975.

E. Beloglazova, *Sovershenno Sekretnyi General (Top Secret General)*, Geroi Otechestva, Moscow, 2005 (in Russian).

V.S. Belous, A.A. Greshilov, N.D. Egupov, V.P. Zhabchuk, V.N. Ivanov, G.V. Kononenko, V.M. Kraskovsky, A.F. Kulakov, V.M. Kutsenko, S.N. Lyutikov, V.V. Mal'tsev, V.S. Matlashov, A.M. Matushchenko, N.K. Ostapenko, V.K. Panyukhin, K.A. Pupkov, N.K. Sokolov, Yu.N. Tret'yakov, and I.S. Shal'nov, *Shchit Rossii: Sistemy Protivoraketnoi Oborony (Shield of Russia: Systems of Missile Defense)*, Izdatel'stvo MGTU im. N.E. Baumana (Publishing House of N.E. Bauman Moscow State Technical University), Moscow, 2009 (in Russian).

A. Ben David, "Under Cover," *Aviation Week and Space Technology*, vol. 176, no. 30, pp. 20, 21, 30 Sept. 2014.

B. Berdichevsky, *Traektoriya Zhizni. Lyudi, Samolety, Rakety (Trajectory of Life. People, Airplanes, Rockets)*, AGRAF, Moscow, 2005 (in Russian).

S. Beria, *Moi Otets—Lavrentii Beria (My Father—Lavrentii Beria)*, Sovremennik, Moscow, 1994 (in Russian).

B. Berkowitz, *The National Reconnaissance Office at 50 Years: A Brief History*, Center for the Study of National Reconnaissance, Chantilly, VA, 2011.

C.E. Bohlen, *Witness to History. 1929–1969*, W.W. Norton and Co., New York, NY, 1973.

D.C. Brown, "On the Trail of Hen House and Hen Roost," *Studies in Intelligence*, vol. 13, no. 2, pp. 11–19, 1969.

D.A. Brugioni, "The Tyuratam Enigma," *Air Force Magazine*, pp. 108, 109, March 1984.

D.A. Brugioni, *Eyes in the Sky: Eisenhower, the CIA, and Cold War Aerial Espionage*, Naval Institute Press, Annapolis, MD, 2010.

J.W. Bullard, *History of the Field Army Ballistic Missile Defense System Project, 1959–1962*, U.S. Army Missile Command, Redstone Arsenal, AL, 1963.

M. Cagle, *Development, Production, and Deployment of the Nike Ajax Guided Missile System, 1945–1959*, Historical Monograph, U.S. Army Ordinance Missile Command, Redstone Arsenal, AL, 1959.

M. Cagle, *History of the Nike Hercules Weapons System*, Historical Monograph, Project No. AMC 75 M, U.S. Army Missile Command, Redstone Arsenal, AL, 1973.

Central Intelligence Agency, Special Intelligence Estimate SIE-5, *The Scale and Nature of the Soviet Air Defense Effort 1952–1954*, 8 Nov. 1952.

Central Intelligence Agency, National Intelligence Estimate NIE 11-5-55, *Soviet Capabilities and Probable Programs in the Guided Missile Field*, 12 March 1957.

Central Intelligence Agency, National Intelligence Estimate NIE 11-5-59, *Soviet Capabilities in Guided Missiles and Space Vehicles*, 3 Nov. 1959.

Central Intelligence Agency, *Preliminary Report of Mission ODE—8009*, 14 Feb. 1960.

Central Intelligence Agency, PAR-582–60, ODE Mission 4155, 16 April 1960.

Central Intelligence Agency, National Intelligence Estimate NIE 11-3-65, *Soviet Strategic Air and Missile Defenses*, 18 Nov. 1965.

Central Intelligence Agency, *Launch Complex G. Tyuratam Missile Test Center, USSR*, Photographic Intelligence Report CIA/PIR 61074, Dec. 1965.

Selected Bibliography

Central Intelligence Agency, National Intelligence Estimate NIE 11-3-67, *Soviet Strategic Air and Missile Defenses*, 9 Nov. 1967.

Central Intelligence Agency, National Intelligence Estimate NIE 11-13-82, *Soviet Ballistic Missile Defense, Vol. 1: Key Judgments and Summary*, 13 Oct. 1982.

Central Intelligence Agency, Imagery Analysis Division, *Hen Roost Antennas. Sary Shagan Missile Antimissile Test Center, USSR*, CIA-RDP78T05161A000400010043-8, PIR 6133, 1 April 1966.

Central Intelligence Agency, Office of Scientific Intelligence, SC-02164-58, *Electronic Aspects of the Soviet Air Defense System*, Scientific Intelligence Report CIA/SI/HTA/SIR-2/58, 3 March 1958.

Central Intelligence Agency, Office of Special Activities (OSA), *Accomplishments of the U-2 Program*, TCS-7519-60, 27 May 1960. (Pedlow and Welzenbach, 1998, p. 316, provide a quote from this document that allows its attribution to Director of Central Intelligence Allen Dulles; they also note that James Q. Weber probably wrote the draft of this document.)

Central Intelligence Agency, Photographic Intelligence Center, PIC/B-1/60, *Radar Sites Associated with Moscow Air Defense System*, Photographic Intelligence Brief, 25 Feb. 1960.

Central Intelligence Agency, Photographic Intelligence Center, PIC/B-2/60, *Radar Sites Associated with Moscow Air Defense System*, Photographic Intelligence Brief, 26 Feb. 1960.

Central Intelligence Agency, Photographic Intelligence Center, PIC/R-4/60, *Gage-Patty Cake Radar Sites Probably Associated with Moscow Air-Defense System*, Photographic Intelligence Report, May 1960.

Central Intelligence Agency, Photographic Intelligence Center, PIC/JR-1010/61, *Antimissile Complex, Sary Shagan, USSR*, Joint Photographic Intelligence Report, April 1961.

B.E. Chertok, *Rakety i Lyudi* (*Rockets and People*), Mashinostroenie, Moscow, 1994 (in Russian). (English translation: B. Chertok, *Rockets and People,* vol. I, NASA, Washington, DC, 2005.)

B.E. Chertok, *Rakety i Lyudi. Goryachie Dni Kholodnoi Voiny* (*Rockets and People. Hot Days of the Cold War*), Mashinostroenie, Moscow, 1997 (in Russian). (English translation: B. Chertok, *Rockets and People. Hot Days of the Cold War,* vol. III, NASA, Washington, DC, 2010.)

K.-S. Chiang, *Soviet Russia in China*, Farrar, Straus and Giroux, New York, 1967.

"Chronicle. On the 80th Birthday of G.V. Kisun'ko," *Journal of Communications Technology and Electronics*, vol. 43, no. 12, p. 1431, 1998. (The journal is a cover-to-cover translation of a leading Russian, formerly Soviet, scientific journal *Radiotekhnika i Elektronika*, vol. 43, no. 12, p. 1531, 1998.)

J.D. Clark, *Ignition! An Informal History of Liquid Rocket Propellants*, Rutgers University Press, New Brunswick, NJ, 1972.

S. Courtois, N. Werth, J.-L. Panné, A. Paczkowski, K. Bartošek, and J.L. Margolin, *The Black Book of Communism: Crimes, Terror, Repression*, Harvard University Press, Cambridge, MA, 1999.

M.V. Davydov, *Gody i Lyudi* (*Years and People*), Radio i Svyaz', Moscow, 2009 (in Russian).

D.A. Day, J.M. Logsdon, and B. Latell (editors), *Eye in the Sky: The Story of the Corona Spy Satellites*, Smithsonian Institution Press, Washington, DC, 1998.

S.D. Drell, "Physics and U.S. National Security," *Reviews of Modern Physics*, vol. 71, no. 2, pp. S460–S470, 1999.

N.D. Drozdov, *Iz Istorii Sozdaniya Sistemy Protivoraketnoi Oborony v SSSR* (*From the History of Development of the Missile Defense System in the USSR*), 1998; manuscript (Drozdov-Fromhistory.doc) posted on the website of the Moscow Aviation Institute, http://www.mai.ru/colleges/war/ballist/books/; accessed on 28 April 2006 (in Russian).

D. Dyer, *TRW. Pioneering Technology and Innovation since 1900*, Harvard Business School Press, Boston, MA, 1998.

F.J. Dyson, "Defense Against Ballistic Missiles," *Bulletin of the Atomic Scientists*, no. 6, pp. 12–18, June 1964.

G.A. Efremov, L.E. Makarov, A.O. Degtyarev, A.V. Blagov, L.A. Bondarenko, A.I. Burgansky, N.E. Dementieva, V.S. Kudryashov, Yu.V. Mel'nikov, V.I. Nikitenko, A.A. Perevezentsev, V.A. Polyachenko, G.F. Resh, A.N. Strakhov, and L.M. Shelepin, *Tvortsy i Sozdateli. Oda Kollektivu (Creators and Makers. Ode to the Collective)*, Bedretdinov i Ko., Moscow, 2009 (in Russian).

V. Egorov, *Gody, Otdannye Protivoraketnoi Oborone: Vospominaniya Razrabotchikov (Years Given to Missile Defense: Recollections of Developers)*, Moskovskie uchebniki—SiDiPress, Moscow, 2009 (in Russian).

F. Eliot, "Moon Bounce ELINT," *Studies in Intelligence*, vol. 11, no. 2, pp. 59–65, 1967.

Yu.N. Erofeev, *Aksel' Ivanovich Berg. Zhizn' i Deyatel'nost' (Aksel' Ivanovich Berg. His Life and Work)*, Goryachaya Liniya—Telekom, Moscow, 2007 (in Russian).

M.D. Evtif'ev, *Iz Istorii Sozdaniya Zenitno-Raketnogo Shchita Rossii (From the History of Creation of the Antiaircraft-Missile Shield of Russia)*, Vuzovskaya Kniga, Moscow, 2000 (in Russian).

O. Falichev and E. Sukharev, "Askania v KB-1 (Askania in Design Bureau No. 1)," *Strela*, vol. 3, no. 110, March 2012 (in Russian).

O.V. Golubev, Razrabotka, sozdanie i ispytaniya sistem navedeniya protivoraket na ballisticheskie tseli v otechestvennoi PRO. Vospominaniya rukovoditelya rabot (Development, Building, and Testing of Antimissile Guidance Systems for Ballistic Targets in the National Missile Defense. Recollections of the Head of the Development), in N.G. Zavaly, *Rubezhi Oborony—v Kosmose i na Zemle. Ocherki Istorii Raketno-Kosmicheskoi Oborony (Defensive Lines—in Space and on Earth. Stories of the History of Missile and Space Defense)*, pp. 570–583, Veche, Moscow, 2003 (in Russian).

O.V. Golubev, Yu.A. Kamensky, M.G. Minasyan, and B.D. Pupkov, "Zadachi upravleniya i otsenki effektivnosti v razrabotkakh otechestvennoi sistemy PRO. I. Ekkperimental'nyi poligonnyi kompleks PRO (Sistema A) [Control Problems and Efficiency Assessment in the Development of the National Missile Defense System. I. Experimental Test-Range Missile Defense Complex (System A)]," *Tekhnicheskaya Kibernetika*, no. 6, pp. 166–174, 1992 (in Russian).

O.V. Golubev, Yu.A. Kamensky, M.G. Minasyan, and B.D. Pupkov, "Zadachi upravleniya i otsenki effektivnosti v razrabotkakh otechestvennoi sistemy PRO. II. Sistema PRO g. Moskvy (Sistema A-35). [Control Problems and Efficiency Assessment in the Development of the National Missile Defense System. II. Missile Defense System of Moscow (System A-35)]," *Tekhnicheskaya Kibernetika*, no. 6, pp. 186–192, 1993 (in Russian).

O.V. Golubev, Yu.A. Kamensky, M.G. Minasyan, and B.D. Pupkov, *Rossiyskaya Sistema Protivoraketnoi Oborony. Proshloe i Nastoyashchee—Vzglyad Iznutri (Russian System of Missile Defense. Past and Present—A View from the Inside)*, Tekhnokonsalt, Moscow, 1994 (in Russian).

O.V. Golubev, Yu.A. Kamensky, and B.D. Pupkov, "Zadachi upravleniya i otsenki effektivnosti v razrabotkakh otechestvennoi sistemy PRO. III. Razvitie sistemy PRO g. Moskvy. Dal'neishee sovershenstvovanie sistemy PRO. (Control Problems and Efficiency Assessment in the Development of the National Missile Defense System. III. Development of the Missile Defense System of Moscow. Further Improvement of the Missile Defense System)," *Teoriya i Sistemy Upravleniya*, no. 2, pp. 175–180, 1995 (in Russian).

Ye. Gordon and V. Rigmant, *Tupolev Tu-16 Badger. Versatile Soviet Long-Range Bomber*, Aerofax—Ian Allan Publishing, Leicester, England, 2004.

Government Accounting Office, GAO/NSIAD-94–219 Report, *Ballistic Missile Defense. Records Indicate Deception Program Did Not Affect 1984 Test Results*, Washington, DC, 1994.
A.A. Greshilov, N.D. Egupov, and A.M. Matushchenko, *Yadernyi Shchit (Nuclear Shield)*, Logos, Moscow, 2008 (in Russian).
I. Groettrup, *Rocket Wife*, Andre Deutsh Ltd., London, 1959.
M. Gruntman, *Blazing the Trail: The Early History of Spacecraft and Rocketry*, American Institute of Aeronautics and Astronautics (AIAA), Reston, VA, 2004.
M. Gruntman, *From Astronautics to Cosmonautics*, BookSurge, Charleston, NC, 2007.
M. Gruntman, *Enemy among Trojans. A Soviet Spy at USC*, Figueroa Press, Los Angeles, CA, 2010.
M. Gruntman, "Socks for the First Cosmonaut of Planet Earth," *Quest*, vol. 18, no. 1, pp. 44–48, 2011; downloadable from http://astronauticsnow.com/astrosocks/index.html or directly from http://astronauticsnow.com/astrosocks/gruntman_quest_v_18_n_1_2011.pdf.
B.I. Gubanov, *Triumf i Tragediya "Energii." Razmyshleniya Glavnogo Konstruktora. Tom 3. "Energiya"—"Buran." (Triumph and Tragedy of "Energia." Thoughts of the Chief Designer. Volume 3. "Energia"—"Buran."*), Nizhegorodskii Institut Ekonomicheskogo Razvitiya (NIER), Nizhnii Novgorod, 1998 (in Russian). Electronic version at http://www.buran.ru/htm/gubanov3.htm; accessed on 27 March 2011.
A.A. Gubenko, Tochka Otshcheta—Saryshagan (Reference point—Saryshagan) in N.G. Zavaly, *Rubezhi Oborony—v Kosmose i na Zemle. Ocherki Istorii Raketno-Kosmicheskoi Oborony (Defensive Lines—in Space and on Earth. Stories of the History of Missile and Space Defense)*, pp. 665–680, Veche, Moscow, 2003 (in Russian).
Guided Missiles and Astronautics Intelligence Committee, SH-0288/60, *Soviet Surface-to-Surface Missile Deployment*, 1 Sept. 1960.
Guided Missile and Astronautics Intelligence Committee, TH 1869-60, *Memorandum*, 15 Sept. 1960.
A. Gusarov, *Ne Sluzhboi Edinoi Zhil Poligon (Not Only Service in the Life of the Range)*; the manuscript in 11 parts and conclusion posted at the website of the veterans of the Saryshagan range at http://veteran.priozersk.com/articles/xxxx, where *xxxx* stands for *2198* (part 1), *2213* (part 2), *2226* (part 3), *2300* (part 4), *2356* (part 5), *2377* (part 6), *2428* (part 7), *2460* (part 8), *2471* (part 9), *2487* (part 10), *2538* (part 11), *2551* (conclusion); accessed 5 May 2014; fl. C. 2012 (in Russian).
C. Hansen, *U.S. Nuclear Weapons: The Secret History*, Aerofax, Arlington, TX, 1988.
O. Hays, Jr., *Home from Siberia: The Secret Odysseys of Interned American Airmen in World War II*, Texas A&M University Press, College Station, TX, 1990.
D.E. Hoffman, *The Dead Hand: The Untold Story of the Cold War Arms Race and Its Dangerous Legacy*, Doubleday, New York, 2009.
H.M. Hua, "The Black Cat Squadron," *Air Power History*, vol. 49, no. 1, 2002.
H.M. Hua, *Lost Black Cats: Story of Two Captured Chinese U-2 Pilots*, Authorhouse, Bloomington, IN, 2005.
Investigation of National Defense Missiles. Hearings before the Committee of Armed Services, House of Representative, 85th Congress, Second Session, pursuant to H. Res. 67, Government Printing Office, Washington, DC, 1958.
V.M. Ivantsov, Sozdanie radiolokatsionnykh stantsii nadgorizontnogo polya SPRN [Development of Above-the-Horizon Radar Coverage for SPRN (Ballistic Missile Attack Early Warning System)], in N.G. Zavaly, *Rubezhi Oborony—v Kosmose i na Zemle. Ocherki Istorii Raketno-Kosmicheskoi Oborony (Defensive Lines—in Space and on Earth. Stories of the History of Missile and Space Defense)*, pp. 473–491, Veche, Moscow, 2003 (in Russian).
V.I. Ivkin and G.A. Sukhina, compilers, Zadacha Osoboi Gosudarstvennoi Vazhnosti. Iz Istorii Sozdaniya Yadernogo Oruzhiya i Raketnykh Voisk Strategicheskogo Naznacheniya

(1945–1959) [Task of Special State Importance. From the History of Development of Nuclear-Missile Weapons and Strategic Rocket Forces (1954–1959)], *Rossiiskaya Politicheskaya Entsiklopediya ROSSPEN (Russian Political Encyclopedia)*, Moscow, 2010 (in Russian).

N.L. Johnson, E. Stansbery, J.-C. Liou, M. Horstman, C. Stokely, and D. Whitlock, "The Characteristics and Consequences of the Break-up of the Fengyun-1C Spacecraft," *Acta Astronautica*, vol. 63, pp. 128–135, 2008.

P.I. Kachur and A.V. Glushko, *Valentin Glushko. Konstruktor Raketnykh Dvigatelei i Kosmicheskikh Sistem (Valentin Glushko. Designer of Rocket Engines and Space Systems)*, Politekhnika, Saint Petersburg, 2008 (in Russian).

N.P. Kamanin, *Skrytyi Kosmos (Hidden Space)*, Infortekst, Moscow, 1995 (in Russian).

Yu. Kamensky, interview by E. Zhirov, "Noch'yu kto-to pomochilsya na ostatki golovnoi chasti (At Night Somebody Relieved Himself on the Remnants of the Warhead)," *Kommersant Vlast'*, no. 21 (474), 4 June 2002, web version, http://www.kommersant.ru/doc.aspx?DocsID=325435&print=true, accessed on 31 July 2010 (in Russian).

Yu.A. Kamensky, Porazhenie BG BR i yadernaya bezopasnost' v otechestvennykh sistemakh PRO (Destruction of Warheads of Ballistic Missiles in National Systems of Missile Defense), in N.G. Zavaly, *Rubezhi Oborony—v Kosmose i na Zemle. Ocherki Istorii Raketno-Kosmicheskoi Oborony (Defensive Lines—in Space and on Earth. Stories of the History of Missile and Space Defense)*, pp. 599–611, Veche, Moscow, 2003 (in Russian).

V.V. Kavel'kina, M.A. Golov, I.V. Shumakova, A.A. Glushkov, A.B. Gordienko, and V.I. Shovkunov, *60 Let v Stroyu. Poligon Kapustin Yar, 1946–2006 (60 Years in Operation. Test Site Kapustin Yar, 1946–2006)*, GTsMP "Kapustin Yar," Znamensk, 2006 (in Russian).

G.P. Kennedy, *Vengeance Weapon 2*, Smithsonian Institution Press, Washington, DC, 1983.

L.L. Kerber, *A Delo Shlo k Voine (And Things Were Moving Toward War)*, Izobretatel-Ratsionalizator, 2008–2010; http://militera.lib.ru/memo/russian/kerber/index.html; accessed on 11 Jan. 2015 (in Russian).

M. Khodarenok, "Tochka otshcheta v istorii PRO (Starting Point in the History of Missile Defense)," *Vozdushno-Kosmicheskaya Oborona (Air and Missile Defense)*, no. 50, issue 1, pp. 54–73, 2010.

N.S. Khrushchev, *Velichestvennaya Programa Kommunisticheskogo Stroitel'stva v Nashei Strane. Rech' na Prieme Vypusknikov Voennykh Akademii 14 Noyabrya 1958 goda (The Majestic Program of Communism Building in Our Country. A Speech at the Reception for Graduates of Military Academies on 14 November 1958)*, Gospolitizdat, Moscow, 1958 (in Russian).

N.S. Khrushchev (S. Talbott, editor), *Khrushchev Remembers: The Last Testament*, Little, Brown, and Co., Boston, 1974.

S. Khrushchev, *Rozhdenie Sverkhderzhavy: Kniga ob Ottse (Birth of a Superpower: The Book about the Father)*, Vremya, Moscow, 2000 (in Russian).

L.G. Khvatov, Sozdanie radiolokatorov tochnogo navedeniya i ikh ispytaniya v sostave otechestvennykh sistem PRO (Development of Precise Guidance Radars and Their Testing in National Missile Defense Systems), in N.G. Zavaly, *Rubezhi Oborony—v Kosmose i na Zemle. Ocherki Istorii Raketno-Kosmicheskoi Oborony (Defensive Lines—in Space and on Earth. Stories of the History of Missile and Space Defense)*, pp. 584–598, Veche, Moscow, 2003 (in Russian).

G.V. Kisun'ko, "On the Theory of the Excitation of Radiowaves Guide," *Comptes Rendus (Doklady) de l'Academie des Sciences de l'URSS*, (cover-to-cover translation into English; later known as *Soviet Physics—Doklady*), vol. 51, no. 3, pp. 199–202, 1946a.

G.V. Kisun'ko, "K teorii vozbuzhdeniya volnovodov (On the Theory of Excitation of Radiowave Guides)," *Bulletin de l'Academie des Sciences de l'URSS*, Série Physique, vol. 10, no. 2, pp. 217–224, 1946b (in Russian).

G.V. Kisun'ko, *Elektrodinamika Polykh Sistem* (*Electrodynamics of Hollow Systems*), VKAS, Leningrad, 1949 (in Russian).

G.V. Kisun'ko, *Sekretnaya Zona. Ispoved' General'nogo Konstruktora* (*Secret Zone. Confession of the Chief Designer*), Sovremennik, Moscow, 1996 (in Russian).

A.I. Kokurin and N.V. Petrov (compilers), R.G. Pikoya (editor), *Lubyanka. VChK–OGPU–NKVD–NKGB–MGB–MVD–KGB, 1917–1960, Reference Book*, International Fund "Democracy," Moscow, 1997 (in Russian).

Yu.P. Kornilov, "Neizvestnyi 'Polyus' (Unknown 'Polyus')," *Zemlya i Vselennaya* (*Earth and Universe*), no. 4, pp. 18–23, 1992 (in Russian).

S.N. Konyukhov (editor), *Prizvany Vremenem. Ot Protivostoyaniya k Mezhdunarodnomu Sotrudnuchestvu* (*Called by the Time. From Confrontation to International Cooperation*), 2nd edition, ART-PRESS, Dnepropetrovsk, 2009 (in Russian).

D. Kosmachev, O nekotorykh voprosakh inzhenernogo obespecheniya razrabotok (Some Problems of Engineering Support of Development Projects), in V.D. Anisimov et al. (editors), *Korporatsiya "Vympel." Sistemy Raketno-Kosmicheskoi Oborony* (*Corporation "Vympel." Systems of Missile and Space Defense*), pp. 121–125, Oruzhie i Tekhnologii, Moscow, 2005 (in Russian).

G. Kostyrchenko, *Out of the Red Shadows: Anti-Semitism in Stalin's Russia*, Prometheus Books, New York, 1995.

G.V. Kostyrchenko (editor), *Gosudarstvennyi Anti-semitizm v SSSR: Ot Nachala Do Kul'minatsii, 1938–1953* (*State Anti-Semitism in the USSR: From the Beginning to Culmination, 1938–1953*), Mezhdunarodnyi Fond Demokratiya: Materik, Moscow, 2005 (in Russian).

L. Kuchma, *Ukraina—Ne Rossiya* (*Ukraine—Not Russia*), Vremya, Moscow, 2003 (in Russian).

A.F. Kulakov, *Balkhashskii Poligon* (*Balkhash Test Site*), ZAO Moskovskie Uchebniki—SiDiPress, Moscow, 2006 (in Russian).

A.L. Kun (editor), *Turkestanskii Al'bom. Po Rasporyazheniyu Turkestanskogo General-Gubernatora General-Adjiutanta K.P. von Kaufmana I-go. Chast' Etnograficheskaya. Tuzemnoe Naselenie v Russkikh Vladeniyakh Srednei Azii, 1871–1872. Sostavil A.L. Kun* (*Turkestan Album. By the Order of Turkestan General-Governor General-Adjutant K.P. Kaufman I. Ethnographic Part. Indigenous Population in Russian Possessions in Central Asia.* Compiled by A.L. Kun), 1872 (in Russian).

A. Kuriksha, Sozdanie SPRN i SKKP [Creation of SPRN (Ballistic Missile Attack Early Warning System) and SKKP (System for Control of Space)], in V.D. Anisimov et al. (editors), *Korporatsiya "Vympel." Sistemy Raketho-Kosmicheskoi Oborony* (*Corporation "Vympel." Systems of Missile and Space Defense*), pp. 101–111, Oruzhie i Tekhnologii, Moscow, 2005 (in Russian).

J.-F. Lagrot, "Le Saiga, Miracule des Steppes (The Saigas, the Miracle of the Steppes)," *Sciences et Avenir*, pp. 64–67, Dec. 2009 (in French).

K. Lantratov, "'Zvezdnye voiny', kotorykh ne bylo (The 'Star Wars' That Never Happened)," 2005; manuscript posted at the websites http://www.buran.ru/htm/str163.htm and http://www.buran.ru/other/skif-lan.pdf, accessed on 27 March 2011 (in Russian). English translation: K. Lantratov, "The 'Star Wars' That Never Happened: The True Story of the Soviet Union's Polyus (Skif-DM) Space-Based Laser Battle Stations." Part 1, *Quest*, vol. 14, no. 1, pp. 5–14, 2007. Part 2: *Quest*, vol. 14, no. 2, pp. 5–18, 2007.

C.G. Lasby, *Project Paperclip*, Atheneum, New York, 1971.

V.V. Litvinov (editor), *Korporatsiya "Vympel." Sistemy Raketno-Kosmicheskoi Oborony* (*Corporation "Vympel." Systems of Missile and Space Defense*), Oruzhie i Tekhnologii, Moscow, 2005 (in Russian).

R.M. Lloyd, *Physics of Direct Hit and Near Miss Warhead Technology*, American Institute of Aeronautics and Astronautics (AIAA), Reston, VA, 2001.

M.M. Lobanov, *Razvitie Sovetskoi Radiolokatsionnoi Tekhniki* (*Development of Soviet Radar Technology*), Voennoe Izdatel'stvo (Military Publishing House), Moscow, 1982 (in Russian).

H.S. Lowenhaupt, "On the Soviet Nuclear Scent," *Studies in Intelligence*, vol. 11, no. 4, pp. 13–29, Fall 1967.

H.S. Lowenhaupt, "Mission to Birch Woods," *Studies in Intelligence*, vol. 12, no. 4, pp. 1–12, Fall 1968.

V. Lukashevich, *Istoriya Sozdaniya Polyusa* (*History of Development of Polyus*), manuscript posted at the website http://www.buran.ru/htm/cargo.htm; accessed on 27 March 2011 (in Russian).

O. Lutz, "Some Special Problems of Power Plants," in *History of German Guided Missiles Development*, AGARDograph No. 20, AGARD/NATO, Verlag E. Appelhans & Co., Brunswick, Germany, pp. 238–252, 1957.

K. Magnus, *Raketensklaven. Deutsche Forscher hinter rotem Stacheldraht* (*Rocket Slaves. German Scientists Behind Red Barbed Wire*), Deutsche Verlags-Anstalt, Stuttgart, 1993 (in German).

V. Markov, Istoriya sozdaniya i razvitiya Tsentral'nogo Nauchno-Proizvodstvennogo Ob"edineniya "Vympel" (History of Creation and Development of the Central Scientific-Industrial Association "Vympel"), in V.D. Anisimov et al. (editors), *Korporatsiya "Vympel." Sistemy Raketho-Kosmicheskoi Oborony* (*Corporation "Vympel." Systems of Missile and Space Defense*), pp. 30–38, Oruzhie i Tekhnologii, Moscow, 2005 (in Russian).

V.I. Markov, Razrabotka i sozdanie tekhnicheskikh sredstv RKO predpriyatiyami TsNPO "Vympel" (Development of Technical Means of Missile and Space Defense by Organizations of TsNPO "Vympel"), in N.G. Zavaly, *Rubezhi Oborony—v Kosmose i na Zemle. Ocherki Istorii Raketno-Kosmicheskoi Oborony* (*Defensive Lines—in Space and on Earth. Stories of the History of Missile and Space Defense*), pp. 399–432, Veche, Moscow, 2003 (in Russian).

R.A. McDonald and S.K. Moreno, *Raising the Periscope . . . Grab and Poppy: America's Early ELINT Satellites*, Center for the Study of National Reconnaissance, Chantilly, VA, 2005.

N. Mikhailov, Razmyshleniya of bylom (Thoughts about the Past), in V.D. Anisimov et al. (editors), *Korporatsiya "Vympel." Sistemy Raketho-Kosmicheskoi Oborony* (*Corporation "Vympel." Systems of Missile and Space Defense*), pp. 39–47, Oruzhie i Tekhnologii, Moscow, 2005 (in Russian).

Ministry of Radio Industry, Design Bureau No. 1, *Sistema Zenitnogo Upravlyaemogo Raketnogo Oruzhiya S-225, Kratkie Svedeniya, Po Materialam Eskiznogo Proekta* (*System of Antiaircraft Guided Rocket Weapon S-225, Brief Description, Based on Preliminary Design*), 54 pages, 1965 (in Russian); http://narod.ru/disk/53029046001.3b55a9ce263fafd1db0bb10cf6f11cb8/%D0%B0%D0%B7%20%D0%BE%D0%B1%D1%89%20%D1%81%D0%B2%D0%B5%D0% B4.djvu.html; accessed on 19 July 2013.

M.F. Moynihan, "The Scientific Community and Intelligence Collection," *Physics Today*, vol. 53, no. 12, pp. 51–56, 2000.

National Photographic Interpretation Center, NPIC/B-20/61, *Probable SSM Launch Area near Makat, USSR*, Photographic Interpretation Brief, Aug. 1961.

National Photographic Interpretation Center, TH 0747-62KH, *Soviet Surface-to-Surface Missile Deployment*, Guided Missiles and Astronautics Intelligence Committee, 1 Jan. 1962.

National Photographic Interpretation Center, NPIC/R-21/62, *Antimissile Test Center Sary Shagan, USSR*, Photographic Interpretation Report, Feb. 1962.

National Photographic Interpretation Center, NPIC/R-81/64, *SA-1 SAM Sites, Moscow, USSR, Northwest Quadrant*, Photographic Interpretation Report, Feb. 1964a.

National Photographic Interpretation Center, NPIC/R-89/64, *Moscow Air Defense Radar Sites, USSR*, Photographic Interpretation Report, Feb. 1964b.

M. Nemirovskaya and V. Shnitser, "Frol," *Alef*, 2011 (in Russian); http://www.alefmagazine.com/pub1745.html, accessed 8 Feb. 2013.

F.C.E. Oder, J.C. Fitzpatrick, and P.E. Worthman, *The Gambit Story*, National Reconnaissance Office, Chantilly, VA, 1991.

A. Orlov, Testimony, The Scope of Soviet Activity in the United States, Hearings before the Subcommittee to Investigate the Administration of the Internal Security Act and other Internal Security Laws of the Committee of the Judiciary, United States Senate, Eighty-Fifth Congress, First Session, 14 and 15 Feb. 1957, Part 51, pp. 3421–3473, U.S. Government Printing Office, Washington, DC, 1957.

A. Orlov, "The U-2 Program: A Russian Officer Remembers," *Studies in Intelligence*, pp. 5–14, Winter 1998/1999.

G.W. Pedlow and D.E. Welzenbach, *The CIA and the U-2 Program, 1954–1974*, Central Intelligence Agency, Washington, DC, 1998.

R.L. Perry, *A History of Satellite Reconnaissance: The Perry Gambit History*, Center for the Study of National Reconnaissance, Chantilly, VA, 2012 (originally published in 1973).

M. Pervov, *Zenitnoe Raketnoe Oruzhie Protivovozdushnoi Oborony Strany (Antiaircraft Missile Weapons of the Country Air Defense)*, Aviarus-XXI, Moscow, 2001 (in Russian).

M. Pervov, *Sistemy Raketno-Kosmicheskoi Oborony Rossii Sozdavalis' Tak (Systems of Missile and Space Defense of Russia Were Created This Way)*, Aviarus-XXI, Moscow, 2003 (in Russian).

Pravda, Rech' tovarishcha I.V. Stalina na predvybornom sobranii izbiratelei Stalinskogo izbiratel'nogo okruga g. Moskvy 9 fevralya 1946 g. (Speech by Comrade I.V. Stalin at the electoral meeting of voters of the Stalin Electoral District of Moscow on 9 Feb. 1946), pp. 1 and 2, 10 Feb. 1946 (in Russian).

Pravda, Rech' tovarishcha R. Ya. Malinovskogo (Speech by Comrade R. Ya. Malinovsky), pp. 4 and 5, 25 Oct. 1961 (in Russian).

Pravda, Beseda tovarishcha N.S. Khrushcheva s gruppoi amerikanskikh zhurnalistov 13 iulya 1962 goda (Conversation of Comrade N.S. Khrushchev with a Group of American Journalists on 13 July 1962), pp. 1 and 2, 18 July 1962 (in Russian).

Pravda, Vazhnye problemy. Vystuplenie A.N. Kosygina pered angliiskimi i inostrannymi zhurnalistami (Important Issues. Remarks of A.N. Kosygin to English and Foreign Journalists), pp. 1 and 3, 11 Feb. 1967 (in Russian).

Pravda, Obrashchenie ko vsem uchenym mira (Appeal to All Scientists of the World), p. 4, 10 April 1983 (in Russian).

B.V. Raushenbakh (editor), *S.P. Korolev i Ego Delo: Svet i Teni v Istorii Kosmonavtiki (S.P. Korolev and His Cause: Light and Shadows in the History of Cosmonautics)*, compiled by G.S. Vetrov, Nauka, Moscow, 1998 (in Russian).

J. Raymond, "The Army Rocket Makes First Kill of a Ballistic Missile," *New York Times*, p. 3, 12 Feb. 1960.

R. Reagan, Address to the Nation on Defense and National Security, 23 March 1983, in *Public Papers of the Presidents of the United States: Ronald Reagan, 1983* (in two books), Book I (1 Jan to 1 July 1983), pp. 437–443, Government Printing Office, Washington, DC, 1984.

T.C. Reed, *At the Abyss: An Insider's History of the Cold War*, Ballantine Books, New York, 2004.

V. Repin, Osnovnye etapy sozdaniya raketno-kosmicheskoi oborony (Main Stages of Creation of National Missile and Space Defense), in V.D. Anisimov et al. (editors), *Korporatsiya "Vympel." Sistemy Raketho-Kosmicheskoi Oborony (Corporation "Vympel." Systems of Missile and Space Defense)*, pp. 50–72, Oruzhie i Tekhnologii, Moscow, 2005 (in Russian).

V.G. Repin, Sobytiya i lyudi (Events and People), in N.G. Zavaly, *Rubezhi Oborony—v Kosmose i na Zemle. Ocherki Istorii Raketno-Kosmicheskoi Oborony (Defensive Lines—in*

Space and on Earth. Stories of the History of Missile and Space Defense), pp. 433–472, Veche, Moscow, 2003 (in Russian).

J.T. Richelson, *The Wizards of Langley: Inside the CIA's Directorate of Science and Technology*, Perseus Book Group, Boulder, CO, 2002.

N. Riehl and (introduction, commentary, and translation by) F. Seitz, *Stalin's Captive: Nicolaus Riehl and the Soviet Race for the Bomb*, American Chemical Society and the Chemical Heritage Foundation, Washington, DC, 1996.

K.C. Ruffner (editor), *CORONA: America's First Satellite Program*, Center for the Study of Intelligence, Central Intelligence Agency, Washington, DC, 1995.

M.B. Saukke, *Neizvestnyi Tupolev* (*Unknown Tupolev*), Fond "Russkie Vityazi," Moscow, 2006 (in Russian).

K.H. Schirrmacher, Guidance of Surface-to-Air Missiles by Means of Radar, in *History of German Guided Missiles Development*, AGARDograph No. 20, AGARD/NATO, Verlag E. Appelhans & Co., Brunswick, Germany, pp. 187–200, 1957.

E. Schuyler, *Turkistan: Notes of a Journey in Russian Turkistan, Khokand, Bukhara, and Kuldja*, Scribner, Armstrong, and Co., New York, 1876.

M. Selinger, "The Road to Iron Dome," *Aerospace America*, vol. 50, no. 4, pp. 26–31, 2013.

S.M. Semenov (chief editor), *Raspletin. 100-letiyu so Dnya Rozhdeniya Posvyashchaetsya* (*Raspletin. Dedicated to 100th Birthday Anniversary*), Mezhdunarodnyi Ob"edinennyi Biograficheskii Tsentr, Moscow, 2008 (in Russian).

S.M. Semenov (chief editor) and V.N. Korovin, *General'nyi Konstruktor B.V. Bunkin* (*General Designer B.V. Bunkin*), Mezhdunarodnyi Ob"edinennyi Biograficheskii Tsentr, Moscow, 2012 (in Russian).

Yu.P. Semenov (chief editor), *Raketno-Kosmicheskaya Korporatsiya "Energia" Imeni S.P. Koroleva: 1946–1996* (*S.P. Korolev Rocket-Space Corporation "Energia": 1946–1996*), RKK Energia, Korolev, Russia, 1996 (in Russian).

A.B. Shirokorad, *Entsiklopediya Otechestvennogo Raketnogo Oruzhiya, 1817–2002* (*Encyclopedia of National Rocket Weapons, 1817–2002*), editor A.E. Taras, ACT, Moscow, and Kharvest, Minsk, 2003 (in Russian).

Sistema–25 Protivovozdushnoi Oborony Moskvy (*Sistema-25 of Air Defense of Moscow*), Album (photographs and schematics), 66 pages; (in Russian); http://narod.ru/disk/187015 93000.86bfb67a896783a5c7678829c33a884f/%D0%B0%D0%BB%D1%8C%D0%B1% D0%BE%D0%BC%20%D1%8125.djvu.html; accessed on 19 July 2013.

W.B. Smith, *My Three Years in Moscow*, J.B. Lippincott, New York, 1950.

V.D. Sokolovsky (editor), *Military Strategy: Soviet Doctrine and Concepts*, Frederick A. Praeger, New York, 1963; the original Russian edition was published in Sept. 1962.

A. Solzhenitsyn, *The Gulag Archipelago 1918–1956*, Harper & Row, New York, 1973.

G.P. Sutton, *History of Liquid Propellant Rocket Engines*, American Institute of Aeronautics and Astronautics (AIAA), Reston, VA, 2006.

V. Svetlov, "V-1000—pervaya protivoraketa (V-1000—The First Antimissile Missile)," *Vozdushno-Kosmicheskaya Oborona* (*Air and Missile Defense*), no. 4, issue 41, pp. 52–63, 2008 (in Russian).

N. Talensky, "Anti-missile Systems and Disarmament," *International Affairs*, no. 10, pp. 15–19, Oct. 1964.

E. Teller, *Better a Shield than a Sword, Perspectives on Defense and Technology*, Macmillan, New York, 1987.

L.P. Temple, III, *Shades of Gray. National Security and the Evolution of Space Reconnaissance*, American Institute of Aeronautics and Astronautics (AIAA), Reston, VA, 2005.

A. Timberlake, *A Reference Grammar of Russian*, Cambridge University Press, New York, 2004.

A.A. Tolkachev, Mif o nevidimosti i neuyazvimosti GCh BR v pustyne Betpak-Dala (Myth about Invisibility and Invincibility of Warheads of Ballistic Missiles in Betpak Dala

Desert), in N.G. Zavaly, *Rubezhi Oborony—v Kosmose i na Zemle. Ocherki Istorii Raketno-Kosmicheskoi Oborony (Defensive Lines—in Space and on Earth. Stories of the History of Missile and Space Defense)*, pp. 612–634, Veche, Moscow, 2003 (in Russian).

Toward New Horizons: Science, the Key to Air Supremacy (commemorative edition), Headquarters Air Force Systems Command, 1992.

Yu.K. Tsukov, Goryachii vozdukh poligona—zapiski ispytatelya (Hot Air of the Range —Notes of a Test Officer), in N.G. Zavaly, *Rubezhi Oborony—v Kosmose i na Zemle. Ocherki Istorii Raketno-Kosmicheskoi Oborony (Defensive Lines—in Space and on Earth. Stories of the History of Missile and Space Defense)*, pp. 367–385, Veche, Moscow, 2003 (in Russian).

U.S. Army, *History of Strategic Air and Ballistic Missile Defense*, vol. 1 and vol. 2, U.S. Army, Center of Military History, 2009.

U.S. Department of State, Alleged Violations of Soviet Territory, *U.S. Department of State Bulletin*, vol. 35, no. 892, pp. 191–192, 30 July 1956.

O. Veideman, *Vospominaniya o S-25. Chast' 1–8 (Recollections about S-25, Parts 1–8)*, 2007 (in Russian); manuscript posted by Oleg Veideman under the screenname *veideol* at the website http://veideol.livejournal.com in 2007. The URLs are: http://veideol.livejournal.com/519.html (part 1); http://veideol.livejournal.com/771.html and http://veideol.livejournal.com/2745.html (part 2); http://veideol.livejournal.com/1247.html (part 3); http://veideol.livejournal.com/1433.html (part 4); http://veideol.livejournal.com/1676.html (part 5); http://veideol.livejournal.com/1916.html (part 6); http://veideol.livejournal.com/2294.html (part 7); and http://veideol.livejournal.com/2450.html (part 8); accessed on 20 July 2010. Veideman resides in Tallinn, Estonia.

Yu.G. Venediktov, Voennye stroitely mesto sluzhby ne vybirayut (Military Builders Do Not Choose Locations of Service), in N.G. Zavaly, *Rubezhi Oborony—v Kosmose i na Zemle. Ocherki Istorii Raketno-Kosmicheskoi Oborony (Defensive Lines—in Space and on Earth. Stories of the History of Missile and Space Defense)*, pp. 681–692, Veche, Moscow, 2003 (in Russian).

V.M. Vinogradov, A.G. Smirnov, B.I. Zhaivoronok, Yu.V. Nastenko, V.V. Zub, A.V. Zhdanov, A.Ya. Kostin, K.I. Popov, M.V. Ryabov, V.S. Logachev, I.V. Pribylov, G.V. Goryachkin, I.T. Saenko, V.V. Zakharov, V.S. Yakushev, A.I. Mitrokhin, and V.B. Ivanov, *Grif "Sekretno" Snyat. Kniga ob Uchastii Sovetskikh Veonnosluzhashchikh v Arabo-Israil'skom Konflikte (Classification "Secret" Removed. A Book about Participation of Soviet Military Personnel in the Arab-Israeli Conflict)*, Moscow, 1997 (in Russian).

K.A. Vlasko-Vlasov, *Ot "Komety" do "Oko" (From "Kometa" to "Oko")*, Ol'ga, Moscow, 2002 (in Russian).

K.A. Vlasko-Vlasov, Front v kosmose (Front Line in Space), in N.G. Zavaly, *Rubezhi Oborony—v Kosmose i na Zemle. Ocherki Istorii Raketno-Kosmicheskoi Oborony (Defensive Lines—in Space and on Earth. Stories of the History of Missile and Space Defense)*, pp. 502–550, Veche, Moscow, 2003 (in Russian).

F. von Hippel, "Gorbachev's Unofficial Arms-Control Advisers," *Physics Today*, vol. 66, no. 9, pp. 41–47, 2013.

H. von Zborowski, "BMW-Developments," in *History of German Guided Missiles Development*, AGARDograph No. 20, AGARD/NATO, Verlag E. Appelhans & Co., Brunswick, Germany, pp. 297–324, 1957.

Yu.V. Votintsev, "Neizvestnye voiska ischeznuvshei sverkhderzhavy (Unknown Troops of the Vanished Superpower)," *Voenno-Istoricheskii Zhurnal (Military-Historical Journal)*, Part 1, no. 8, pp. 54–61, Aug. 1993; Part 2, no. 9, pp. 26–38, Sept. 1993; Part 3, no. 10, pp. 32–42, Oct. 1993; Part 4, no. 11, pp. 12–27, Nov. 1993 (in Russian).

Yu.V. Votintsev, Voiska Protivoraketnoi i Protivokosmicheskoi Oborony, 1967–1986 (Missile Defense and Space Defense Troops, 1967–1986), in N.G. Zavaly, *Rubezhi Oborony—v Kosmose i na Zemle. Ocherki Istorii Raketno-Kosmicheskoi Oborony (Defensive Lines—in*

Space and on Earth. Stories of the History of Missile and Space Defense), pp. 12–70, Veche, Moscow, 2003 (in Russian).

J. Walker, L. Bernstein, and S. Lang, *Seize the High Ground: The U.S. Army in Space and Missile Defense*, Department of the Army, 2005.

J.A. Walker, F. Martin, and S.S. Watkins, *Strategic Defense: Four Decades of Progress*, Historical Office, U.S. Army Space and Strategic Defense Command, 1995.

Wall Street Journal, "The Arms Control Illusion," editorial, p. A12, 11 Aug. 2014a.

Wall Street Journal, "The Parliament of Palestine," editorial, p. A12, 18–19 Oct. 2014b.

A.D. Wheelon, "Corona: The First Reconnaissance Satellite," *Physics Today*, vol. 50, no. 2, pp. 24–30, 1997.

A.D. Wheelon, "Technology and Intelligence," *Technology and Society*, vol. 26, nos. 2–3, pp. 245–255, 2004.

J.B. Wiesner and H.F. York, "National Security and the Nuclear-Test Ban," *Scientific American*, vol. 211, no. 4, pp. 27–35, Oct. 1964.

F.H. Winter and R. van der Linden, "Out of the Past. An Aerospace Chronology," *Aerospace America*, vol. 49, no. 3, pp. 44–45, 2011.

V.N. Yakovlev, *Raketnyi Shchit Otechestva (Rocket Shield of the Fatherland)*, TsIPK RVSN, Moscow, 1999 (in Russian).

S.G. Zabetakis and J.F. Peterson, "The Diyarbakir Radar," *Studies in Intelligence*, vol. 8, no. 4, pp. 41–47, 1964.

A.G. Zakharov, *Kak Eto Bylo. Vospominaniya Nachal'nika Kosmodroma Baikonur (How It Was. Recollections of the Commander of Cosmodrome Baikonur)*, [s.n.], 1996 (in Russian).

N.G. Zavaly, *Rubezhi Oborony—v Kosmose i na Zemle. Ocherki Istorii Raketno-Kosmicheskoi Oborony (Defensive Lines—in Space and on Earth. Stories of the History of Missile and Space Defense)*, Veche, Moscow, 2003 (in Russian).

E.V. Zhadeiko, Gospoda ofitsery, vasha zhizn' pod pritselom (Gentlemen Officers, Your Life in the Crosshairs), in Yu.V. Tret'yakov (author-compiler), *Sorok Pyat' Sorok Pyatomu (Forty-Five to the Forty-Fifth)*, pp. 636–669, Znanie, Moscow, 2005 (in Russian).

E.V. Zhadeiko, *Russkii Karavan (Russian Caravan)*, (in Russian); manuscript posted at the website of the veterans of Saryshagan, http://priozersk.com/zd/731 (PDF file URL http://www.priozersk.com/books/EVZRK.pdf), accessed on 20 March 2014.

D.B. Zimin, Tridtsat' Let v RTI (Thirty Years in RTI [Radiotechnical Institute]), in N.G. Zavaly, *Rubezhi Oborony—v Kosmose i na Zemle. Ocherki Istorii Raketno-Kosmicheskoi Oborony (Defensive Lines—in Space and on Earth. Stories of the History of Missile and Space Defense)*, pp. 635–647, Veche, Moscow, 2003 (in Russian).

INDEX

2P24, 21
3M8, 21
5N77, 243
8A11. *See* R-1
8A62. *See* R-5
8K63. *See* R-12
8K64. *See* R-16
8K65. *See* R-14
8K67. *See* R-36
8K71. *See* R-7
8K75. *See* R-9A
8K81. *See* UR-200
8K84. *See* UR-100
8Zh38. *See* R-2
11K67. *See* Tsyklon-2A
11K69. *See* Tsyklon-2
32-B, 54, 55, 58
205, 50, 52. *See also* V-300
207, 50. *See also* V-300
207A, 50. *See also* V-300
215, 50, 66, 72. *See also* V-300
307-144ss, 30
697-355ss/op, 30
720-435, 71
893-533, 71
1017-419ss, 15
1059. *See* Model 1059
1323, post box, 21
2837-1349, 32
2838-1201, 58
3140-1028, 19
3389-142bss/op, 32
12866-525, 15
A-4, 92–94, 107, 146. *See also* V-2

A-35, 3, 73, 87, 105, 108–109, 111, 117, 141, 149, 150, 166, 178, 189, 180, 216–220, 223–227, 229–234, 247
A-35M, 3, 109, 150, 166, 217, 224–226, 228, 232–234
A-100, 41–42, 61–64, 99
A-135, 3, 112, 117, 220, 231–232, 234, 252
A-350, 217, 223, 227
Abkhaziya, 56
ABM Treaty, 229, 248
Academy of Sciences, USSR, 4, 6, 23, 42, 87, 98, 105, 115, 141, 200, 220, 244–245, 250
advanced science degrees, USSR, 36, 37, 87
Aerospace America, AIAA, 2
AIAA. *See* American Institute of Aeronautics and Astronautics
Air Force Engineering Academy, 91
Aldan, 117, 141, 217, 223
Algeria, 122
All-Union Scientific-Research Institute of Radioelectronic Systems, 246
Almaz, 21, 36, 38, 59, 252
Almaz-Antei Corporation, 38, 59, 78, 252. *See also* Almaz; Antei
Al'perovich, Karl S., 36, 37, 39, 40, 41, 45, 54, 56, 58, 65, 75, 78, 79, 84, 88
American Institute of Aeronautics and Astronautics, 2
Amur-P, 117, 220, 231
AN/APQ-13, 25
Andropov, Yuri V., 6

297

AN/FPS-17, 118, 132
AN/FPS-79, 118, 132
Angara, 237
Antei, 21, 38, 59, 78, 252
Antimissile Missile Initial Guidance Radar. See RSVPR
antisatellite weapons, 3, 9, 69, 96, 103, 107, 122, 134, 138, 200, 220, 222, 234, 236–242
anti-Semitism, 9, 36, 52, 84
Antwerp, 91, 93
Aral Sea, 81, 147, 148
Aral'sk, 138
Argon. See KH-5
Argun', 141, 150, 217, 219, 228
Arkharov, M.A., 254
Armenia, 83
Arrow 2, 10
Arrow 3, 10
Article Channel Radar. See RKI
Arzamas-16, 15, 65
Arzanov, Levon S., 128, 129
AS-1, 26, 27. See also Kometa
Atlas, 263
Atomic Energy Commission, 258
Atyrau. See Gur'yev
author of this book. See Gruntman, Mike
Avrora, 228, 230, 233–234
Azerbaijan, 17
Azov, 217

B-29, 24, 25
B-52, 78
B-200, 41, 42, 45–48, 56, 65, 66, 69, 72, 84, 145, 147, 176
Badger, 27, 130. See also Tu-16
Baidukov, Georgii F., 72
Baikonur, 15, 55, 80, 118, 120, 121, 133, 222, 238, 242, 249, 251. See also Tyuratam missile range
Balkhash-9, 122, 238
Balkhash, lake, 119, 122, 125, 133, 135, 150, 155–156, 164–165, 168, 169
Balkhash, town, 123, 127, 128, 154
ballistic missile attack early warning, 3, 42, 107, 109, 112, 122, 229, 231, 234, 236, 237, 238, 242. See also SPRN
Balqash. See Balkhash, lake
Baranovichi, 74, 75

Barmin, Vladimir P., 53
Basistov, Anatolii G., 231, 232
Bazalt, Scientific-Industrial Conglomerate. See SKB-47
Beagle. See IL-28
Beer Sheba, Israel, 10
Behsher, 118, 132
Belarus, 75, 78
Belaya, 237
Belgium, 93
Bell Telephone Laboratories, Inc., 256, 260
Berg, Aksel' I., 23, 33–36, 111, 207, 163, 207
Beria, Lavrentii P., 16–17, 19, 30, 31, 56, 57, 83, 85, 89, 95
Beria, Sergo, 16–20, 24, 26, 31, 33, 34, 37, 92, 195, 244, 246, 248
Berkut, 29–30, 32–34, 37–43, 45, 49–57, 61, 77, 85, 86, 94, 101–103. See also S-25
Berlin, 13, 14, 35, 95
betonka, 65–66, 75
Betpak Dala, 117, 120, 127, 131
Bison. See Myasishchev-4
Black Sea, 56, 82, 123
Bocharov, Vladimir, 131
Bohlen, Charles (Chip) E., 46
Bolshevik, factory, 182
Bolshevo, 91
bomber gap, 7, 78, 132
Borovsk, 225
Breitbart, Anton Ya., 91–93
Brest, 74
Brezhnev, Leonid I., 33, 96, 222
Brucker, Wilber N., 260
Brugioni, Dino A., 133
Bruk, Isaak S., 141
Bull. See Tu-4
Bunkin, Boris V., 36, 58, 59
Buran, space vehicle, 126, 249, 250
Burlakov, Yuri G., 228
Burshtein, Iliya L., 36
Burtsev, Vsevolod S., 115
Burya, 54

Cab. See Li-2
California, 172, 263
Caspian Sea, 1, 131
Cat House, 224, 225. See also Dunai-3U
Caucasus, 16, 17

Center of Military History, U.S. Army, 2
Central Air Club, 25
Central Asia, 81, 117, 118, 120, 132–134
Central Committee, CPSU, 4, 38, 71, 80, 94, 98, 103, 107, 110, 112, 185, 207, 212, 216, 220, 236, 246, 248
Central Computing Station, 185, 187
Central Design Bureau Luch, 244. *See also* TsSKB Luch
Central Finding Station—Preliminary. *See* TsSO-P
Central Intelligence Agency. *See* CIA
Central Scientific-Industrial Association Vympel. *See* Vympel, TsNPO
Central Scientific-Research Institute of Chemistry and Mechanics. *See* NII-6
Central Scientific-Research Radar Institute. *See* NII-108
Central System Indicator. *See* TsIS
Chekhov, 224, 225
Chelkar, 147, 148, 194, 195
Chelomei, Vladimir N., 105, 197, 220, 222, 223, 236, 239, 246–247
Chelyabinsk-70, 15, 159
Chernobyl', 242
Chertok, Boris E., 18, 19, 36
Chesapeake Bay, Maryland, 171, 172
Chiang Kai-shek, 14
Chief of Naval Operations, U.S., 9
China, People's Republic of, PRC, 9, 14, 82, 118, 146
China, Republic of, Taiwan, 14, 82
CIA, 45, 46, 75, 81, 118, 132, 138, 139, 141, 143, 148, 168, 170, 171–173, 175, 184, 217, 224
Commission on Military-Industrial Matters. *See* VPK
Committee No. 1, 15
Committee No. 2, 15
Communist Party of the Soviet Union, 80, 94, 98, 103, 112, 185, 207, 212, 220, 236, 248
Comprehensive Test Ban Treaty, 8
construction troops, 118, 124, 126–129, 131, 141, 147
Cornell Aeronautical Laboratories, 261
Corona, 64, 73, 78, 137, 168
Corporal, 7, 257, 258, 260, 261
Council for Radars, 15, 35

Council of Ministers, USSR, 5, 15–17, 19, 30, 32, 54, 55, 58, 71, 72, 96, 101, 103, 107, 110, 117, 129, 184, 185, 207, 212, 216, 220, 235, 236
countermeasures, 58, 200, 211, 212, 228. *See also* penetration aids
CPSU. *See* Communist Party of the Soviet Union
Crimea, 78, 81, 82
CTBT. *See* Comprehensive Test Ban Treaty
Dal', 54, 56, 117, 141
Daryal, 42, 112, 232
Daugava, 112, 232
David's Sling, 10
DC-3, Douglas, 130. *See also* Li-2

decoys, 105, 149, 212, 219, 221, 226, 228–230, 233. *See also* penetration aids
Defense Forces, Israel, 10
Dementiev, Petr V., 80
Democratic People's Republic of Korea. *See* Korea, North
Department of State, United States, 74
Design Bureau of Chemical Machine Building. *See* KB Khimmash
Design Bureau of General Machine Building. *See* KB OM
Detachment B, CIA, 118
Directorate of Engineering Works No. 32. *See* UIR-32
Directorate of Engineering Works No. 130. *See* UIR-130
divizion, 66
Diyarbakir, 118, 132
Dnepr, 42, 112, 232
Dnepropetrovsk, 148, 200, 201, 211, 239
Dnestr, 42, 112, 122, 141, 236, 237, 238, 242
Dnestr-M, 112
Dog House, 172, 217, 225. *See also* Dunai-3
Dolgoprudny, 63, 64
Don, 42
Don-2N, 112
Don-N, 228
Dorokhov, Stepan D., 129, 204, 254
Douglas Aircraft Corp., 256, 261
Dragon Returnee Program, 46

Drozdov, N.D., 223, 244
D.Sc. *See* advanced science degrees, USSR
Duga, 166, 231
Dulles, Allen, 136
Dunai-1, 111
Dunai-2, 111, 135, 137–142, 156, 161–173, 178, 179, 185, 186, 191, 193, 194, 199, 200, 203, 217, 231
Dunai-3, 111, 166, 172, 179, 217, 218, 224–225
Dunai-3M, 225
Dunai-3U, 224, 225
Dzhezkazgan, 135

E-05, 224, 225, 227
E-24, 224, 225
E-31, 224, 225
E-33, 224, 225, 226
early warning of ballistic missile attack. *See* ballistic missile attack early warning
Egypt, 27
Eisenhower, Dwight D., 73, 75, 78, 82, 95, 132, 133
ELINT, U.S., 132
Elizarenkov, 246
Emba, 252
Embassy in Moscow, U.S., 74, 75
Energia-Buran, 54
Energia, NPO, 52, 249. *See also* OKB-1
Energia, space launcher, 54, 126, 248, 249
Energomash, 114
Enisei, 224, 225
Equipment Board, War Department, 255
ESL, Inc., 172
Experimental Radar. *See* RE

FABMDS, 264
Fagot. *See* MIG-15
Fakel, 59, 112, 114, 181, 182, 217, 223
Far East, Soviet, 24, 242
Fédération Aéronautique Internationale, 80
Field Army Ballistic Missile Defense System. *See* FABMDS
Fifth Directorate, 109, 110
Fili, 75, 77
First Chief Directorate, 30
First Special Army of Air Defense, 41, 71
Fourth Chief Directorate, 72, 109

Fourth Directorate, 41
France, 22
Fresco. *See* MIG-17
Fryazino, 20

Gagarin, Yuri A., 2, 55, 80–81, 235
Galactic Radiation Background, satellite. *See* GRAB
Galosh, 217. *See also* A-350
Gambit-1. *See* KH-7
Gegechkori, Sergei A., 57. *See also* Beria, Sergo
General Combat Code. *See* OBP
General Staff, Chief of, USSR, 4, 94, 129
Georgia, 16, 17, 57
geostationary satellites, 137
German specialists, in USSR, 21, 38, 39, 40, 46, 56, 86
Germany, 14, 21, 22, 35, 40, 43, 52, 57, 74, 75, 86, 91, 95
Glavspetsmash, 58, 96, 102
Glavspetsmontazh, 96
GL MkII, 20, 93
Glushko, Valentin P., 23, 75, 114, 223
GNIIP-8, 43, 55, 117
GNIIP-10, 129, 131, 135, 138, 141, 145, 151, 153, 159, 168, 187, 188, 190, 192, 204, 252, 254
Goa. *See* SA-3
Golubev, Oleg V., 23, 24, 109, 216, 217, 233, 234, 236
Gonor, Lev R., 94
Goodyear Aircraft Co., 261
Gorbachev, Mikhail S., 8, 251
Gorky, 61
GPU, 16. *See also* KGB
GRAB, 137
Great Britain, 22
Gruntman, Alexander Yu., 126
Gruntman, Mike, 80
Grushin, Petr D., 58, 59, 60, 61, 112, 113, 181, 197
Gubanov, Boris I., 251
Gubenko, Alexander A., 117, 118, 120, 123, 124, 127, 128, 144, 252
Guideline. *See* SA-2
Guild. *See* SA-1
GULAG, 22
Gulf War, First, 10
Gulshad, 141

INDEX

Gurevich, Mikhail I., 24
Gur'yev, 194
Guryev. *See* Gur'yev

Hamas, 10
Hammaguir, 122
Hawk, 256–259
Hen House, 138–141, 164, 170–172. *See also* TsSO-P
Hen Roost, 138–139, 141, 164, 165, 170, 172–173. *See also* Dunai-2
Hero of Socialist Labor, 83, 88, 94, 101, 247
Hero of the Soviet Union, 72
Hezbollah, 10
Hills, Lee, 4
Hoch, Johannes, 38–40
HOE. *See* Homing Overlay Experiment
Homing Overlay Experiment, 7
Honest John, 7, 256, 258–259
Hound. *See* Mi-4
Hungary, 14
hypergolic combination, 52

ICBM, 2, 4, 7, 18, 53, 54, 96, 98, 100, 105, 110, 149–150, 179, 211–212, 220, 239, 256, 257, 263
Il-14, 130
IL-28, 72, 130
Ile, river, 122
Incirlik, 118
India, 9
Indonesia, 27
INEUM, 141, 142
Institute of Control Computers. *See* INEUM
Institute of History of Natural Sciences and Technology, 245
Institute of Precision Mechanics and Computer Engineering. *See* ITMVT
intercontinental ballistic missile. *See* ICBM
intermediate range ballistic missile. *See* IRBM
Iran, 9, 118, 132, 171
Iran, Shah of. *See* Shah of Iran
Iraq, 10
IRBM, 1, 7, 55, 96, 105, 148–149, 195, 199–201, 204, 205, 257, 263
Irkutsk, 236, 237, 242
Iron Dome, Israel, 10

IS, 237, 239, 240, 241
Isaev, Alexei I., 129
Isaev, Alexei M., 50, 51, 182
Israel, 9, 10
Istra. *See* RKTs-35TA
ITMVT, 115, 186, 187, 188, 190, 203

Japan, 9, 24, 157
Jian Zhongzheng. *See* Chiang Kai-shek

Kaktus, 211
Kaliningrad, Moscow region, 18, 250
Kaliningrad (Western USSR), 74
Kalmykov, Valerii D., 83, 84, 85, 89, 96, 206, 215, 216, 232, 246
Kama, 62
Kamchatka peninsula, 149
Kamensky, Yuri A., 22, 232, 244, 245
Kapustin Yar, 117
Kapustin Yar test range, 15, 30, 43, 54–56, 58, 59, 72, 84, 97, 103, 117, 118, 120, 132, 133, 138, 140, 142, 147, 148, 185, 194, 195, 199, 212, 252
Kap Yar. *See* Kapustin Yar test range
Karaganda, 152, 184
Karamursel, 118, 132
Kármán, Theodore von, 255
Katyusha. *See* M-13
Kaunas, 74
Kazakhstan, 1, 2, 15, 43, 80, 107, 108, 115, 117–118, 122, 123, 131, 133, 148, 184, 194, 195, 238, 252
KB-1, 23, 32, 33, 36–43, 54–59, 72, 73, 75, 77, 78, 81, 83–85, 89, 98–99, 101–106, 112, 114, 186, 195, 216, 239, 244, 252
KB-11, 15
KB Khimmash, 52, 182
KB OM, 54
Keldysh, Mstislav V., 4
Kennel, 27. *See also* Kometa
KGB, 16, 21–22, 24, 37, 40, 56, 85
KH-4, 219, 227, 243
 Mission 1114-2, 243
 Mission 1116-2, 227
 Mission 1117-2, 219
KH-5, 119, 122, 162, 238
 Mission 9058A, 119, 162
 Mission 9066A, 122, 238

KH-7, 39, 114, 125, 126, 136, 143,
    144, 145, 148, 150, 152, 154, 156,
    164, 177, 187, 188, 202, 218, 219,
    226, 237, 238
  Mission 4027, 125, 126, 136, 150, 152,
    154, 156, 164, 188, 219, 226, 238
  Mission 4029, 148
  Mission 4030, 39, 114, 202
  Mission 4032, 143, 144, 145, 177, 187, 237
  Mission 4038, 218
Khimki, 22, 49, 50, 51, 55, 75, 77, 112, 114
Khrushchev, Nikita S., 1, 4, 33, 41, 57, 81,
    96, 97, 123, 134, 149, 197, 204,
    206, 222, 223, 239, 246–248
Khrushchev, Sergei N., 41, 246, 247
Khvatov, Leonid G., 153, 220, 245, 247
Kiev, 16, 57, 75, 77, 78
kiloton, 157
kinetic kill vehicle, 1
Kisun'ko, Grigorii V., 17, 36, 38, 58, 72,
    83–89, 99–101, 103, 105–112,
    114, 115, 118, 120, 127, 137, 138,
    142, 149, 151, 155, 157, 173,
    178–181, 184, 187, 190, 193, 197,
    199, 200, 204, 206, 207, 211,
    215–217, 219, 223, 226, 228–234,
    236, 244–247, 252, 254
Klin, 224–226
Kobalt-M, 25
Koktas, 152
Kolosov, Andrei A., 103, 239
Kometa, 19, 24–27, 31, 32, 58, 92, 102,
    103, 112
Kometa, TsNII. *See* TsNII Kometa
Komsomolsk-on-Amur, 242
Konev, Ivan S., 94
Korea, 29
Korea, North, 9
Korea, South, 9, 14
Korenev, Georgii V., 24
Korolev, Sergei P., 4, 18, 52, 55, 75, 94, 100,
    103, 115, 146
Korolev, town, 18, 250
Koshlyakov, Nikolai S., 23
Kosygin, Alexei N., 5
Kozlov, Frol A., 222
Kozorezov, Konstantin I., 181, 195, 197,
    198, 199, 204
Krasnyi Kavkaz, cruiser, 25
Krasnyi Oktyabr', 182. *See also* Plant No.
    466

Krona, 166, 231
Krot, 211
Krug. *See* SA-4
Krylov, Nikolai I., 222
KS-1. *See* Kometa
KST-60, 207, 208
KT-50, 144
Kubinka, 66, 68
Kuchma, Leonid D., 222, 242
Kuksenko, Pavel N., 17, 19, 23–24, 26, 27,
    29, 32, 33, 36, 37, 38, 40, 45, 54,
    56, 86, 101
Kulakov, Alexander F., 120, 145–146, 158,
    178, 193, 194, 206, 215
Kuleshov, Pavel N., 72
Kunaev, Dinmukhamed A., 184
Kuntsevo, 32, 179
Kupol-10, 174
Kwajalein, 263

La-5, 49
La-7, 49
Lake Seliger, 38, 57
laser space battle station. *See* Polyus
laser weapons, 117, 138, 141, 150, 219,
    244, 250
Latvia, 242
Lavochkin, Semen A., 25, 43, 49, 51, 54, 56,
    72, 112, 195
Lebedev, Sergei A., 115, 186
Leningrad, 16, 19, 38–39, 42, 45, 49, 61, 72,
    73, 78, 103, 146, 182
Lenin Hills, 78
Lenin Prize, 247
Lenin, Vladimir I., 189
Leonov, Leonid V., 42
Li-2, 130
Lipsman, Frol P., 113, 185
List, 212
Lithuania, 75, 78
Livshits, Nakhim A., 38, 101, 102
London, 91
Lozhki, 63
Lukin, Fedor V., 101
Luneburg lens, 228
Luzhniki, 75, 76, 78

M-1, 141
M-2, 141

INDEX

M-3, 141
M-4, 141
M-8, 53
M-13, 53
M-40, 187, 188, 189, 191, 192, 193, 203
M-50, 187, 189
Magnus, Kurt, 39
Makat, 194, 195, 199, 202
Malenkov, Georgii M., 35
Malinovsky, Rodion Ya., 3
Manchuria, 24
Mao Tse-tung. *See* Mao Zedong
Mao Zedong, 14
maps, Moscow, 75
Markov, Vladimir I., 56, 105, 228, 231–234, 242
Marx, Karl, 189
Maryland, 171
Matters of the Rocket Weapons, 15
McElroy, Neil, 255
Meshed, 118, 132
Mi-4, 130
Michurinsk, 63, 64
Middle East, 9
MiG-15, 79, 130
MiG-17, 79, 130
MiG aircraft, 24
Mikhailov, Nikolai V., 215, 251, 254
Mikoyan, Anastas I., 24
Mikoyan, Artem I., 24, 25, 80
Military Academy of Communications, 16, 17, 38, 103
Military-Industrial Commission. *See* VPK
Minister of Armaments, 20, 32
Minister of Defense, 3, 71, 216, 233, 245, 251, 254
Minister of Defense Industry, 103
Minister of Radio Industry, 56, 233, 246
Ministry of Armaments, 19, 20, 32, 42, 95
Ministry of Defense, 4, 31, 41, 43, 58, 72, 91, 106, 109, 110, 112, 117, 123, 179, 197, 212, 223, 228, 250, 252
Ministry of Defense Industry, 112, 179, 197
Ministry of General Machine Building, 96. *See also* MOM
Ministry of Internal Affairs, 37, 38, 40, 65
Ministry of Means of Communications, 42
Ministry of Middle Machine Building, 96
Ministry of Radio Industry, 96, 212, 221, 231, 233

Minsk, 74, 75
Minsredmash. *See also* Ministry of Middle Machine Building
Mints, Alexander L., 23, 42, 72, 98–100, 102, 103, 105, 106, 111, 112, 138, 142, 149, 232
Missile and Space Defense Forces, 3, 80, 227, 233
missile defense, politicization of, 8
missile defense, strategic, 2
missile defense, theater, 9
missile gap, 78
Mixed Aviation Test Division, 60th, 129
Model 1059, 147
MOM, 96, 251
Moon, permanent base on, 54
Moon, scattering of radar signals from, 137, 171, 172
Moscow river, 77, 78
Moscow Scientific-Research Radiotechnical Institute, 113
Moscow State University, 75, 78
Mozharovsky, Georgii M., 91
Mukachevo, 242
Murav'ev, Konstantin K., 18
Murmansk, 179, 242
Musatov, Alexander N., 224
mutually assured destruction, 8, 9
MVD, 37, 65. *See also* KGB
Myasishchev-4, 75, 77
Mykolaiv. *See* Nikolaev
Mymrin, Mikhail G., 110
Mytishchi, 103

Naro-Fominsk, 224, 225
National Photographic Interpretation Center. *See* NPIC
NATO, 9, 149
Navaho, 54
Naval Research Laboratory. *See* NRL
Nedelin, Mitrofan I., 94
Neman, 166, 228
New Mexico, 256
N.E. Zhukovsky Air Force Engineering Academy. *See* Air Force Engineering Academy
NIEMI, 21, 33, 42, 252. *See also* NII-20
NII-2, 223
NII-4, 91
NII-6, 32, 195

NII-20, 19–21, 32, 33, 42, 92, 93, 94, 112, 179, 181, 195, 252
NII-20, Ministry of Means of Communications, 42
NII-37, 166, 231, 242
NII-88, 18, 19, 30, 38, 50, 52, 94, 182
NII-108, 20, 33, 35, 36, 42, 84, 87, 111, 163, 164, 165, 207
NII-129, 113, 185
NII-160, 20
NII-244, 42, 185, 228. *See also* NII-20, Ministry of Means of Communications
NII-885, 94
NII-1011, 15
NIIDAR, 165, 166, 224, 231, 242
NII Istok, 20. *See also* NII-160
NIIP-5, 15. *See also* Tyuratam missile range
NIIRP, 216, 231, 244, 245. *See also* OKB-30
Nike, 256–257
Nike-Ajax, 256, 257
Nike-Hercules, 7, 256–258, 260, 261, 263
Nike-Zeus, 256, 258, 262, 263
Nikolaev, 242, 243
Nilovsky, Sergei F., 43, 117
NIRTI, 166. *See also* NIIDAR
NKGB, 16. *See also* KGB
NKVD, 16. *See also* KGB
Novodevich'e cemetery, 113, 254
NPIC, 133
NPO Astrofizika, 244, 246. *See also* TsKB Luch
NRL, 172
nuclear explosion, S-25, 72
nuclear explosions, Saryshagan. *See* Operation K
nuclear intercept, 7, 109, 217, 230
Nudol', 224, 225

OBP, 189–191, 193, 203
Odessa, 78, 123
Oganov, Nikolai I., 42
OGPU, 16. *See also* KGB
OKB-1, 52. *See also* Energia, NPO
OKB-2, 58. *See also* Fakel
OKB-30, 111, 114, 216, 244. *See also* NIIRP
OKB-301, 43, 49–52, 54, 55, 56, 112

OKB-586, 200. *See also* Yuzhnoe Design Bureau
Omega, 69
Omega-2, 69
Omel'chenko, Ivan D., 233
Operation K, 157, 159, 170, 172, 178, 179, 230
Order of Lenin, 26, 72
Order of the Red Banner, 23
Orenburg, 133
Orlov, Alexander, 61
OS-1, 236, 237, 240
OS-2, 141, 236, 237, 238, 239, 240
Osa, 59
Ostapenko, Nikolai K., 39–40, 85, 98, 99, 178, 184
over-the-horizon radar, 3, 56, 107, 231, 236, 242

Pacific Ocean, 263
Pakistan, 9, 118
Paperclip, 22
Patriot, 247, 256
penetration aids, 35, 104, 149, 200, 207–212. *See also* countermeasures
People's Republic of China. *See* China, People's Republic of, PRC
Peshawar, 118
Pivovarov, Anatolii V., 56, 87
plant, airframe, Fili, 75, 77
Plant No. 37, 165, 166
Plant No. 293, 55, 58
Plant No. 301, 50
Plant No. 465, 20, 32, 33, 35, 93
Plant No. 466, 182. *See also* Krasnyi Oktyabr'
Plato, Project, 256
Pleshakov, Petr S., 207, 211
Pluton, 93
Podlipki, 18, 20, 22, 30, 50, 52, 75, 77, 91, 94, 103, 146, 250
Poland, 75
Polet-1, 239
Polet-2, 239
Polyus, 7, 8, 248, 249, 251
Polyus, laser space battle station. *See* Polyus
Pomaznev, M., 32
Potsdam Conference, 14
Powers, Francis Gary, 59, 73, 80

Precise Guidance Radar. *See* RTN
Priozersk, 119, 122, 124, 130, 135, 136, 141, 143, 150–158, 161–162, 236, 238
prisoners, specialists, Soviet, 23–24, 40
PRO, 234, 242
Proton, 54, 220, 250

Quemoy, 14

R-1, 18, 32, 146, 147
R-2, 32, 146, 147
R-3, 32
R-5, 147, 148, 167, 194, 195, 198, 206
R-5M, 147
R-7, 53, 149, 239
R-12, 1, 147–149, 194–195, 198–201, 206, 208, 209–210, 257
R-16, 239
R-36, 212
R-101, 30
R-105, 30
R-110, 30
Rabinovich, Samuil P., 112, 179, 181
radar cross-section. *See* RCS
Radio Corp. of America, 261
RALAN, 42–43, 98–103, 106, 107, 111. *See also* RTI
Raspletin, Alexander A., 33, 34–38, 40, 43, 44–45, 57–58, 72, 83–89, 98, 101, 103, 104, 105, 216, 220, 221
Raytheon Co., 257
RCS, 97, 100
RE, 142, 175
RE-1, 142–145, 147, 186. *See also* RE
RE-2, 142, 143, 147, 149, 186. *See also* RE
RE-3, 150
RE-4, 141, 150, 219
Reagan, Ronald, 5, 6, 8, 251
reconnaissance, 4, 39, 59, 62, 64, 66–69, 71, 73, 75, 78, 82, 96, 132
reconnaissance, space. *See* satellite reconnaissance
Redstone Arsenal, 260
Repin, Vladislav G., 229–232, 242
repressions, 16
Republic of China, ROC. *See* China, Republic of, Taiwan
Rheintochter, 30
Riga, 179

ring road, 65, 107–108
RKI, 217, 219, 224, 227
RKI-35, 224, 227
RKI-35TA, 150, 217, 219. *See also* RKI
RKTs, 217, 219, 224, 225, 227
RKTs-35, 224–225
RKTs-35TA, 150, 217, 219. *See also* RKTs
RS-10, 145, 173–176
RS-11, 173–174, 176
RSVPR, 112, 161, 162, 179–181, 185, 196, 204
RTI, 42, 99, 111–112, 114, 137–139, 141, 164, 171, 176, 231, 232, 236, 238, 242. *See also* RALAN
RTN, 111, 119, 127, 139, 142, 146–147, 153, 158, 161–162, 173–179, 186–187, 191, 193–194, 198–199, 204, 211, 216, 253
RTN-1, 131, 141, 185, 203
RTN-2, 141–144, 185, 200, 203
RTN-3, 141, 185, 203
Ryabikov, Vasilii M., 30, 31

S-25, 56, 58–59, 61–63, 65, 66–73, 75, 77, 79, 88, 97, 99, 101–102, 117–119, 142, 145, 176, 225–226. *See also* Berkut
S-50, 72–73
S-75, 58–59, 73, 80–83, 101, 102, 117, 134, 136, 141, 150, 156, 181
S-125, 59, 73
S-200, 59, 117, 141
S-225, 105, 117, 141, 217, 220, 221
S-300, 59, 117
S-400, 59
SA-1, 50, 68, 102, 225, 226–227. *See also* S-25
SA-2, 59, 73, 81–83, 101–102, 136, 150, 181. *See also* S-75
SA-3, 59, 73. *See also* S-125
SA-4, 20–21, 33
SA-5, 59. *See also* S-200
SA-8, 20, 33, 59
SA-10, 59. *See also* S-300
SA-12, 21. *See also* S-300
SA-12a, 21
SA-12b, 21
SA-15, 21, 33, 59
SA-20, 59. *See also* S-300
SA-21, 59. *See also* S-400

SA-23, 21
Saddler. *See* SS-7
Safeguard, missile defense system, 7
SALT, 229
Samsun, 118, 132
Sandal. *See* SS-4
Sapwood. *See* SS-6
Sarov, 15. *See also* Arzamas-16
Sarymsek, Peninsula, 122
Saryshagan, 117
Saryshagan, railroad station, 110, 117, 119, 122–127, 133, 135, 150, 156, 162
Saryshagan test range, 1, 2, 3, 4, 7, 55, 103, 117, 122, 123, 124, 128–130, 133–137, 139, 143, 156, 159, 162, 193, 252, 254
    1st directorate, 132
    2nd directorate, 132
    3rd directorate, 132
    4th directorate, 132
    Launch Complex A, 139, 141
    Launch Complex B, 139, 141, 184
    radar site 1, 135, 141
    radar site 2, 135, 141, 168, 170
    Site 1, 131, 135, 141, 146, 185
    Site 2, 119, 123, 127, 135, 141, 142–150, 158, 177, 185, 253
    Site 3, 141, 153, 185
    Site 4, 124, 125, 127, 141, 152, 153, 154, 156
    Site 4p, 124, 125, 141, 151, 152
    Site 4v, 123, 125, 128, 141, 150, 151, 153
    Site 6, 141, 161, 162, 180, 181, 182, 184, 185, 196, 204
    Site 7, 130, 141, 181
    Site 8, 140, 141, 164
    Site 9, 141
    Site 14, 140, 141, 164, 165, 167–169
    Site 15, 140, 141, 164, 168
    Site 16, 141
    Site 17, 141
    Site 20, 141
    Site 21, 141
    Site 22, 141
    Site 35, 141, 153, 253
    Site 38, 141, 150, 156, 217, 219, 244
    Site 40, 141, 151–154, 161–162, 185, 187–188, 190–192
    Site 51, 141, 150
    Site 52, 141
    Site 53, 141
    Site 54, 141
site designations, 138, 141
Saryshagan test range, nuclear explosions, 155–159. *See also* Operation K
Sasin. *See* SS-8
Satellite Destroyer, 237, 239, 241. *See also* IS
Satellite Finder node, 236. *See also* OS-1, OS-2
satellite killers, 237. *See also* IS
satellite reconnaissance, 39, 73, 119, 122, 126, 136, 143–145, 152, 154, 156, 164, 177, 226–227, 237–238
Savin, Anatolii I., 103, 239
SB-1, 13, 19, 20–26, 27, 31–33, 244. *See also* KB-1
Scarp. *See* SS-9
Schmetterling, 30
Scientific-Production Association Astrofizika. *See* NPO Astrofizika
Scientific-Research Electromechanical Institute. *See* NIEMI
Scientific-Research Institute of Long-Range Radio Communications. *See* NIIDAR
Scientific-Research Institute of Radio Instrument Building. *See* NIIRP
Scientific-Research Radiotechnical Institute. *See* NIRTI
Scientific-Research Test Range No. 5. *See* NIIP-5
Scrag. *See* SS-10
Scud, missile, 10
Scunner. *See* SS-1
SDI. *See* Strategic Defense Initiative
Sego. *See* SS-11
Semipalatinsk, 118, 133
Sena, Lev A., 23
Sevastopol, 82, 242
Sevruk, Dominik D., 23, 50, 182
Shah of Iran, 118
Sharashka, 22–23
Shchukin, Alexander N., 31, 35–36, 58, 72, 84, 98, 99
Shokin, Alexander I., 35
short-range ballistic missile. *See* SRBM
Shyster. *See* SS-3
Siberia, 24, 118, 133, 236, 237, 242
Sibling. *See* SS-2
"Silicon Valley," Soviet, 35
Sinop, 118, 132

INDEX

Sistema A, 1
SKB-1, 231
SKB-2, 231
SKB-3, 231
SKB-30, 103, 106, 107, 142, 147, 153, 181, 187, 247
SKB-31, 103
SKB-41, 103, 239
SKB-47, 197
Skean. *See* SS-5
Skif-DM, 248. *See also* Polyus
SKKP, 234
SL-11, 240, 241. *See also* Tsyklon-2
Sliozberg, Mikhail L., 19, 20
SM-71P, 182, 196
Smirnov, Yu. L., 245
Smirnov, Leonid V., 96, 245
Smirnov, Sergei M., 23
Smith, Walter Bedell, 13, 17
Smolensk, 63, 75
SOFT TOUCH, 133
Sokolovsky, Vasilii D., 4, 5, 94, 95
SON-2ot, 20, 93
Sorge, Christian, 46
Sosul'nikov, Vladimir P., 111, 163, 164, 165, 217, 224
Space Control System, 234, 236. *See also* SKKP
Special Advisory Group, 255
Sperry Gyroscope Co., 261
SPK, 179
SPRN, 234, 242, 244
Sputnik, 4, 18, 55, 236
Sputnik-3, 149
SS-1 32, 146. *See also* R-1
SS-2, 32, 146. *See also* R-2
SS-3, 147, 194, 195. *See also* R-5
SS-4, 54, 147, 149, 194, 195, 199, 200, 201, 204–205, 257. *See also* R-12
SS-5, 54. *See also* R-14
SS-6, 53, 149. *See also* R-7
SS-7, 239. *See also* R-16
SS-8, 54. *See also* R-9A
SS-9, 212. *See also* R-36
SS-11, 54, 220. *See also* UR-100
Stalingrad, 15
Stalin, Joseph V., 13–16, 19, 24, 29–32, 34, 35, 36, 41, 56–60, 83, 85, 94–96
Stalin Prize, 27
Stanford, California, 171, 172
Stanford Research Institute, 261

State Central Interservice Test Range, 252
State Central Test Range No. 4, 15. *See also* Kapustin Yar test range
State Committee for Defense, 35
State Committee on Radioelectronics, 206, 215
State Scientific-Research Test Range No. 10, 129. *See also* GNIIP-10
Station for Command Transmission. *See* SPK
Stepanov, Andrei M., 203
Stilwell, Joseph W., 255
Stilwell report, 255
Stockman, Harvey, 73
Strategic Arms Limitation Talks. *See* SALT
Strategic Defense Initiative, 6, 8, 251
Strategic Rocket Forces, 16, 148, 194–195, 200, 220, 222, 252
Strela, computer, 185
Subbotin, Vladimir L., 129
Sukhumi, 56
Suslov, Mikhail A., 245, 246
Suslov, Revolii M., 245, 246
Sverdlovsk, 56, 57, 59
Syr Darya, 121

Taifun, 30
Taiwan. *See* China, Republic of, Taiwan
TALENT, 136. *See also* U-2, overflights
Taran, 105, 220, 222, 223, 247
Target Channel Radar. *See* RKTs
Tashkent, 61, 133
Teller, Edward, 5, 6, 255, 263
Temnikov, 63
Terra, 150, 219. *See also* laser weapons
TG-02, 52
TGU, 30, 31, 35–36, 38, 39, 41, 42, 55, 58, 83, 85, 96, 98–101, 102
Third Chief Directorate. *See* TGU
Thor-Able, 137
Titan I, 263
TNT, 7, 157, 263
Tobol, 224, 225
Tomashevich, Dmitrii L., 23, 54, 58, 59
Tonka. *See* TG-02
Tor, 59
Trabzon, 118
Transit-2A, 137
trinitrotoluene. *See* TNT
Trofimchuk, Mikhail I., 129

Tsentr Upravleniya Poletami. *See* TsUP
TsIS, 191–194
TsKB-20, 20
TsKB Luch, 244, 246
TsKKP, 71, 242
TsNII-108. *See* NII-108
TsNII Kometa, 103
TsNIIMash, 53
TsSO-P, 111, 135, 137–142, 149, 156, 164, 166, 170–172, 178–179, 236, 238
TsUP, 53
Tsyklon-2, 239, 240–242
Tsyklon-2A, 239, 240
Tu-4, 24, 25, 56
Tu-16, 27, 130
Tu-104, 130. *See also* Tu-16
Tupolev, Andrei N., 23, 24, 25
Turkestan, 80, 81, 133, 194
Turkestan military district, 80, 81
Turkey, 9, 82, 118, 132
Tushino, 22, 25
Tyuratam, 120
Tyuratam missile range, 15, 55, 80, 118, 121, 122, 123, 126, 128, 131–133, 147, 149, 222, 238, 241, 247, 249, 251
  Launch Complex G, 238
  Site 90, 244
Tyuratam, railroad station, 133

U-2, overflights, 3, 46, 59, 62, 63, 73–83, 132–136, 137, 140, 151, 152, 168, 169, 184, 188
  Mission 2013, 76, 78
  Mission 2014, 68, 69, 73, 76–78
  Mission 2020, 78
  Mission 2021, 78
  Mission 2023, 78
  Mission 4035, 133
  Mission 4155, 81, 128, 130, 134, 136, 140, 169, 188
  Mission 8009, 64, 81, 82
U-2, overflights, of People's Republic of China, 82
UIR-32, 117, 123, 127, 129, 151, 155
UIR-130, 126
Ukraine, 16, 57, 75, 78, 79, 123, 148, 200, 201, 239, 242, 243
United Nations, 9, 149

Upravlenie Inzhenernykh Rabot No. 32. *See* UIR-32
Upravlenie Inzhenernykh Rabot No. 130. *See* UIR-130
UR-100, 220, 222
UR-200, 239
Ural mountains, 15, 56, 57, 73, 80, 159
Ur'ev, Naum I., 212
U.S. Air Force, 118, 132
U.S. Army, 2, 7, 255, 256, 257–258, 261, 263
U.S. Navy, 137, 172, 255
U.S. Senate, 61
USSR, military expenditures, 1, 3, 234
Ustinov, Dmitrii F., 19, 32, 33, 96, 112, 222, 244
Ustinov, Nikolai D., 244, 246

V-1, 39
V-2, 15, 16, 18, 39, 91–93. *See also* A-4
V-300, 43, 49–56, 58, 70, 72
V-301, 49
V-750, 50, 181, 182
V-1000, 50, 51, 52, 59, 112, 141, 161, 162, 178, 181–185, 191, 194–200, 203–204, 206, 207, 208
Vandenberg Air Force Base, 263
Vasilevsky, Alexander M., 94
v/ch 03080, 129. *See also* GNIIP-10
v/ch 19313, 123. *See also* UIR-32
v/ch 29139, 43. *See also* GNIIP-8
v/ch 46180, 242
v/ch 77969, 41. *See also* 4th Directorate
Veideman, Oleg, 66, 69
Veisbein, Mikhail M., 101, 103, 138
Venediktov, Yuri G., 155
Verba, 211
Vershynin, Konstantin A., 94
Vetoshkin, Sergei I., 30, 31, 55, 72
Vietnam, 14
Vilnius, 74
Vito, Carmine, 73
Vlasko-Vlasov, Konstantin A., 240
VNIIRT, 42. *See also* NII-20, Ministry of Means of Communications
Vnukovo, 130
Volga, 166
Volga, river, 15
Volgograd. *See* Stalingrad
Vol'man, Iosif I., 85, 86
Vorob'evy Hills. *See* Lenin Hills

Voronov, Alexander V., 181, 197–198
Votintsev, Yuri V., 3, 80, 112, 228, 233, 235
Voznyuk, Vasilii I., 55
V.P. Chkalov Central Air Club. *See* Central Air Club
VPK, 55, 72, 96, 102, 109, 167, 176, 192, 195, 197, 204, 208–210, 245–246
Vympel, Central Scientific-Industrial Association. *See* Vympel, TsNPO
VympelKom, 245
Vympel, OKB, 112, 216, 231. *See also* NIIRP
Vympel, TsNPO, 112, 114, 231–234, 251

W-50, 258
Wasserfall, 15, 19, 30, 49
Watkins, James D., 9
Western Electric Co., 260–261
Wheelon, Albert "Bud" D., 157, 170, 171–173
White Book, 212
White Sands Missile Range. *See* WSMR
Woomera, 122
World War II, 13, 14, 15, 17, 18, 20–22, 24, 30, 31, 33–34, 49, 52–53, 72, 93–95, 107, 185, 244, 255

WSMR, 7, 256–259, 261
Wuerzburg, 35

Yakovlev, Alexander S., 25
Yakovlev, Nikolai D., 94
Yangel, Mikhail K., 148, 199, 200, 201, 239
Yekaterinburg. *See* Sverdlovsk
Yerevan, 83
Yo-Yo, 45, 46, 47, 48. *See also* B-200
Yuzhnoe Design Bureau, 148, 199, 200, 201, 211, 239, 241. *See also* OKB-586

Zagorsk, 61, 224, 225, 227
Zahedan, 118
Zaikov, Lev N., 251
Zakharov, Alexander G., 247
Zakson, Mikhail B., 36, 84, 85, 89
Zelenograd, 35
Zhadeiko, Evgenii V., 120, 131, 146, 184
Zhukov, Georgii K., 94
Zimin, Dmitrii B., 245
ZUR-205. *See* 205

SUPPORTING MATERIALS

A complete listing of titles in the Library of Flight series is available from AIAA's electronic library, Aerospace Research Central (ARC), at arc.aiaa.org. Visit ARC frequently to stay abreast of product changes, corrections, special offers, and new publications.

AIAA is committed to devoting resources to the education of both practicing and future aerospace professionals. In 1996, the AIAA Foundation was founded. Its programs enhance scientific literacy and advance the arts and sciences of aerospace. For more information, please visit www.aiaafoundation.org.